Unified Symmetry
In the Small and in the Large 2

Unified Symmetry

In the Small and in the Large 2

Edited by

Behram N. Kursunoglu

Global Foundation, Inc.
Coral Gables, Florida

Stephen Mintz

Florida International University
Miami, Florida

and

Arnold Perlmutter

University of Miami
Coral Gables, Florida

Plenum Press • New York and London

Library of Congress Cataloging-in-Publication Data

Unified symmetry : in the small and in the large 2 / edited by Behram
N. Kursunoglu, Stephen Mintz, and Arnold Perlmutter.
    p.   cm.
   "Proceedings of the Twenty-third Coral Gables Conference on
Unified Symmetry: In the Small and in the Large, held February 2-5,
1995, in Coral Gables, Florida"--T.p. verso.
   Includes bibliographical references and index.
   ISBN 0-306-45189-1
   1. Unified field theories--Congresses. 2. Supersymmetry-
-Congresses. 3. Cosmology--Congresses. I. Kurşunoğlu, Behram,
1922-   . II. Mintz, Stephen. III. Perlmutter, Arnold, 1928-   .
IV. Coral Gables Conference on Unified symmetry: In the Small and in
the Large (23rd : 1995)
QC173.68.U555
530.1'42--dc20   95-39325
                                                                 CIP

Proceedings of the Twenty-third Coral Gables Conference on Unified Symmetry: In the Small and in the
Large, held February 2–5, 1995, in Coral Gables, Florida

This volume was taken from a series of conferences sponsored by Global Foundation, Inc.,
Coral Gables, Florida

ISBN 0-306-45189-1

© 1995 Plenum Press, New York
A Division of Plenum Publishing Corporation
233 Spring Street, New York, N. Y. 10013

10 9 8 7 6 5 4 3 2 1

All rights reserved

No part of this book may be reproduced, stored in a retrieval system, or transmitted in any form or by any
means, electronic, mechanical, photocopying, microfilming, recording, or otherwise, without written
permission from the Publisher

Printed in the United States of America

PREFACE

The Twenty-third Coral Gables conference on Unified Symmetry in the Small and in the Large was convened February 2-5, 1995. The shift of the traditional conference time from the last part of January was caused by the 1995 Superbowl's choice of our preferred date for their game. The conference was dedicated to reminiscences of Julian Schwinger. The death of Eugene P. Wigner in the early part of January 1995 was observed with a deep sorrow during the conference. At about that time the news of Asim Barut's death made 1995 an inauspicious year for physicists. In the meantime physics at the frontiers marched on as it did before. There were no path-breaking discoveries, but hope and persistence were still there. In 1964 (the first Coral Gables conference) if we had asked a physicist to give us a sincere opinion on what is "hot" in physics we would have expected him or her to point out the narrow area of their own research. The answer to this question in 1995 is still the same as it would have been in 1964. The mindset is a human quality and even in physics the physicist can respond like a religious believer.

The diversity of topics in the 1995 conference, ranging from monopole condensation, CMBR temperatures, dark matter to gravity waves, the top quark, the Higgs boson, the neutrino mass, and quark structure, made it unique in that the solution of any of these topics should trigger a revolution in physics. Yes, a revolution is what is needed to focus once again, after the physicist's tired atomic era, public attention on physics and physicists. The editors do not know what that desired revolution could be but they would hope that for the sake of the good old days of the Coral Gables conferences that the commencement of that revolution in physics should take place in one of the newly reinstated Orbis Scientiae before the arrival of the twenty-first century.

This conference was in part supported by grants from National Science Foundation and U.S. Department of Energy. The conference also received support from Martin-Marietta Astronautics and the Northrop Grumman Aerospace Corporation. Finally, the Trustees and the Chairman of the Global Foundation wish to extend special thanks to Edward Bacinich of Alpha Omega Research Foundation for his generous support of this conference.

Editors
Coral Gables, Florida
May 19, 1995

CONTENTS

SECTION I - JULIAN SCHWINGER IN RETROSPECT

Recollections of Julian Schwinger 3
    M. Hamermesh

Schwinger in Coral Gables 7
    Behram N. Kursunoglu

**Nonequilibrium Problems in Quantum Field
Theory and Schwinger's Closed Time Path Formalism** 11
    Fred Cooper

SECTION II - DIRECTIONS IN PARTICLE PHYSICS AND COSMOLOGY

Second Phase of the General Theory of Relativity 29
    Behram N. Kursunoglu

**Exact Solutions for Confinement of Electric Charge
via Condensation of a Spectrum of Magnetic Charges** 33
    Behram N. Kursunoglu

**Calculation of Cosmological Parameters and Their
Approximations in the Standard Big Bang Model** 53
    Ralph A. Alpher and Robert Herman

Estimating the Energy Spectrum of Cold Dark Matter Signal 87
    Yun Wang

**Moment and Wavelet Analysis of Correlations in Multihadron
and Galaxy Distributions** 97
    Peter Caruthers

Dilaton-Driven Inflation in String Cosmology 111
    Ram Brustein

Gravity Driven Inflation 119
    Janna J. Levin

SECTION III - CURRENT STATUS OF GRAVITY WAVE DETECTION

The Search for Gravitation Waves 125
    Barry C. Barish

**Reducing Thermal Noise in Interferometric Detectors
of Gravitational Waves** 133
    Peter R. Saulson

SECTION IV - NEUTRINOS AND MUONS

A Search for $\nu_\mu \to \nu_e$ Oscillation at LAMPF 139
   Jeremy Margulies

Neutrino Reactions in Nuclei in the Large and in the Small 151
   S. L. Mintz and M. Poukaviani

Physics Interest in $\mu^+ \mu^-$ Colliders 165
   V. Barger

SECTION V - STRINGS AND SUPERSTRINGS

Spin Field Vertices and Gauge Symmetry 175
   L. Dolan

Dynamical Supersymmetry Breaking:
Some Recent Developments 187
   Michael Dine

Identification as Black Holes of All Massive Superstring States 197
   Paul H. Frampton

Effects of the Top Landau Pole on Electroweak Physics in
SUSY Unification 203
   Pran Nath and R. Arnowitt

SECTION VI - PROGRESS IN SOME NEW AND OLD IDEAS

Integrals of Motion for the Sine-Gordon Model with
Boundary at the Free Fermion Point 213
   Luca Mezincescu and Rafael I. Nepomeceche

Reflection Matrices and Polymers at a Surface 221
   Murray T. Batchelor and C. Ming Yung

The Classical Space-Time from the Chern Simons Gauge
Theory of Gravity 229
   Freydoon Mansouri

Theorems on Estimating Perturbative Coefficients in Quantum
Field Theory and Statistical Physics 241
   Mark A. Samuel and Stephen D. Druger

SECTION VII - SPIN PHYSICS AT HIGH ENERGY

Highlights of the Spin '94 Meeting 267
   A. D. Krisch

Accelerating Polarized Beams at the AGS 287
   Thomas Roser

Polarized Photon Beams at Fermilab 293
   R. A. Phelps

Index 301

SECTION I

JULIAN SCHWINGER IN RETROSPECT

RECOLLECTIONS OF JULIAN SCHWINGER

M. Hamermesh

University of Minnesota
School of Physics
116 Church Street, S. E.
Minneapolis, MN 55455

When I entered CCNY in 1932, I majored in mathematics, but I also took many physics courses. At City College I met Julian Schwinger, who had graduated from Townsend Harris High School in 1934. There he had Irving Lowen as his physics teacher . Even then Julian was known as a prodigy who read Dirac and articles in the physics journals. His first published paper was jointly with Otto Halpern in 1935 on ""Polarization of Electrons by Double Scattering."

Julian entered CCNY in 1934. Our acquaintance developed because we were in the same class for Modern Geometry in 1934 (5 ?). The text for this course in projective geometry was by Graustein and was taught by Professor Reynolds, who was head of the Math Department. The book was not very exciting, and Reynolds was a somewhat foggy instructor. Julian paid little attention to the course, but had to appear occasionally to satisfy the attendance requirements. He was spending most of his time on research and reading in journals, and devoted minimal time to courses. At this time he also helped some of the instructors at City with their thesis research. So his time spent on all the other courses was also little, which led to problems in other departments. In the geometry class, Julian would ask me what the subject was for this day's class (having not looked at the text). Reynolds must have realized that Julian was not prepared, and would call on him to solve problems on the blackboard. Like most problems in projective geometry, these were fairly easy, but did require seeing a picture and using geometrical reasoning. But Julian believed in the Cartesian edict that analytic methods are the thing, and would struggle till the bell rang. Julian's life was difficult. It appeared likely that he would flunk out of CCNY.

Fortunately I.I. Rabi, who had taken an interest in Julian, arranged for him to transfer to Columbia, with the understanding that he would work simultaneously for the MS and PhD degrees. In 1936 I graduated from CCNY, (transferring from math to physics because that was the condition for my getting a job as tutor at CCNY). I entered graduate school at New York University. Irv Lowen was my teacher in a dynamics course, and later I had Halpern as my thesis adviser. By 1937 I was impatient with courses and looking for a chance to do serious research. I used to attend the joint NYU-Columbia theoretical physics seminar that met once a week. Among the regular attendees were Halpern, Gene Feenberg and Fermi. I would see Julian regularly at the seminars. I asked Julian about possibly working at Columbia where he was the theorist for a group of experimenters working in nuclear physics (V.Cohen, H.Goldsmith and J. Manley).

The system at Columbia was to use unpaid people in research, which made it a heaven for all the many people working at colleges in New York City (for example S.Millman and J.Zacharias). My first publication was a note about albedo for neutrons in a paper by Manley, Goldsmith and Schwinger .

Julian and I began to work together almost daily. After my teaching during the day, I would meet Julian at 4 PM after his waking around 2 PM. He had developed the habit of working only at night, partly because he wanted to avoid the various big shots and famous visitors who would

Unified Symmetry: In the Small and in the Large 2
Edited by B. N. Kursunoglu *et al.*, Plenum Press, New York, 1995

come and try to see and talk to him. I recall that we once stood for hours outside Pupin (the physics building at Columbia) under an overhang during a rain, so he could avoid George Placzek. Julian's group did their work on the 13'th floor of Pupin. He and I would talk about physics in between measurements (which consisted mainly of running with samples to be counted in proportional counters at the other end of the corridor). This was the period of my education by Julian. He would ask me questions and see that I needed instructuion. He would then proceed to teach me about the subject. I remember his teaching me about Bessel functions. This was all done with no notes . It ended with his going through most of the material in Watson's book. He was surprised that I knew nothing about group theory and proceeded to go through Wigner's book with me. I forgot most of this until, much later. when I needed to use it for my own research. This was a difficult time for me. After my dinner (his breakfast) we would go back to Pupin, and work until around midnight. We would then go to the Sandwich Shop nearby and have a snack. At about 1 or 2 AM I would take the subway home to East New York, while he would walk home. At times, when we were deep in calculations, we would work all night . I remember that we once went next morning to the theoretical seminar, and both fell asleep.

In 1937 a paper by Julian and Ed Teller was published entitled "Scattering of Neutrons by Ortho- and Parahydrogen." Much later the paper was reprinted in Julian's selected papers (1978), with Julian's comment, "because I, not my distinguished colleague, wrote it." Julian and I worked on improving and extending this paper. We did some brutal calculations on scattering by ortho- and paradeuterium.We began to write up the material for publication . Julian was to go in the fall to Berkeley to be Oppenheimer's assistant. We would sit every night for hours trying to get a suitable opening paragraph. Julian would read the few lines but nothing was satisfactory. I began to get thoroughly annoyed, and he left for the West leaving me with all the material. He was lucky to escape because toward the end I contemplated murdering him. In 1939 I received an urgent telegram from Julian. It appeared that Alvarez and Pitzer were doing measurements on ortho- and parahydrogen and threatening to have someone at Berkeley repeat our work. I sent all the material to Julian and shortly afterward letters by Julian and by A and P appeared in the Physical Review. I felt therefore that all our work on deuterium was for naught.

In 1943 Julian went to the MIT Radiation Lab. At the same time I came to Harvard from Stanford. Julian spent most of the war working on waveguide theory with Harold Levine and Nate Marcuvitz ,while I worked on operations analysis, antenna design and radar reflection. We began to see each other regularly. Julian had found a way of deriving our results for deuterium much more simply, and we finished the job quickly. The papers on D appeared in 1946 and on H in 1947. We also had many social contacts. Julian was courting Clarice Carroll, and they would meet at our apartment in Cambridge. At the end of the war I went to NYU as professor of physics.

At Harvard, I had worked very closely with John VanVleck. Since Harvard was considering offering a professorship to Julian, Van asked me whether I felt that Julian could meet classes at reasonable times. I tried to reassure him. Julian assumed his new post, and he and Clarice were married in Boston in 1947. My wife and I came from New York for the great event. I have a photo of Bernie Feld and me as ushers sporting top hats.

At NYU in 1947 I taught a course in nuclear physics, using notes written by Julian . I wrote notes extending his work, and he proposed that we sign a book contract. We did, but in 1948 I went to the Argonne Lab and couldn't continue. Then Herman Feshbach signed a similar contract, but this too was dropped. The final outcome of all this effort was the series of books by Herman many years later. In the course of about ten years of close contact, I learned something of Julian's work habits. He had file drawers of as yet unpublished papers, which would often appear long after they were written,e.g.,"Neutron Polarization from Resonance Scattering by He" (69,681 (1946)) and an "unpublished" paper for the New York office of the AEC , NYO-3071, entitled " On Angular Momentum," written in 1952, which appeared much later in a collection of papers edited by Biedenharn and Van Damm. I also remember that he would become interested in problems, work extensively on them, and then lose interest and drop the subject. In particular I recall some interesting work on absorption of sound in gases.

In 1972 Julian left Harvard and went to UCLA, and we saw each other only on rare occasions. Julian was a very private person, who was immersed in his work in physics. There was always only a handful of people with whom he was comfortable. I know that at UCLA he would lunch with Seth Putterman and Bob Finkelstein. He and Clarice lived a quiet life with few close friends. Still I always found it fun to be with them. Julian was his own man. He never wanted to be under someone's direction. He often chose methods and problems that he liked, even though most physicists weren't interested in them. A good example is his work on Source Theory and also his work in later years on attempting to find an explanation for claims about cold fusion.

He often chose problems like the Fermi model and made them a thing of beauty. In 1994 he was working on a theory of sonoluminescence, just before his death.

Julian had several hobbies. He loved fine cars, starting with big Buicks and later he came to a Coral Gables Conference sporting a new Lamborghini. He was also a tennis enthusiast. Once he was invited to give the Erikson lecture at Minnesota. He called and asked me to find him a tennis partner during his visit. I did, but unfortunately the partner was far above Julian's class, to Julian's obvious displeasure.;

I remember with pleasure my long association with Julian. I miss him -- a great theoretical physicist, a pleasant companion and a good friend.

SCHWINGER IN CORAL GABLES

Behram N. Kursunoglu
Global Foundation, Inc., Coral Gables, Florida

Julian Schwinger, as his quasi-dated letter below recorded, has come to Coral Gables for the first time in January of 1964 to attend to the inauguration of the first Coral Gables Conference on "Symmetry Principles at High Energy" held during January 30-31. I was told by those who knew him well, that I must be congratulated for succeeding to talk to Schwinger over the phone when I invited him to our first conference. This telephone encontre was arranged by Clarice Schwinger.

**INSTITUT
DES HAUTES ÉTUDES SCIENTIFIQUES**
BURES-SUR-YVETTE (S.-&-O.)
928-48-83

Cambridge Jan 6

Dear Prof. Kursunoglu:

This is a long delayed affirmative response to your kind invitation to attend the conference of Symmetry Principles at High Energy. My wife and I plan to arrive in Miami somewhat earlier and accordingly we shall try to find our own accommodations. We are both looking forward to our first visit to Miami.

Very sincerely yours,

Julian Schwinger

P.S. The address is still Harvard University.

His presentation in the conference was titled "A Ninth Baryon". His motivation can best be explained by the following exert from his speech as it appeared in the proceedings of our conference "Symmetry Principles at High Energy", January 30-31, 1964, W.H. Freeman and Co., San Francisco and London.

A NINTH BARYON?

Julian Schwinger

Harvard University, Cambridge, Massachusetts

Let me begin by urging you to ignore the deliberately provocative title of this lecture and to appreciate that what I want to talk about is not some modification of existing classification schemes, but rather a fundamental field theory of matter, i.e., everything.
First let me tell you what I mean by a field theory. I have in mind the idea that fields are the fundamental dynamical variables in terms of which general dynamical statements must be made. Furthermore, fields are not directly observable in any immediate sense as particles, but rather that the function of a dynamical theory is to interpret the structure of nature in terms of these fundamental field variables. As a consequence of the dynamics, what emerges are various excitations which are stable or quasi-stable, namely, the physical particles. The function of dynamics then is to establish the contact between the underlying substratum of dynamical variables, the fields, and what is actually observed, the particles. Stated in this way, it would seem that we have an impossible task, because it is no secret that theoretical physics lacks the tools to establish this contact.
It is here, however, that I would like to point to the fact that the leptons may represent the essential clue to breaking this barrier. Namely, the leptons, with their essentially weak (as far as we know) interactions, are such that the dynamical chain between fields and particles may be particularly short. It may indeed be possible (and I will provisionally assume that it is possible) to establish a one-to-one correspondence between the particles that we actually see and the corresponding fundamental fields. It would appear that there is nothing particularly subtle in that correspondence, so that quantum electrodynamics, at least, has been on the right track. In other words, we have a clue for the corresponding fundamental fields that are to be associated at least with <u>that</u> aspect of nature.
Now that by itself will not get us very far; but what I want to do also is to appeal to a symmetry, to essentially explore the unobservable substratum of the strong interactions of the nucleonic world with the aid of an

A Nassettin Hoja (a Turkish folklore hero who lived in the middle Anatolis in 13th century) story was chosen by the editors to precede Schwinger's paper, relates that

> One night Hoja had a dream that a man gave him eight gold liras and that he began to haggle by asking him for nine.
>
> At this point he woke up, and finding that there was nothing in his hand, shut his eyes tight, stretched out his hand, saying, "Very well, bring them here. I'll take eight.

Schwinger has also served, along with J. Robert Oppenhimer, as a member for the Scientific Council of the Center for Theoretical Studies. In 1965 after receiving his Nobel prize in December of 1965 he arrived in Coral Gables with his new Lamborghini with a license plate bearing the famous number 137 which, of course, behooved one of the makers of the new quantum electrodynamics. His annual month-long visits with Clarice to South Florida continued until 1977 when they had already moved to Los Angeles. During those years Coral Gables was one of the Mecca's of sciences. The visitors to the Center for Theoretical Studies of the University of Miami included P.A.M. Dirac, Lars Onsager (then a distinguished professor at CTS), Francis Crick, John Eccles, Munfred Eigen, Willis Lamb, Edward Teller, Robert Hofstadter, Murray Gellmann, Lord C.P. Snow, Eugene P. Wigner, Abraham Pais, Abdus Salam, John Bardeen, Nikolai Basov, Hans A. Bethe, Nikolai Bogolubov, Norman Ramsey, Robert Mullikan, Wendell Stanley, and many other distinguished visitors.

Julian Schwinger liked the combination of South Florida's pleasant winters and the masters, like himself, in the sciences. He came to the campus in late afternoon to attend the daily tea meetings in the CTS Fifteen Chair Conference room and give his lecture on sourcery. At one occasion some of the CTS postdoctoral fellows were raising strong objections to Schwinger's most cherished new theory. He was quite annoyed and bitterly complained about their attitudes. We were all very surprised at his serious concern with the postdoctorals viewpoints against his theory. In the evening we drove Schwinger and Clarice to Dirac's Apartment to dine with them and enjoy Margaret Dirac's Hungarian chicken.

NONEQUILIBRIUM PROBLEMS IN QUANTUM FIELD THEORY AND SCHWINGER'S CLOSED TIME PATH FORMALISM

Fred Cooper,

Theoretical Division, Los Alamos National Laboratory
Los Alamos, NM 87545

INTRODUCTION

We review the closed time path formalism of Schwinger using a path integral approach. We apply this formalism to the study of pair production from strong external fields as well as the time evolution of a nonequilibrium chiral phase transition.

In 1961 in his classic paper "Brownian Motion of a Quantum Particle"[1], Schwinger solved the formidable technical problem of how to use the action principle to study initial value problems. Previously, the action principle was formulated to study only transition matrix elements from an earlier time to a later time. The elegant solution of this problem was the invention of the closed time path (CTP) formalism. This formalism was first used to study field theory problems by Mahanthappa and Bakshi[1].

With the advent of supercomputers, it has now become possible to use this formalism to numerically solve important field theory questions which are presented as initial value problems. Two of these problems we shall review here. They are

1. The time evolution of the quark-gluon plasma[2].

2. Dynamical evolution of a non-equilibrium chiral phase transition following a relativistic heavy ion collision[3].

The basic idea of the CTP formalism is to take a diagonal matrix element of the system at a given time $t = 0$ and insert a complete set of states into this matrix element at a different (later) time t'. In this way one can express the original fixed time matrix element as a product of transition matrix elements from 0 to t' and the time reversed (complex conjugate) matrix element from t' to 0. Since each term in this product is a transition matrix element of the usual or time reversed kind, standard path integral representations for each may be introduced. If the same external source operates in

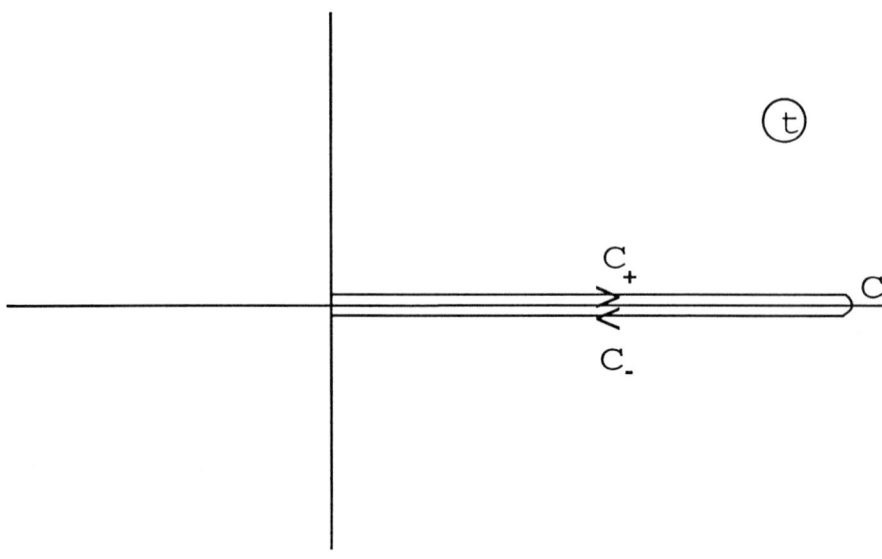

Fig. 1

the forward evolution as the backward one, then the two matrix elements are precisely complex conjugates of each other, all dependence on the source drops out and nothing has been gained. However, if the forward time evolution takes place in the presence of one source J_+ but the reversed time evolution takes place in the presence of a *different* source J_-, then the resulting functional is precisely the generating functional we seek.

Complex time contour $\mathcal{C}$ for the closed time path propagators.

Indeed

$$\begin{aligned} Z_{in}[J_+, J_-] &\equiv \int [\mathcal{D}\Psi]\langle in|\psi\rangle_{J_-} \langle \psi|in\rangle_{J_+} \\ &= \int [\mathcal{D}\Psi]\langle in|\mathcal{T}^* exp\left[-i\int_0^{t'} dt J_-(t)\phi(t)\right] |\Psi, t'\rangle \times \\ &\quad \langle \Psi, t'|\mathcal{T} exp\left[i\int_0^{t'} dt J_+(t)\phi(t)\right] |in\rangle \end{aligned} \quad (1)$$

so that, for example,

$$\left.\frac{\delta W_{in}[J_+, J_-]}{\delta J_+(t)}\right|_{J_+=J_-=0} = -\left.\frac{\delta W_{in}[J_+, J_-]}{\delta J_-(t)}\right|_{J_+=J_-=0} = \langle in|\phi(t)|in\rangle \quad (2)$$

is a true expectation value in the given time-independent Heisenberg state $|in\rangle$. Here $\phi(t) = \phi(x,t)$ and we are supressing the coordinate dependence and the integration over the spatial volume in what follows for notational simplicity.

Since the time ordering in eq. (1) is forward (denoted by $\mathcal{T}$) along the time path from 0 to t' in the second transition matrix element, but backward (denoted by $\mathcal{T}^*$) along the path from t' to 0 in the first matrix element, thus the name: *closed time path generating functional*. If we deform the backward and forward directed segments of the path slightly in opposite directions in the complex t plane, the symbol $\mathcal{T}_\mathcal{C}$ may be introduced for path ordering along the full closed time contour, $\mathcal{C}$ depicted in Fig.1.

This deformation of the path corresponds precisely to opposite $i\epsilon$ prescriptions along the forward and backward directed segments, which we shall denote by $C_\pm$ respectively in the following.

If we have an arbitrary initial density matrix ρ then we have instead:

$$Z[J_+, J_-, \rho] \equiv \text{Tr}\left\{\rho\left(\mathcal{T}^* \exp\left[-i\int_0^{t'} dt\, J_-(t)\phi(t)\right]\right)\left(\mathcal{T}\exp\left[i\int_0^{t'} dt\, J_+(t)\phi(t)\right]\right)\right\}$$

$$= \int [\mathcal{D}\varphi][\mathcal{D}\varphi'][\mathcal{D}\psi]\, \langle\varphi|\rho|\varphi'\rangle\, \langle\varphi'|\mathcal{T}^* exp\left[-i\int_0^{t'} dt\, J_-(t)\phi(t)\right]|\psi\rangle$$

$$\times \langle\psi|\mathcal{T} exp\left[i\int_0^{t'} dt\, J_+(t)\phi(t)\right]|\varphi\rangle . \tag{3}$$

Variations of this generating function will yield Green's functions in the state specified by the initial density matrix, i.e. expressions of the form,

$$\text{Tr}\{\rho\phi(t_1)\phi(t_2)\phi(t_3)...\}. \tag{4}$$

Introducing the path integral representation for each transition matrix element in eq. (3) results in the expression,

$$Z[J_+, J_-, \rho] = \int [\mathcal{D}\varphi][\mathcal{D}\varphi']\, \langle\varphi|\rho|\varphi'\rangle \int [\mathcal{D}\psi] \int_\varphi^\psi [\mathcal{D}\phi_+] \int_{\varphi'}^\psi [\mathcal{D}\phi_-] \times$$

$$\exp\left[i\int_0^\infty dt\, (\, L[\phi_+] - L[\phi_-] + J_+\phi_+ - J_-\phi_-)\right], \tag{5}$$

where L is the classical Lagrangian functional, and we have taken the arbitrary future time at which the time path closes $t' \to \infty$.

The double path integral over the fields ϕ_+ and ϕ_- in (5) suggests that we introduce a two component contravariant vector of field variables by

$$\phi^a = \begin{pmatrix} \phi_+ \\ \phi_- \end{pmatrix} ; \quad a = 1, 2 \tag{6}$$

with a corresponding two component source vector,

$$J^a = \begin{pmatrix} J_+ \\ J_- \end{pmatrix} ; \quad a = 1, 2 . \tag{7}$$

Because of the minus signs in the exponent of (5), it is necessary to raise and lower indices in this vector space with a 2×2 matrix with indefinite signature, namely

$$c_{ab} = diag\,(+1, -1) = c^{ab} \tag{8}$$

so that, for example

$$J^a c_{ab} \Phi^b = J_+\phi_+ - J_-\phi_- . \tag{9}$$

These definitions imply that the correlation functions of the theory will exhibit a matrix structure in the 2×2 space. For instance, the matrix of connected two point functions in the CTP space is

$$G^{ab}(t, t') = \frac{\delta^2 W}{\delta J_a(t)\delta J_b(t')}\bigg|_{J=0} . \tag{10}$$

Explicitly, the components of this 2×2 matrix are

$$\begin{aligned}
G^{21}(t,t') &\equiv G_>(t,t') = i\text{Tr}\{\rho\,\Phi(t)\overline{\Phi}(t')\}_{con} \, , \\
G^{12}(t,t') &\equiv G_<(t,t') = i\text{Tr}\{\rho\,\overline{\Phi}(t')\Phi(t)\}_{con} \\
G^{11}(t,t') &= i\text{Tr}\{\rho\,\mathcal{T}[\Phi(t)\overline{\Phi}(t')]\}_{con} = \theta(t,t')G_>(t,t') + \theta(t',t)G_<(t,t') \\
G^{22}(t,t') &= i\text{Tr}\{\rho\,\mathcal{T}^*[\Phi(t)\overline{\Phi}(t')]\}_{con} = \theta(t',t)G_>(t,t') + \theta(t,t')G_<(t,t')
\end{aligned} \quad (11)$$

Notice that
$$G^{11}(t,t) = G^{22}(t,t) \quad (12)$$

with the usual convention that
$$\theta(t,t) = \tfrac{1}{2} \, . \quad (13)$$

The 2×2 matrix notation originated with Schwinger's classic article in 1960[1].

In what follows we will use an alternative generating functional[4] using the Complex Path Ordered Form:

$$\int [\mathcal{D}\psi]\langle\varphi'|\mathcal{T}^*\exp\left[-i\int_0^\infty dt\,J_-(t)\phi(t)\right]|\psi\rangle\langle\psi|\mathcal{T}\exp\left[i\int_0^\infty dt\,J_+(t)\phi(t)\right]|\varphi\rangle$$
$$= \langle\varphi'|\mathcal{T}_C\exp\left[i\int_C dt\,J(t)\phi(t)\right]|\varphi\rangle \quad (14)$$

so that (3) may be rewritten more concisely in the CTP complex path ordered form,

$$\begin{aligned}
Z_C[J,\rho] &= \text{Tr}\left\{\rho\left(\mathcal{T}_C\exp\left[i\int_C dtJ(t)\phi(t)\right]\right)\right\} \\
&= \int [\mathcal{D}\varphi^1]\int [\mathcal{D}\varphi^2]\,\langle\varphi^1|\rho|\varphi^2\rangle \int_{\varphi^1}^{\varphi^2}[\mathcal{D}\phi]\exp\left[i\int_C dt\,(L[\phi]+J\phi)\right] \, . \quad (15)
\end{aligned}$$

This is identical in structure to the usual expression for the generating functional in the more familiar in-out formalism. Only difference — path ordering according to the complex time contour $\mathcal{C}$ replacing the ordinary time ordering prescription along only $\mathcal{C}_+$. For example, the propagator function becomes

$$\begin{aligned}
G(t,t') &= \theta_C(t,t')G_>(t,t') + \theta_C(t',t)G_<(t,t') \\
&\equiv \theta_C(t,t')G^{21}(t,t') + \theta_C(t',t)G^{12}(t,t')
\end{aligned} \quad (16)$$

where θ_C is the CTP complex contour ordered theta function defined by

$$\theta_C(t,t') \equiv \begin{cases} \theta(t,t') & \text{for } t,t' \text{ both on } \mathcal{C}_+ \\ \theta(t',t) & \text{for } t,t' \text{ both on } \mathcal{C}_- \\ 1 & \text{for } t \text{ on } \mathcal{C}_- \, , \, t' \text{ on } \mathcal{C}_+ \\ 0 & \text{for } t \text{ on } \mathcal{C}_+ \, , \, t' \text{ on } \mathcal{C}_- \end{cases} \quad (17)$$

With this definition of $G(t,t')$ on the closed time contour, the Feynman rules are the ordinary ones, and matrix indices are not required. In integrating over the second half of the contour $\mathcal{C}_-$ we have only to remember to multiply by an overall negative sign to take account of the opposite direction of integration, according to the rule,

$$\int_C dt = \int_{0\,\mathcal{C}_+}^\infty dt - \int_{0\,\mathcal{C}_-}^\infty dt \, . \quad (18)$$

A second simplification is possible in the form of the generating functional of (15), if we recognize that it is always possible to express the matrix elements of the density matrix as an exponential of a polynomial in the fields[5]

$$\langle \varphi^1|\rho|\varphi^2 \rangle = \exp\left[R + R_a(t_0)\varphi^a(t_0) + R_{ab}(t_0)\varphi^a(t_0)\varphi^b(t_0) + \ldots\right]. \quad (19)$$

Since any density matrix can be expressed in this form, there is no loss of generality involved in expressing ρ as an exponential. If we add this exponent to that of the action in (15), and integrate over the two endpoints of the closed time path φ^1 and φ^2, then the only effect of the non-trivial density matrix ρ is to introduce source terms into the path integral for $Z_C[J,\rho]$ with support *only* at the endpoints. This means that the density matrix can only influence the boundary conditions on the path integral at $t = 0$, where the various coefficient functions R_a, R_{ab}, etc. have the simple interpretations of initial conditions on the one-point (mean field), two-point (propagator), functions *etc.* It is clear that the equations of motion for $t \neq 0$ are not influenced by the presence of these terms at $t_0 = 0$. In the special case that the initial density matrix describes a thermal state, $\rho_\beta = \exp\{-\beta H\}$ then the trace over ρ_β may be represented as an additional functional integration over fields along the purely imaginary contour from $t = -i\beta$ to $t = 0$ traversed before $\mathcal{C}_-$ in Fig. 1. In this way the Feynman rules for real time thermal Green's functions are obtained[6]. Since we consider general nonequilibrium initial conditions here we have only the general expression for the initial ρ above and no contour along the negative imaginary axis in Fig. 1.

To summarize, we may take over all the results of the usual scattering theory generating functionals, effective actions, and equations of motion provided only that we

1. substitute the CTP path ordered Green's function(s) (16) for the ordinary Feynman propagators in internal lines;

2. integrate over the full closed time contour, $\mathcal{C}$, according to (18); and

3. satisfy the conditions at $t = 0$ corresponding to the initial density matrix ρ.

Closed time path contour and causality

Rules for evaluating the time integrals using the closed time path [CTP] contour shown in Fig. (1).

The integration path is given by

$$\int_\mathcal{C} dt = \int_{0:c_+}^\infty dt - \int_{0:c_-}^\infty dt . \quad (20)$$

The *causal* Green's functions are given by functions of the form,

$$A(t,t') = \Theta_c(t,t')A_>(t,t') + \Theta_c(t',t)A_<(t,t') , \quad (21)$$

where $\Theta_c(t,t')$ is defined in eq.(17) These causal Green's functions are symmetric.

$$A_>(t,t') = A_<(t',t)$$

To prove causality of any graph we need two lemmas.

1. Lemma 1- a loop of two causal functions,such as self energy graph, is another causal function.

$$B(t,t') = \Theta_c(t,t')B_>(t,t') + \Theta_c(t',t)B_<(t,t')$$
$$C(t,t') = \Theta_c(t,t')C_>(t,t') + \Theta_c(t',t)C_<(t,t'), \quad (22)$$

the self energy graph

$$A(t,t') = iB(t,t')C(t,t'),$$

then

$$A(t,t') = \Theta_c(t,t')A_>(t,t') + \Theta_c(t',t)A_<(t,t') \quad (23)$$

where

$$A_{>,<}(t,t') = iB_{>,<}(t,t')C_{>,<}(t,t') \quad (24)$$

2. Lemma 2 -Matrix product of two causal functions is causal

$$A(t_1,t_3) = \int_c dt_2\, B(t_1,t_2)C(t_2,t_3), \quad (25)$$

we find then

$$A(t,t') = \Theta_c(t,t')A_>(t,t') + \Theta_c(t',t)A_<(t,t'), \quad (26)$$

where

$$A_{\gtrless}(t_1,t_3) = -\int_0^{t_3} dt_2\, B_{\gtrless}(t_1,t_2)[C_>(t_2,t_3) - C_<(t_2,t_3)]$$
$$+ \int_0^{t_1} dt_2\, [B_>(t_1,t_2) - B_<(t_1,t_2)]C_{\gtrless}(t_2,t_3). \quad (27)$$

Now consider the product of three causal functions:

$$A(t_1,t_4) = \int_c dt_2 \int_c dt_3\, B(t_1,t_2)C(t_2,t_3)D(t_3,t_4). \quad (28)$$

We can work this case out by applying the second lemma from left to right. That is, we can let

$$E(t_1,t_3) = \int_c dt_2\, B(t_1,t_2)C(t_2,t_3). \quad (29)$$

Then $E(t_1,t_3)$ is causal. We are then left with:

$$A(t_1,t_4) = \int_c dt_3\, E(t_1,t_3)D(t_3,t_4), \quad (30)$$

and so A is also causal.

After doing the integrals sequentially one is left with

$$f(t) = \int_c dt_1\, F(t,t_1) = \int_0^t [F_>(t,t_1) - F_<(t,t_1)], \quad (31)$$

which explicitly displays the causality (dependence only on earlier times).

To see how these rules work in practice, consider the terms contributing to order $1/N$ to the induced current determining the backreaction on an initially strong electric field[4]. The current is just $\text{tr}\{\gamma^\mu \tilde{G}\}$. Where $\tilde{G}$ is the full fermion propagator to order $1/N$.

Using the above lemmas we obtain that the Maxwell eqs. of motion take the form,

$$\partial_\nu F^{\mu\nu}(x) = \langle j^\mu(x)\rangle = -\frac{ie^2}{2}\text{tr}\left\{\gamma^\mu[G_>(x,x) + G_<(x,x)]\right\}$$
$$+ \frac{2e^2}{N}\int_0^t dt_1 d^3\vec{x}_1 \int_0^{t_1} dt_2 d^3\vec{x}_2 \mathcal{I}m\,\text{tr}\left\{\gamma^\mu\left[G_>(x,x_1) - G_<(x,x_1)\right]\times\right.$$
$$\left.[\Sigma_<(x_1,x_2)G_>(x_2,x) - \Sigma_>(x_1,x_2)G_<(x_2,x)]\right\}. \quad (32)$$

Here:

$$\Sigma_<(x_1,x_2) = i\gamma^\mu G_<(x_1,x_2)\gamma^\nu D_{\nu\mu <}(x_1,x_2). \quad (33)$$

This immediately displays the causality of the result: that is only previous times contribute to the space-time integrations.

TIME EVOLUTION OF THE QUARK GLUON PLASMA

Our model for the production of the quark-gluon plasma begins with the creation of a flux tube containing a strong color electric field. We assume the kinematics of ultrarelativistic high energy collisions results in a boost invariant dynamics in the longitudinal (z) direction where all expectation values are functions of the proper time $\tau = \sqrt{t^2 - z^2}$. We therefore introduce the light cone variables τ and η, which will be identified later with fluid proper time and rapidity. These coordinates are defined in terms of the ordinary lab-frame Minkowski time t and coordinate along the beam direction z by

$$z = \tau\sinh\eta \quad , \quad t = \tau\cosh\eta. \quad (34)$$

The Minkowski line element in these coordinates has the form

$$ds^2 = -d\tau^2 + dx^2 + dy^2 + \tau^2 d\eta^2. \quad (35)$$

Hence the metric tensor is given by

$$g_{\mu\nu} = \text{diag}(-1,1,1,\tau^2). \quad (36)$$

For simplicity, here we discuss pair production from an abelian Electric Field and the subsequent quantum back-reaction on the Electric Field. The physics of the problem can be understood for constant electric fields as a simple tunneling process. If the electric field can produce work of at least twice the rest mass of the pair in one compton wavelength, then the vacuum is unstable to tunnelling. This condition is:

$$\frac{eE\hbar}{mc} \geq 2mc^2. \quad (37)$$

The problem of pair production from a constant Electric field (ignoring the back reaction) was studied by J. Schwinger in 1951 [7]. The WKB argument is as follows: One imagines an electron bound by a potential well of order $|V_0| \approx 2m$ and submitted to an additional electric potential eEx. The ionization probability is proportional to the WKB barrier penetration factor:

$$\exp[-2\int_0^{V_0/e} dx\{2m(V_0 - |eE|x)\}^{1/2}] = \exp(-\frac{4}{3}m^2/|eE|) \quad (38)$$

In his classic paper Schwinger was able to analytically solve for the effective Action in a constant background electric field and determine an exact pair production rate:

$$w = [\alpha E^2/(2\pi^2)] \sum_{n=1}^{\infty} \frac{(-1)^{n+1}}{n^2} \exp(-n\pi m^2/|eE|). \tag{39}$$

By assuming this rate could be used when the Electric field was slowly varying in time, the first back reaction calculations were attempted using semi classical transport methods. Here we use the CTP formalism and perform the field theory calculation. The lagrangian density for QED in curvilinear coordinates gives rise to the action

$$S = \int d^{d+1}x \, (\det V) \left[\frac{-i}{2} \bar{\Psi} \tilde{\gamma}^\mu \nabla_\mu \Psi + \frac{i}{2}(\nabla_\mu^\dagger \bar{\Psi}) \tilde{\gamma}^\mu \Psi - im\bar{\Psi}\Psi - \frac{1}{4}F_{\mu\nu}F^{\mu\nu} \right], \tag{40}$$

where[8]

$$\nabla_\mu \Psi \equiv (\partial_\mu + \Gamma_\mu - ieA_\mu)\Psi \tag{41}$$

From the action (40) we obtain the Heisenberg field equation for the fermions,

$$(\tilde{\gamma}^\mu \nabla_\mu + m)\Psi = 0, \tag{42}$$

which takes the form

$$\left[\gamma^0 \left(\partial_\tau + \frac{1}{2\tau} \right) + \gamma_\perp \cdot \partial_\perp + \frac{\gamma^3}{\tau}(\partial_\eta - ieA_\eta) + m \right] \Psi = 0, \tag{43}$$

Variation of the action with respect to A_ν yields the Maxwell equations: If the electric field is in the z direction and a function of τ only, we find that the only nontrivial Maxwell equation is

$$\frac{1}{\tau}\frac{dE(\tau)}{d\tau} = \frac{e}{2}\langle [\bar{\Psi}, \tilde{\gamma}^\eta \Psi] \rangle = \frac{e}{2\tau}\langle [\Psi^\dagger, \gamma^0 \gamma^3 \Psi] \rangle. \tag{44}$$

We expand the fermion field in terms of Fourier modes at fixed proper time τ,

$$\Psi(x) = \int [d\mathbf{k}] \sum_s [b_s(\mathbf{k})\psi^+_{\mathbf{k}s}(\tau)e^{ik\eta}e^{i\mathbf{P}\cdot\mathbf{x}} + d_s^\dagger(-\mathbf{k})\psi^-_{-\mathbf{k}s}(\tau)e^{-ik\eta}e^{-i\mathbf{P}\cdot\mathbf{x}}]. \tag{45}$$

The $\psi^\pm_{\mathbf{k}s}$ then obey

$$\left[\gamma^0 \left(\frac{d}{d\tau} + \frac{1}{2\tau} \right) + i\gamma_\perp \cdot \mathbf{k}_\perp + i\gamma^3 \pi_\eta + m \right] \psi^\pm_{\mathbf{k}s}(\tau) = 0, \tag{46}$$

We square the Dirac equation by introducing

$$\psi^\pm_{\mathbf{k}s} = \left[-\gamma^0 \left(\frac{d}{d\tau} + \frac{1}{2\tau} \right) - i\gamma_\perp \cdot \mathbf{k}_\perp - i\gamma^3 \pi_\eta + m \right] \chi_s \frac{f^\pm_{\mathbf{k}s}}{\sqrt{\tau}}. \tag{47}$$

The spinors χ_s are chosen to be eigenspinors of $\gamma^0 \gamma^3$,

$$\gamma^0 \gamma^3 \chi_s = \lambda_s \chi_s \tag{48}$$

with $\lambda_s = 1$ for $s = 1, 2$ and $\lambda_s = -1$ for $s = 3, 4$. They are normalized,

$$\chi_r^\dagger \chi_s = 2\delta_{rs}. \tag{49}$$

The sets $s = 1, 2$ and $s = 3, 4$ are two different complete sets of linearly independent solutions of the Dirac equation Inserting (47) into the Dirac equation (46) we obtain the quadratic mode equation

$$\left(\frac{d^2}{d\tau^2} + \omega_{\mathbf{k}}^2 - i\lambda_s \dot{\pi}_\eta\right) f_{\mathbf{k}s}^\pm(\tau) = 0, \qquad (50)$$

where now

$$\omega_{\mathbf{k}}^2 = \pi_\eta^2 + \mathbf{k}_\perp^2 + m^2. \qquad (51)$$

We obtain

$$\frac{1}{\tau}\frac{dE(\tau)}{d\tau} = -\frac{2e}{\tau^2}\sum_{s=1}^{4}\int [d\mathbf{k}](\mathbf{k}_\perp^2 + m^2)\lambda_s |f_{\mathbf{k}s}^+|^2, \qquad (52)$$

We would like to compare the field theory calculation with a semiclassical transport approach which uses as a source of particle production Schwinger's production rate for time independent fields.

Assuming boost-invariant initial conditions for f, invariance of the Boltzmann-Vlasov assures that the distribution function is a function only of the boost invariant variables $(\tau, \eta - y)$ or (τ, p_η). The kinetic equation reduces to

$$\frac{\partial f}{\partial \tau} + eF_{\eta\tau}(\tau)\frac{\partial f}{\partial p_\eta} = \pm[1 \pm 2f(\mathbf{p},\tau)]e\tau|E(\tau)|$$
$$\times \ln\left[1 \pm \exp\left(-\frac{\pi(m^2 + \mathbf{p}_\perp^2)}{e|E(\tau)|}\right)\right]\delta(p_\eta). \qquad (53)$$

Turning now to the Maxwell equation, we have that

$$-\tau\frac{dE}{d\tau} = j_\eta = j_\eta^{cond} + j_\eta^{pol}, \qquad (54)$$

where j^{cond} is the conduction current and j_μ^{pol} is the polarization current due to pair creation[9].

Thus in (1+1) dimensions we have

$$j_\eta^{cond} = 2e\int\frac{dp_\eta}{2\pi\tau p_\tau}p_\eta f(p_\eta, \tau)$$
$$j_\eta^{pol} = \frac{2}{F^{\tau\eta}}\int\frac{dp_\eta}{2\pi\tau p_\tau}p^\tau\frac{Df}{D\tau}$$
$$= \pm[1 \pm 2f(p_\eta = 0, \tau)]\frac{me\tau}{\pi}\text{sign}[E(\tau)]\ln\left[1 \pm \exp\left(-\frac{\pi m^2}{|eE(\tau)|}\right)\right]. \qquad (55)$$

In figure two we compare the results of a numerically solving the back reaction equations in the field theory with the semiclassical transport approach. We find that for this approximation (no rescattering of quarks), the semiclassical method does reasonable well when compared to a coarse grained in time and momentum field theory calculation.

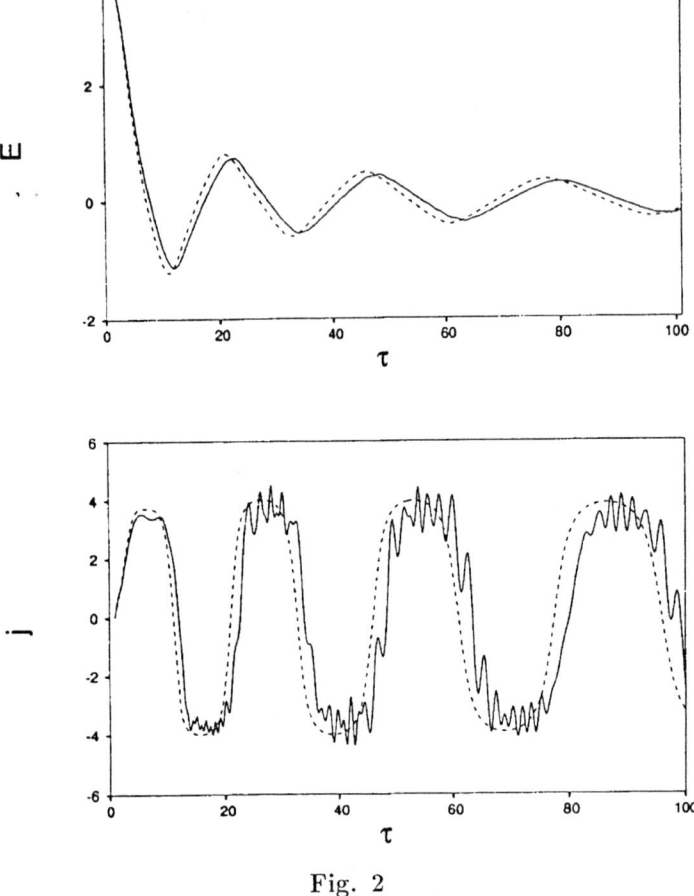

Fig. 2

Proper time evolution of the electric field $E(\tau)$ and the fermion current $j_\eta(\tau)$ for initial conditions $E(\tau = 1) = 4.0$. The field theory calculation is compared to the semiclassical transport approach with a Schwinger source term.

DYNAMICAL EVOLUTION OF A NON-EQUILIBRIUM CHIRAL PHASE TRANSITION

Recently there has been speculation that disoriented chiral condensates (DCC's) can be formed following a heavy ion collision, and these condensates, formed during a quenched phase transition from the unbroken high temperature phase, could lead to events with a nonequilibrium distribution of charged to neutral pions. To see whether these ideas made sense we studied numerically the time evolution of pions produced following a heavy ion collision using the linear sigma model, starting from the unbroken phase. The quenching (if present) in this model is due to the expansion of the initial Lorentz contracted energy density by free expansion into vacuum. Starting from an approximate equilibrium configuration at an initial proper time τ in the disordered phase we study the transition to the ordered broken symmetry phase as the system expands and cools. We determined[3] the proper time evolution of the effective pion mass, the order parameter $<\sigma>$ as well as the pion two point correlation function. We studied the phase space of initial conditions that lead to instabilities (exponentially growing long wave length modes) which can lead to disoriented chiral condensates. The model we used to study the chiral phase transition is the σ model in the large N approximation. This model has the correct chiral properties and gives a reasonable description of low energy pion dynamics. The $O(4)$ σ model is described by the Lagrangian:

$$L = \{\frac{1}{2}\partial\Phi \cdot \partial\Phi - \frac{1}{4}\lambda(\Phi \cdot \Phi - v^2)^2 + H\sigma\}. \tag{56}$$

The mesons form an $O(4)$ vector

$$\Phi = (\sigma, \pi_i)$$

Introducing the order parameter:

$$\chi = \lambda(\Phi \cdot \Phi - v^2).$$

we have the alternative Lagrangian:

$$L_2 = -\frac{1}{2}\phi_i(\Box + \chi)\phi_i + \frac{\chi^2}{4\lambda} + \frac{1}{2}\chi v^2 + H\sigma \tag{57}$$

Perform the Gaussian path integral over the Φ field. Evaluate the remaining χ integral at the stationary phase point. Legendre transforming:

$$\Gamma[\Phi, \chi] = \int d^4x[L_2(\Phi, \chi, H) + \frac{i}{2}N\text{tr ln}G_0^{-1}]$$

$$G_0^{-1}(x, y) = i[\Box + \chi(x)]\,\delta^4(x - y)$$

$$[\Box + \chi(x)]\pi_i = 0 \quad [\Box + \chi(x)]\sigma = H$$

$$\chi = -\lambda v^2 + \lambda(\sigma^2 + \pi \cdot \pi) + \lambda N G_0(x, x). \tag{58}$$

As in the quark-gluon plasma problem we consider the kinematics of an ultrarelativistic Heavy Ion Collision which possesses longitudinal Boost invariance as the center of mass energy goes to infinity. Energy densities become function of the proper time only. Natural coordinates are the proper time τ and the spatial rapidity η defined as

$$\tau \equiv (t^2 - x^2)^{1/2}, \quad \eta \equiv \frac{1}{2}\log(\frac{t - x}{t + x}). \tag{59}$$

We assume that the mean (expectation) values of the fields Φ and χ are functions of τ only (homogeneity in the constant τ hypersurface)

$$\tau^{-1}\partial_\tau \tau \partial_\tau \Phi_i(\tau) + \chi(\tau)\Phi_i(\tau) = H\delta_{i1}$$
$$\chi(\tau) = \lambda(-v^2 + \Phi_i^2(\tau) + N < \phi^2(x, \tau) >). \tag{60}$$

On the otherhand the fluctuation fields (which are quantum operators) are functions of both x and t and obey the sourceless equation:

$$\left(\tau^{-1}\partial_\tau \tau \partial_\tau - \tau^{-2}\partial_\eta^2 - \partial_\perp^2 + \chi(x)\right)\phi(x, \tau) = 0. \tag{61}$$

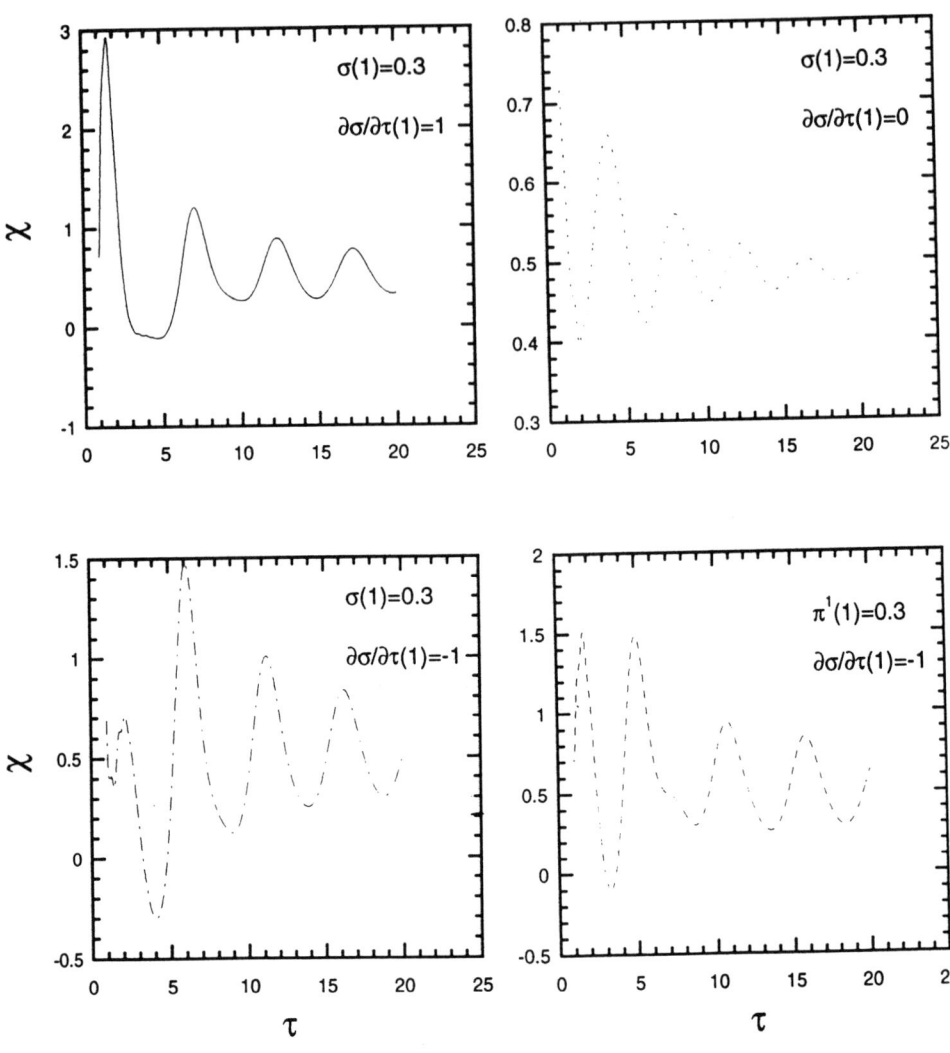

Fig. 3
Proper time evolution of the χ field for four different initial conditions with $f_\pi = 92.5 MeV$.

$$G_0(x, y; \tau) \equiv < \phi(x,\tau)\, \phi(y,\tau) > . \tag{62}$$

We expand this field in an orthonormal basis:

$$\phi(\eta, x_\perp, \tau) \equiv \frac{1}{\tau^{1/2}} \int [d^3 k](\exp(ikx) f_k(\tau)\, a_k + h.c.) \tag{63}$$

where $kx \equiv k_\eta \eta + \vec{k}_\perp \vec{x}_\perp$, $[d^3 k] \equiv dk_\eta d^2 k_\perp/(2\pi)^3$ and the mode functions $f_k(\tau)$ evolve according to (a dot here denotes the derivative with respect to the proper time τ):

$$\ddot{f}_k + (\frac{k_\eta^2}{\tau^2} + \vec{k}_\perp^2 + \chi(\tau) + \frac{1}{4\tau^2}) f_k = 0. \tag{64}$$

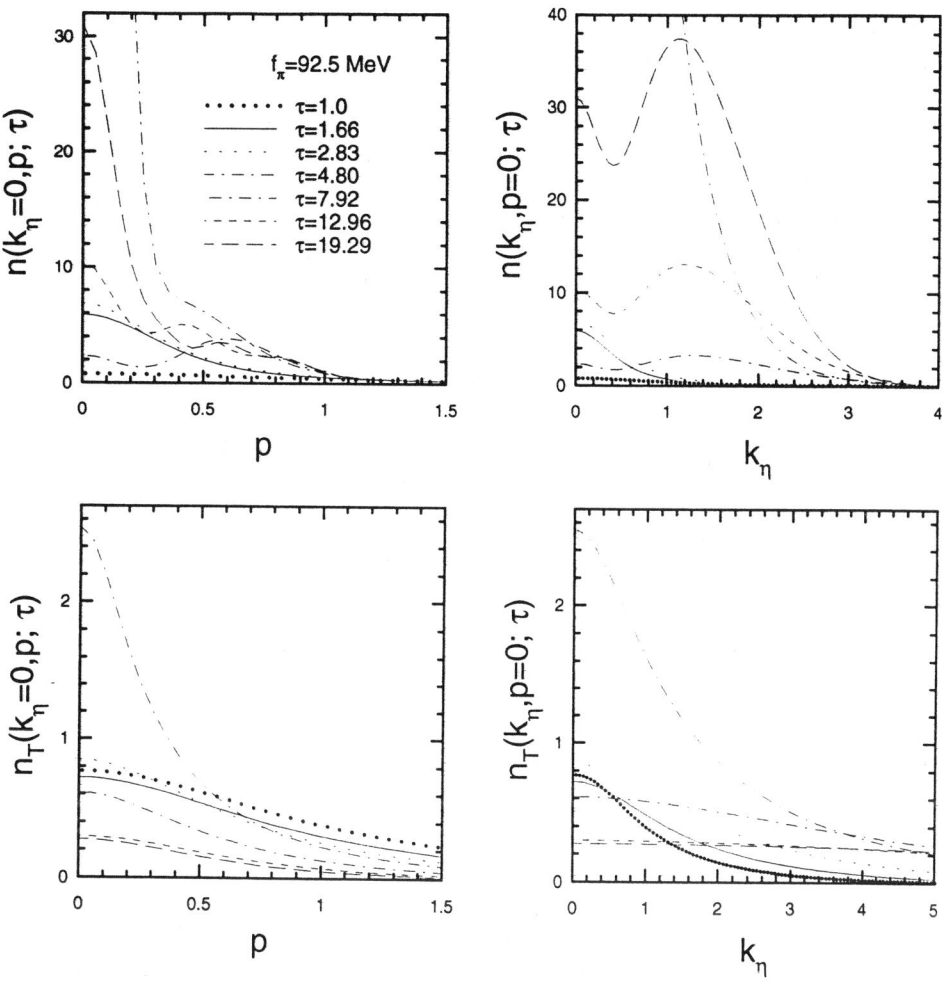

Fig. 4

Slices of $k_\eta = 0$ and $p \equiv |\mathbf{k}_\perp| = 0$ of the proper time evolution of the interpolating phase space particle number density $n(k_\eta, \mathbf{k}_\perp, \tau)$ for $\sigma(1) = \sigma_T$, $\pi^i(1) = 0$ and $\dot\sigma(1) = -1$ compared with the corresponding local thermal equilibrium densities $n_T(k_\eta, \mathbf{k}_\perp, \tau)$.

$$\chi(\tau) = \lambda\left(-v^2 + \Phi_i^2(\tau) + \frac{1}{\tau}N\int [d^3k]|f_k(\tau)|^2\,(1 + 2\,n_k)\right). \tag{65}$$

If we assume the initial density matrix is one of local thermal equilibrium then we have at $\tau = \tau_0$ (the surface of constant energy density and temperature T_0) that:

$$n_k = \frac{1}{e^{\beta_0 E_k^0} - 1}$$

where $\beta_0 = 1/T_0$ and $E_k^0 = \sqrt{k^2 + \chi(\tau_0)}$.

In choosing initial conditions we assumed that the initial value of χ was determined by the equilibrium gap equation for an initial temperature of $200 MeV$. The phase transition in this model occurs at a critical temperature of $160 MeV$. We chose initial σ and π_i expectation values consistent with the constraint

$$\vec{\pi}^2(\tau_0) + \sigma^2(\tau_0) = \sigma_T^2 \tag{66}$$

where σ_T is the equilibrium value of Φ at the initial temperature T, which we choose to be a temperature of $200 MeV$. We varied the value of the initial proper time derivative of the sigma field expectation value and found that there is a narrow range of initial values that lead to the growth of instabilities. Namely

$$.25 < |\dot\sigma| < 1.3 \tag{67}$$

Surprisingly when $|\dot\sigma| > 1.3$ instabilities no longer occur.

Figures 3-4 summarize the results of the numerical simulation for the evolution of the system (65)–(64). We display the auxiliary field χ in units of fm^{-2}, the classical fields Φ in units of fm^{-1} and the proper time in units of fm^{-1} ($1 fm^{-1} = 197 MeV$). In Fig. 3 the proper time evolutions of the auxiliary field χ field is presented for four different initial conditions.

We are interested in knowing how our results differ from the case where the system evolves in local thermal equilibrium which is described by two correlation lengths, the inverse of the effective pion mass associated with χ, and the inverse of the proper time evolving effective temperature $T(\tau) = T_0(\frac{\tau_0}{\tau})^{\frac{1}{3}}$ discussed earlier. We see from Fig. 4 that in the case that $\sigma(1) = \sigma_T$, $\pi^i(1) = 0$ and $\dot\sigma(1) = -1$, where maximum instability exists, complex structures are formed as contrasted to the local thermal equilibrium evolution. The interpolating phase space distribution $n(k_\eta, \mathbf{k}_\perp, \tau)$ obtained numerically, clearly exhibits a larger correlation length in the transverse direction than the equilibrium one and has correlation in rapidity of the order of 1-2 units of rapidity. We notice that in both directions there is structure which does not lend itself to a simple interpretation. On the other hand the local thermal equilibrium evolution is quite regular apart from the normalization of the distributions that are changing with time due to oscillation in the quantity $\chi(\tau)$ which is damped to its equilibrium value once the system expands sufficiently.

Acknowledgments

The work presented here was done in collaboration with Emil Mottola, Yuval Kluger, Salman Habib, Juan Pablo Paz, Ben Svetitsky, Judah Eisenberg, Paul Anderson and John Dawson. This work was supported by the Department of Energy.

References

[1] J. Schwinger, J. Math. Phys. **2**, 407 (1961);
K.T. Mahanthappa, Phys. Rev. **126**, 329 (1962);
P. M. Bakshi and K. T. Mahanthappa, J. Math. Phys. **4**, 1 (1963); **4**, 12 (1963);
L. V. Keldysh, Zh. Eksp. Teo. Fiz. **47**, 1515 (1964) [Sov. Phys. JETP **20**, 1018 (1965)];
G. Zhou, Z. Su, B. Hao and L. Yu, Phys. Rep. **118**, 1 (1985).

[2] Y. Kluger, J. Eisenberg, B. Svetitsky, F. Cooper, and E. Mottola, *Phys. Rev. Lett.* **67**, 2427 (1991), Y. Kluger, J. Eisenberg, B. Svetitsky, F. Cooper, and E. Mottola, *Phys. Rev..* **D48**,190 (1993),

[3] F. Cooper, Y. Kluger, E. Mottola, and J. P. Paz, *Phys. Rev..* **D51**,2377 (1995),

[4] F. Cooper, S. Habib, Y. Kluger, E. Mottola, J. P. Paz, P. R. Anderson, *Phys. Rev.***D50**,2848 (1994).

[5] E. Calzetta and B. L. Hu, *Phys. Rev.* **D35**, 495 (1987).

[6] A. J. Niemi and G. W. Semenoff, *Ann. Phys.* **152**, 105 (1984); *Nucl. Phys.* **B230**, 181 (1984);
R. L. Kobes, G. Semenoff and N. Weiss, *Z. Phys.* **C29**, 371 (1985);
N. P. Landsman and Ch. G. van Weert, *Phys. Rep.* **145**, 141 (1987).

[7] J. Schwinger,*Phys. Rev.* **82**, 664 (1951).

[8] S. Weinberg *Gravitation and Cosmology* (Wiley, New York, 1972).

[9] G. Gatoff, A. K. Kerman, and T. Matsui, *Phys. Rev.***D36**,114 (1987).

SECTION II

DIRECTIONS IN PARTICLE PHYSICS AND COSMOLOGY

SECOND PHASE OF THE GENERAL THEORY OF RELATIVITY

Behram N. Kursunoglu

Global Foundation, Inc., Coral Gables, Florida

INTRODUCTION

Gravity as a universal force is described in Einstein's general theory of relativity as the curvature of the four-dimensional space-time continuum. It is a geometrical description of gravity. The metrical properties of space-time, i.e., measurements of distance and time intervals are included in a ten-component symmetric metric tensor $g_{\mu\nu}$. If, for example, a photon or an electromagnetic wave is emitted then, in principle, the space-time curvature, however small, must change. This change of curvature is due to the fact that the photon or the wave do carry energy and momentum. Thus, an electromagnetic field can act as a source of gravity. When we attempt to discover whether gravity itself can act as a source of the electromagnetic field we are in fact attempting to unify electromagnetic and gravitational fields. The latter statement need not be confused with an often discussed question: Does a charged particle falling in a gravity field emit radiation? If this process is discussed in terms of the *principle of equivalence* then the answer to the charged particle accelerated by gravity, emitting electromagnetic radiation is no.

It is often argued that unification of gravity and electromagnetism alone is not enough since it leaves out the strong and weak interactions. Hence this old fashioned way of doing physics should be abandoned! Yes, only if we attempt to unify just the two long-range forces of electromagnetism and gravitation. However, since electromagnetic field is sourced by the electric charge we need to ask: Why electric charge is *confined* and therefore it cannot, under the action of the electric forces at short distances, explode away? The answer to this question must come from the possible existence of short-range strong attractive forces, generated by charges other than electric charges. In a classical sense the short-range force in question is the part of the total gravitational field that has been missing from the *first phase* of general relativity. Thus, the complete gravitational field consists of two parts: *long-range weak gravitational field* plus *short-range strong gravitational field*. In quantum field theoretical sense strong interactions are carried by a triad of *spin-zero*, *spin-one* and *spin-two* particles. In the same way, in the unified theory, the complete electromagnetic field consists of two parts: *long-range field* plus *short-range weak field*. In the corresponding quantum field theoretical picture weak interactions are carried by *spin-one* particles.

COMPLETION OF GENERAL RELATIVITY

In the presence of the electromagnetic field the field equations of general relativity assume an elegant form where the electromagnetic stress tensor $T_{\mu\nu}$ appears as the source of gravitation. The corresponding Maxwell's equations enter without the electric charge. This theory can be expressed in terms of the first order approximation to the nonsymmetric tensor $\hat{g}^{\mu\nu}$ whose inverse is the covariant nonsymmetric tensor $\hat{g}_{\mu\nu} = g_{\mu\nu} + q_0^{-1}\Phi_{\mu\nu}$ where the constant q_0 has the dimensions of an electric field and where the antisymmetric tensor $\Phi_{\mu\nu}$ is a generalized electromagnetic entity. The approximation to the contravariant tensor $\hat{g}^{\mu\nu}$ (see author's paper following this one) to order $q_0^{-1}\Phi_{\mu\nu}$ yield, the usual general relativity provided we have the *equation of state* $q_0^2 r_0^2 = c^4/2G$ where the constant r_0 represents a fundamental length which spans an interval from zero to the size of the universe. The consequences of the *equation of state* are briefly discussed in the following paper by the author. The second phase of general relativity can thus be obtained by completing the nonsymmetrization of general relativity where not only the metric $\hat{g}_{\mu\nu}$ but the affine connection $\Gamma^\rho_{\mu\nu}$ and, therefore, the corresponding curvature tensor $\hat{R}_{\mu\nu}$ are nonsymmetric entities. The theory in the correspondence limit $r_0 = 0$ or $q_0 = \infty$ reduces to general relativity. The constant parameters r_0 and q_0 are calculated from the theory. For the cosmological time dependence, i.e., initially accelerated expansion of the universe or the so-called *inflation* modification of the Big Bang creation of the universe will be sought in the cosmological solutions of the theory.

THE NATURE OF THE MAGNETIC CHARGE

One of the most fundamental consequences of the second phase of general relativity is the emergence of positive and negative magnetic charge spectrum g_n (n = 0,1,2,3...) confined into a magnetic charge neutral structure, orbiton, i.e., $\sum_0^\infty g_n = 0$. It is also conjectured or proposed as an Ansatz that the infinite sum of the magnetic charges squared is equal to one half of Planck's constant $\hbar$, i.e., $\sum_0^\infty g_n^2 = \frac{1}{2} c\hbar$ revealing the origin of the Planck's constant. We may interpret $\frac{1}{2} g_n^2$ as a *fractional spin* $\frac{1}{2}\hbar_n$ where $\sum_0^\infty \hbar_n = \hbar$. Another neutral distribution of magnetic charge is a dipole consisting of equal amounts but opposite signs of magnetic charges in the two poles of the "magnet" with no limit to the quantity of magnetic charges. However, because of the stability requirement the magnetic charges of the magnet will redistribute themselves into the stable forms of equal number of orbitons in the two poles of the magnet in orbiton and antiorbiton pairs or in quark and antiquark pairs. Thus, predicting a structure for the quarks.

SUPERSYMMETRY

The nonsymmetric structure to order $q_0^{-1}\Phi_{\mu\nu}$ in general relativity remains unchanged if we replace it by the purely imaginary $iq_0^{-1}\Phi_{\mu\nu}$. This, clearly, means that we can choose any one of the two alternative nonsymmetrization where the 16 fundamental field variables are given either by the nonhermitian tensor $\hat{g}_{\mu\nu} = g_{\mu\nu} + q_0^{-1}\Phi_{\mu\nu}$ or by the hermitian tensor $\hat{g}_{\mu\nu} = g_{\mu\nu} + iq_0^{-1}\Phi_{\mu\nu}$. This indifference of general relativity to using hermitian or nonhermitian tensor can best be described as a *supersymmetry degeneracy* where transition from one state to another takes place by the transformations $q_0 \to iq_0$, $r_0 \to ir_0$. The removal of this degeneracy is uniquely achieved by the total nonsymmetrization of general relativity as mentioned above. In this case the nonhermitian branch of the supersymmetric theory describes fermi-Dirac like spin-one-half particles, i.e., quark, and the hermitian branch describes Bose-Einstein like integral-spin particles, i.e., carriers of all forces. The negative energy states in Dirac's theory were not, because of their sign, dropped but were retained and they described antiparticles. In the present case the two supersymmetric states describe particles with half-integral and integral spin systems.

MAGNETIC CHARGE CONDENSATION

The concept of *magnetic charge condensation* was first introduced by this author in 1975 in a physical review paper (see the following paper) to explain the stability of the neutral magnetic charge distribution. Gravitational condensation, prevailing over a long time could have placed or synthesized the magnetic charges g_n, by conservation of magnetic charge currents, into their most stable states that of stratified layers, like in a jigsaw puzzle. This result also contains the confinement of the electric charge through a relationship between electric and magnetic charges. This gave rise to the birth of quark constituents of elementary particles and hence to their short-range interactions evolving into the instant of the Big-Bang creation of the universe.

CLASSICAL EQUATIONS OF MOTION

By substituting the field equations of the generalized theory of gravitation in the action function we obtain its extremum value. By expanding the integrand in powers of $q_o^{-1}\Phi_{\mu\nu}$ and retaining terms to order q_o^{-2} we obtain the action function for a charged particle in an external electromagnetic field. Thus the theory yields, by straightforward treatment of the approximate extremum action function, the Lorentz equations of motion for a charged particle in an external field.

HALF-INTEGRAL SPIN PARTICLES

The success of the classical approximation lays the ground for the derivation of the quantum theory of matter. Our approach entails a path opposite to that of the conventional paradigm of the quantum gravity where general relativity is to be sublimated to the rules of quantum theory to bring it to the "*main stream*" of physics. Instead, the second phase of general relativity, based on the Dirac's way of linearization of the square-root in the extremum action function, provides four kinds of generalized Dirac wave equations comprising all the interactions mediated by massive and massless spins-zero, -one, and -two particles. All these carriers of forces are revealed by decomposing a second rank traceless tensor into its spins zero, one, and two parts. The massive spin-two particles can decay into two spin-one particles within the span of Planck time interval. The usual space-time symmetries and their violation are part of the new wave equations. The nature of the weak interactions along with strong and electromagnetic interactions are, with regard to their symmetries, magnifastly revealed.

To sum up: The second phase of general relativity provides a basis for obtaining quantum formalism for all interactions in place of the difficult or the impossible approach of the quantum gravity.

EXACT SOLUTIONS FOR CONFINEMENT OF ELECTRIC CHARGE VIA CONDENSATION OF A SPECTRUM OF MAGNETIC CHARGES

Behram N. Kursunoglu

Global Foundation, Inc., Coral Gables, Florida, 33146

A spectrum of magnetic charge confinement via *gravitational condensation* was first obtained by this author in a paper, Phy. Rev. 13, March 15, 1976, from his version of the generalized theory of gravitation, Phys. Rev. 88, 1369, (1952), versus the versions by Einstein and Schrödinger which do not yield such a result. At the *dawn of the universe* (i.e. preprimordial era) the material contents of the field could have consisted of the free positive and negative magnetic charge units g_n (n = 0, 1, 2, ...), the *quintessential matter*. This paper contains the earlier proofs mentioned above, that *gravitational condensation*, prevailing over a long time, could have placed or *synthesized* all these units g_n, by conservation of the magnetic charge currents, into their most stable states that of stratified layers, like in a *jigsaw puzzle*. This gave rise to the birth of quark constituents of elementary particles and hence to their short-range interactions evolving into the instant of the Big Bang creation of the universe. According to this theory, the generalized gravitation, which results from removing *supersymmetry degeneracy*, consists of the usual long-range gravity and the short-range strong interaction, while the generalized electromagnetism consists of the usual long-range electromagnetic field and the short-range weak interaction. The theory can also be cast in non-Abelian gauge formulation.

I. Introduction

The *gravitational condensation* of magnetic charges to explain the creation of elementary particles as neutral magnetic charge distribution, was introduced in a 1976 paper[1]. This kind of layered magnetic charge constituency may be assumed to correspond to the structure of a quark of spin $\frac{1}{2}$ where

$$\sum_{0}^{\infty} g_n = 0, \quad g_n = (-1)^n |g_n|, \quad |g_n| > |g_{n+1}|, \quad \lim_{n \to \infty} g_n = 0, \tag{1}$$

$$\frac{1}{c} \sum_{0}^{\infty} g_n^2 = \frac{1}{2} \hbar, \qquad (2)$$

and where the relation (2) is a conjecture or an Ansatz. Thus, the quark structure consists of magnetic charge layers with alternating signs and decreasing magnitudes. The first equation in the relation (1) corresponds to the conservation of the magnetic charge spectrum. The relation between gravitation and magnetic charge was discussed in an earlier paper[2] where an integrable set of equations describing the neutral regions pertaining to neutral surfaces between the opposite signs of magnetic charges were derived without actual solutions of these decoupled set of three differential equations. The integrability of the field equations did, however, demonstrate the condensation of the magnetic charges.

In 1982 I produced a report from the Center for Theoretical Studies titled "New Directions in General Theory of Relativity." Fifty copies of this report of 223 pages were sent to 50 major university physics libraries. The report contained an expanded presentation of the exact solutions of the spherically symmetric field equations of the generalized theory of gravitation pertaining to the confinement of a spectrum of magnetic charges g_n into an elementary particle (i.e., a quark). A copy of the report was sent to the Library of Congress requesting a copyright registration of this work. The December 14, 1982 reply from the Library contained 82-234779 as the card number of my work. I was hoping that this work, based on papers published earlier, may one day turn out to be of interest to other physicists also. Twelve years later the December 6, 1994 issue of the New York Times article by the science reporter Malcolm W. Browne was headlined as "Physicists Say New Math Tool will probe Secrets of Matter" had somewhat similar concepts, but not of the kind presented above.

A few days later a colleague gave me a preprint of a paper "Electric-Magnetic Duality, Monopoles Condensation, and Confinement in N = 2 Supersymmetric Yang-Mills Theory" authored by N. Seiberg and E. Witten. In view of my work 20 years earlier, I was interested in the conceptual part of their paper, especially the part referring to monopole confinement by the process of condensation that was obtained in my paper (Phys. Rev. 13, March 15, 1976, especially page 1551) by an entirely different theory i.e., my own version of the generalized theory of gravitation. It is quite obvious that the authors might not have seen or would not have read this paper or those others that followed it. A detailed account comprising my work on the subject matter covering the period 1950--1991 was published in Physics Essays[3]. However, the trajectories of these entirely different theories may not evolve to a point of intersection. All of these publications were sent to these authors by Federal Express mail at the beginning of February 1995 but I have had no response from them.

II. Nonsmmetric Structure of General Relativity In The Presence of An Electromagnetic Field

The fact that general relativity is the progenitor of the generalized theory of gravitation implies the casting of its Lagrangian in terms of the nonsymmetric quantities. The field equations, for gravity coupled to an electric charge-free electromagnetic field, can be obtained from the variation of the action function

$$S_o = \frac{c^3}{16\pi G} \int L \, d^4 x, \qquad (3)$$

where the Lagrangian L is given by

$$L = \sqrt{(-g)} \, (g^{\mu\nu} + q_o^{-1} \Phi^{\mu\nu}) \, (R_{\mu\nu} - r_o^{-2} q_o^{-1} F_{\mu\nu}) + 2 r_o^{-2} [(1 + \frac{1}{2} q_o^{-2} \Omega) \sqrt{(-g)} - \sqrt{(-g)}], \qquad (4)$$

and where the constants r_o (fundamental length) and q_o (dimensions of an electric field) are related by

$$r_o^2 q_o^2 = \frac{c^4}{2G}. \tag{5}$$

which can be interpreted as an *equation of state*. The relation (5) can be expressed in terms of energy density $\epsilon_o = q_o^2$, and mass density $\rho_o = \frac{c^2}{2Gr_o^2}$ as

$$\epsilon_o = \rho_o c^2. \tag{6}$$

In the Lagrangian (4) the $R_{\mu\nu}$ represents the curvature tensor.

On multiplying the relation (6) by the cubic volume r_o^3 we obtain the gravitational mass

$$M = \frac{c^2}{2G} r_o \simeq N \sqrt{(\frac{e^2+g^2}{2G})} \tag{7}$$

where r_o can be interpreted as the *gravitational size* of a self-gravitating system. Consider the following examples: (a) if r_o is set equal to Planck's length viz.,

$$r_o = \sqrt{(\frac{\hbar G}{c^3})}, \tag{8}$$

then the corresponding mass, as follows from (7), is just half of Planck's mass

$$M_o = \frac{1}{2} \sqrt{(\frac{\hbar c}{G})}; \tag{9}$$

(b)   If we choose for the gravitational size of a proton a value of the order of $r_o \sim 10^{-53}$ or $N \sim 10^{-20}$, then the relation (7) yields the mass of a proton. Thus, the particle mass is determined relative to the total mass of the universe where $r_o = (m_p/m_u)r_u$ is a function of the particle-universe mass ratio and the size r_u of the universe.

(c)   By choosing the value 10^{10} light years for the size of the universe we obtain from (7), total mass of the universe $M_u \sim 10^{33}$ solar masses. Hence we see that the *equation of state* (5) which is the result of the nonsymmetric formulation of general relativity can be qualified as an actual equation of state after the generalization of general relativity into a nonsymmetric theory. Thus, the nonsymmetric property may be regarded as a law of nature. The symmetries involved in this formulation are discussed in the reference (3).

In the Lagrangian the scalar Ω and pseudo-scalar Λ which will be used extensively are defined by

$$\Omega = \frac{1}{2} \Phi^{\mu\nu} \Phi_{\mu\nu}, \quad \Lambda = \frac{1}{4} f^{\mu\nu} \Phi_{\mu\nu}, \tag{10}$$

where

$$f^{\mu\nu} = \frac{1}{2\sqrt{(-g)}} \epsilon^{\mu\nu\rho\sigma} \Phi_{\rho\sigma} \tag{11}$$

is the dual tensor to $\Phi_{\mu\nu}$ and where

$$F_{\mu\nu} = \frac{\partial A_\nu}{\partial x^\mu} - \frac{\partial A_\mu}{\partial x^\nu}. \tag{12}$$

An important symmetry of the Lagrangian (4) and the fundamental relation (5) is the invariance under the transformations

$$r_o \to i r_o, \quad q_o^{-1} \to i q_o^{-1}, \tag{13}$$

which will be referred to as the *supersymmetry transformation*. In fact the Lagrangian (4) with the relation (5) is the same as the usual Lagrangian of general relativity where

$$L = \sqrt{(-g)}\, g^{\mu\nu} R_{\mu\nu} + \frac{2G}{c^4} \sqrt{(-g)}\, \Phi^{\mu\nu} (\tfrac{1}{2} \Phi_{\mu\nu} - F_{\mu\nu}). \tag{14}$$

Thus, the two supersymmetric states, in this case, are the same. This *supersymmetry degeneracy* is removed and the two supersymmetric fields split up by the full nonsymmetric generalization of the Lagrangian (4).

III. Generalized Theory of Gravitation (Second Phase of General Relativity)

The nonsymmetric form as written in (4) with the proviso (5) can be used to obtain the nonsymmetric generalization of general relativity. From (4) we see that the switching on of the electromagnetic field is manifested by the addition of $q_o^{-1} \Phi^{\mu\nu}$ to the gravitational tensor $g^{\mu\nu}$, the latter is the inverse of the symmetric tensor $g_{\mu\nu}$. The constant length r_o has been introduced, like the constant q_o, on the basis of dimensional considerations. The Einstein curvature tensor $R_{\mu\nu}$ has the dimension of (length)$^{-2}$. The nonsymmetric tensor $g^{\mu\nu} + q_o^{-1} \Phi^{\mu\nu}$ can be obtained as a first order approximation, in powers of q_o^{-1}, to the inverse $\hat{g}^{\mu\nu}$ of the covariant nonsymmetric tensor

$$\hat{g}_{\mu\nu} = g_{\mu\nu} + q_o^{-1} \Phi_{\mu\nu}, \tag{15}$$

which satisfy the relation

$$\hat{g}^{\nu\rho} \hat{g}_{\mu\rho} = \delta_\mu^\nu, \tag{16}$$

where, as mentioned before, q_o is either real (i.e., nonhermitian $\hat{g}_{\mu\nu}$), or is pure imaginary (i.e., hermitian $\hat{g}_{\mu\nu}$) which represent the two *supersymmetric fields*. The determinant of the nonsymmetric tensor $\hat{g}_{\mu\nu}$ is given by

$$\hat{g} = \text{Det}(\hat{g}_{\mu\nu}) = g(1 + q_o^{-2} \Omega - q_o^{-4} \Lambda^2), \tag{17}$$

where $g = \text{Det}(g_{\mu\nu})$ and where the sign of the second term is + or − depending on q_o being real or imaginary, respectively. The contravariant tensor $\hat{g}^{\mu\nu}$ is given by

$$\hat{g}^{\mu\nu} = \frac{(1 + \tfrac{1}{2} q_0^{-2} \Omega)\, g^{\mu\nu} - q_0^{-2}\, T^{\mu\nu}}{1 + q_0^{-2} \Omega - q_0^{-4} \Lambda^2} + \frac{q_0^{-1}\, \Phi^{\mu\nu} - q_0^{-3}\, f^{\mu\nu} \Lambda}{1 + q_0^{-2} \Omega - q_0^{-4} \Lambda^2}, \tag{18}$$

where $T^{\mu\nu}$ is of the same form as the energy tensor of the electromagnetic field

$$T^{\mu\nu} = \tfrac{1}{2} g^{\mu\nu}\Omega - \Phi^\mu{}_\rho\, \Phi^{\nu\rho}. \tag{19}$$

The approximation to order q_0^{-1} in the definition (18) of $\hat{g}^{\mu\nu}$ yields the nonsymmetric content $g^{\mu\nu} + q_0^{-1}\, \Phi^{\mu\nu}$ appearing in the Lagrangian (4). Furthermore, the expansion of the square root of the determinant as defined by (18) yields

$$\sqrt{(-\hat{g})} \simeq \sqrt{(-g)}\,(1 + \tfrac{1}{2} q_0^{-2}\, \Omega). \tag{20}$$

Hence, we see that the second term in (4) reduces to, because of the relation (5), $\tfrac{2G}{c^4} \sqrt{(-g)}\, \Omega$. Thus, a straightforward generalization of the Lagrangian (4) is given by

$$\mathcal{L} = \hat{\mathfrak{g}}^{\mu\nu}\,(\hat{R}_{\mu\nu} - r_0^{-2}\, q_0^{-1}\, F_{\mu\nu}) + 2 r_0^{-2}\,[\sqrt{(-\hat{g})} - \sqrt{(-g)}], \tag{21}$$

where

$$\hat{\mathfrak{g}}^{\mu\nu} - \sqrt{(-\hat{g})}\, g^{\mu\nu}, \tag{22}$$

and where the nonsymmetric contracted curvature tensor $\hat{R}_{\mu\nu}$ is given by

$$\hat{R}_{\mu\nu} = -\Gamma^\rho{}_{\mu\nu,\rho} + \Gamma^\rho{}_{\mu\rho,\nu} - \Gamma^\rho{}_{\mu\nu}\Gamma^\sigma{}_{\rho\sigma} + \Gamma^\rho{}_{\mu\sigma}\Gamma^\sigma{}_{\rho\nu}. \tag{23}$$

The components of the nonsymmetric affine connection $\Gamma^\rho{}_{\mu\nu}$ are the algebraic solutions of the 64 *transposition invariant* equations

$$\hat{g}_{\mu\nu\,;\rho} = \hat{g}_{\mu\nu,\rho} - \hat{g}_{\sigma\nu}\Gamma^\sigma{}_{\mu\rho} - \hat{g}_{\mu\sigma}\Gamma^\sigma{}_{\rho\nu} = 0. \tag{24}$$

In the limit of $q_0 \to \infty$ or $r_0 \to 0$ the generalized theory of gravitation reduces to general relativity plus the electromagnetic field.

The field equations of the generalized theory of gravitation are obtained from an action principle

$$\delta S = 0, \tag{25}$$

where

$$S = \frac{q_0^2\, r_0^2}{8\pi c} \int \mathcal{L}\, d^4 x, \tag{26}$$

as

$$\hat{R}_{\mu\nu} = r_o^{-2} (b_{\mu\nu} - g_{\mu\nu}), \tag{27}$$

$$\hat{R}_{\mu\nu} = r_o^{-2} (F_{\mu\nu} - \Phi_{\mu\nu}), \tag{28}$$

$$\hat{g}^{\mu\nu}{}_{,\nu} = 0 \tag{29}$$

The second set of field equations, by cyclic derivatives, can be replaced by

$$\hat{R}_{\mu\nu,\rho} + \hat{R}_{\nu\rho,\mu} + \hat{R}_{\rho\mu,\nu} + r_o^{-2} I_{\mu\nu\rho} = 0, \tag{30}$$

where the axial four-vector

$$I_{\mu\nu\rho} = \Phi_{\mu\nu,\rho} + \Phi_{\nu\rho,\mu} + \Phi_{\rho\mu,\nu}, \tag{31}$$

is the *magnetic charge* current density. In the Einstein and Schrödinger versions of the nonsymmetric theory it was interpreted as the electric charge current. For reasons discussed in the introduction to this paper the interpretation chosen by Einstein and Schrödinger was incorrect. In fact, the electric current density due to an electric charge is defined by

$$J^e_\mu = \frac{1}{4\pi} \frac{\partial}{\partial x^\nu} [\sqrt{(-g)} \, \Phi^{\mu\nu}]. \tag{32}$$

By using the definitions (11) and (31) the magnetic charge current density can also be expressed in the form

$$s^\mu = \frac{1}{4\pi} \frac{\partial}{\partial x^\nu} [\sqrt{(-g)} \, f^{\mu\nu}]. \tag{33}$$

The linearization, by approximation, of the four equations (30), to order q_o^{-2}, yields the wave equation

$$(\nabla^2 - \frac{\partial^2}{c^2 \partial t^2} + \kappa^2) s^\mu = 0, \tag{34}$$

where the corresponding supersymmetric field satisfies the wave equation

$$(\nabla^2 - \frac{\partial^2}{c^2 \partial t^2} - \kappa^2) s^\mu = 0, \tag{35}$$

and where

$$\kappa^2 = 2 \, r_o^{-2}. \tag{36}$$

The wave equations (34) and (35), because of the very large value of κ^2 ($\sim 10^{66}$ cm^{-2}), imply *confinement of the magnetic currents* like $\frac{\sin(\kappa r)}{r}$, and $\frac{\exp(-\kappa r)}{r}$ respectively. The *tachyon* type of solutions in the wave equation (34) do not arise in the exact nonlinear forms of the field equations. From the equation (34) we, further, observe the oscillatory behavior of the magnetic current density s^μ where the distribution of the magnetic charges alternate between positive and negative magnetic charge distributions. An extensive discussion of this and all other currents is included in reference 3.

IV. Spectrum of Magnetic Charges (Half-Integral Spin Particles)

The spherically symmetric field equations for the real field variables $\hat{g}_{\mu\nu} = g_{\mu\nu} + q_o^{-1}\Phi_{\mu\nu}$ are given by[3]

$$\frac{1}{2} r_o^2 \frac{d}{d\beta}[S \exp(\rho_{n\tau})\frac{d\Phi_{ns\tau}}{d\beta}] = R_o^2 \cos\Phi_{ns\tau} + \ell_o^2 \sin\Phi_{ns\tau}, \tag{37}$$

$$\frac{1}{2} r_o^2 \frac{d}{d\beta}[S \exp(\rho_{n\tau})\frac{d\rho_{n\tau}}{d\beta}] = - R_o^2 \sin\Phi_{ns\tau} + \ell_o^2 \cos\Phi_{ns\tau} + \exp(\rho_{n\tau}), \tag{38}$$

$$\frac{1}{2} r_o^2 \frac{d}{d\beta}[\frac{dS}{d\beta}\exp(\rho_{n\tau})] = \exp(\rho_{n\tau})[1 - \frac{\exp(\rho_{n\tau})\sin\Phi_{ns\tau}}{R_o^2 + r_o^2}], \tag{39}$$

$$[\frac{d^2}{d\beta^2} + \frac{1}{4}(\frac{d\Phi}{d\beta})^2]\exp(\frac{1}{2}\rho_{n\tau}) = 0, \tag{40}$$

where

$$S = \frac{\exp(u)}{\cosh^2\Gamma}, \quad \cosh\Gamma = (R_o^2 + r_o^2)\exp(-\rho),$$

$$R_o^2 + r_o^2 = \sqrt{[\exp(2\rho) + \lambda_o^4]}, \quad \ell_o^2 = q_o^{-1} g, \quad \lambda_o^2 = q_o^{-1} e$$

$$\Phi_{ns\tau} - \pm(2n + s + \tau)\pi + (-1)^s \Phi, \quad 0 \le \Phi \le \frac{\pi}{2} \tag{41}$$

$$\rho_{n\tau} = \rho \pm (2n + \tau)i\pi, \quad s = 0, 1, \quad \tau = 0, 1, \quad n = 0, 1, 2, 3....$$

We, further, obtain the relations

$$\sin\Phi_{ns\tau} = (-1)^\tau \sin\Phi, \quad \cos\Phi_{ns\tau} = (-1)^{s+\tau}\cos\Phi, \quad \exp(\rho_{n\tau}) = (-1)^\tau \exp(\rho), \tag{42}$$

$$\exp(\rho_{n\tau})\sin\Phi_{ns\tau} = \exp(\rho)\sin\Phi, \tag{43}$$

and where r and β are related by

$$dr = fd\beta, \quad f = v\cosh\Gamma \tag{44}$$

The field equation (40) represents a harmonic oscillator with a variable frequency $\frac{1}{2}\frac{d\Phi}{d\beta}$.

In general, the existence of magnetic charge layers (or monopoles) with alternating signs of charge densities in traversing through the magnetic charge neutral surfaces are to be obtained by setting the magnetic charge density equal to zero in the field equations. One of the conditions for the existence of layers is the *integrability - which is related to monopole* *condensation of the field equations under the restriction of $s^o=0$ or equivalently*

$$\dot{\rho}_{n\tau} = \Phi_{ns\tau} \tan \Phi_{ns\tau} . \tag{45}$$

It can be integrated as

$$\exp(\rho_{n\tau}) \cos \Phi_{ns\tau} = (-1)^s \ell_o^2 , \tag{46}$$

where the constant of integration is determined from the general solutions[1,2] of the field equations (37) - (40) where

$$\Phi_{ns\tau} \neq N\pi, \quad \Phi_{ns\tau} \neq \frac{1}{2}\pi, \quad N = 0, \pm1, \pm2, \ldots \tag{47}$$

On substituting from (45) in the field equation (40) when written in the form

$$2\ddot{\rho} + \dot{\Phi}^2 + \dot{\rho}^2 = 0, \tag{48}$$

can be integrated by using (45) and obtain

$$\frac{1}{4} \ell_o^2 \dot{\Phi}_{ns\tau}^2 \tan^3 \Phi_{ns\tau} = (-1)^s , \tag{49}$$

where $\frac{1}{4}\ell_o^2$ is a constant of integration.

The equation (49) can easily be integrated by using the substitution

$$\tan \Phi_{ns\tau} = (-1)^s t, \quad t>0, \tag{50}$$

in the equation (49) and integrating we get

$$d\Phi_{ns\tau} = \frac{2}{\ell_o} (-1)^s t^{-\frac{3}{2}} d\beta = \frac{(-1)^s}{1+t^2} dt , \tag{51}$$

where

$$\cos \Phi_{ns\tau} = \frac{(-1)^{s+\tau}}{\sqrt{(1+t^2)}}, \quad \sin \Phi_{ns\tau} = \frac{(-1)^\tau t}{\sqrt{(1+t^2)}}, \tag{52}$$

$$\exp(\rho_{n\tau}) = (-1)^\tau \ell_o^2 \sqrt{(1+t^2)} \tag{53}$$

From the asymptotic limits of the equations (37) - (39) it follows that

$$t = \frac{r^2}{\ell_o^2}, \tag{54}$$

and hence in the regions where magnetic charge density vanishes or in the regions of *monopole condensation* we have the solution

$$\exp(\rho) = \sqrt{(r^4 + \ell_o^4)} . \tag{55}$$

From (51) we obtain

$$\beta = (-1)^\varepsilon \ell_o \left[\sqrt{(t)} - \frac{1}{2\sqrt{2}} [\eta \tan^{-1}[\frac{\sqrt{(2t)}}{1-t}]\eta + \tanh^{-1}[\frac{\sqrt{(2t)}}{1+t}]\eta] \right], \quad (56)$$

where $\eta = \pm 1$ and where we observe the relations

$$\frac{d}{dt}[\tan^{-1}(\frac{\sqrt{(2t)}}{1-t})] = \frac{d}{dt}[\tan^{-1}(\frac{t-1}{\sqrt{(2t)}})], \quad (57)$$

$$\frac{d}{dt}[\tanh^{-1}(\frac{\sqrt{(2t)}}{1+t})] = \frac{d}{dt}[\tanh^{-1}(\frac{1+t}{\sqrt{(2t)}})]. \quad (58)$$

In the result (56) we have for

$$\frac{2t}{(1-t)^2} < 1,$$

$\eta = 1$ and for

$$\frac{2t}{(1-t)^2} > 1,$$

$\eta = -1$.

Now we shall use the above results in the field equations (37) - (39) to reduce them into directly integrable forms. It must be noted that, because of the conditions (47) on the functions $\Phi_{ns\tau}$, and also the fact that $\Phi_{ns\tau}$ are not constant functions of β, all the solutions will describe the interior magnetic charge regions of the field. The field equations pertaining to regions of zero magnetic charge density or the region of *monopole condensation* are now given by

$$\frac{d}{dt}[S \, t^{-\frac{3}{2}}\sqrt{(1+t^2)}] = \frac{1}{2} \frac{t^{\frac{3}{2}}}{(1+t^2)^{3/2}} [\sqrt{(m_g^2 + m_e^2 + m_g^2 t^2)} - 1 + (-1)^s t \, m_g], \quad (59)$$

$$\frac{d}{dt}[S \, t^{-\frac{1}{2}}\sqrt{(1+t^2)}] = \frac{1}{2} \frac{t^{\frac{3}{2}}}{(1+t^2)^{3/2}} [t - t\sqrt{(m_g^2 + m_e^2 + m_g^2 t^2)} + m_g(1+t^2) + (-1)^s m_g], \quad (60)$$

$$\frac{d}{dt}[\frac{dS}{dt} t^{\frac{3}{2}} (1+t^2)^{\frac{3}{2}}] = \frac{1}{2} \frac{m_g}{\sqrt{(1+t^2)}} t^{3/2} [1 - \frac{m_g t}{\sqrt{(m_g^2 + m_e^2 + m_g^2 t^2)}}], \quad (61)$$

where the eigenvalues m_g and m_e are given by

$$m_e^2 = \frac{\lambda_o^4}{r_o^4} = \frac{2G \, e^2}{c^4 \, r_o^2}, \quad m_g^2 = \frac{\ell_o^4}{r_o^4} = \frac{2G \, g^2}{c^4 \, r_o^2}, \quad (62)$$

$$m_e^2 + m_g^2 = \frac{2G}{c^4} \frac{e^2 + g^2}{r_o^2}. \quad (63)$$

A special case where the function Φ is a constant yields the result $m_e^2 + m_g^2 = 1$ or

$$r_o^2 = \frac{2G}{c^4}(e^2 + g^2), \qquad (64)$$

in which case we have

$$m_g^2 = m^2 = \frac{g^2}{e^2 + g^2}. \qquad (65)$$

V. Confinement of the Electric Charge

For the regions of zero magnetic charge density the field equations, as seen from (60) - (61), are decoupled. The solutions for S of the equations (59) - (61) will be represented by the surfaces $S = f_1(t,m)$, $S = f_2(t,m)$, $S = f_3(t,m)$, respectively. The equations (59) - (61) are not compatible for every t and m. The surfaces of compatibility for these equations correspond to the magnetic layers with alternating signs of magnetic charge. From the definition (65) we see that the magnetic and electric charges are related by

$$g^2 = e^2 \frac{m^2}{1 - m^2}. \qquad (66)$$

The relation (66), via the equations (59) - (61), provides, for a given spectrum of m-values, a discrete or "quantized" relationship between the magnetic charge spectrum g_n and the electric charge e. The integrated form of the equations (59) - (61) are given in the Appendix to this paper.

The confinement of the positive or negative electric charge is thus obtained with the confinement of a spectrum of magnetic charges g_n generating a strong short-range force. This follows from the linearized wave equation (34) plus the above exact solutions of the field equations. The special result (64) above reveals more explicitly the short-range force playing a role where in a classical sense a short-range gravitational force is acting as a source of an electromagnetic force by holding the spectrum of magnetic charges together with the electric charge where, as follows from (64), we have

$$\frac{r_o c^4}{2G} = \frac{e^2 + g^2}{r_o} = Mc^2, \qquad (67)$$

or

$$\frac{GM_o^2}{r_o} = \frac{e^2 + g^2}{r_o},$$

where $M_o = \sqrt{2} M$, and where r_o is the range of the forces involved. From (67) we obtain the relation

$$1 - \frac{2GM}{c^2 r_o} = 0, \qquad (68)$$

which is reminiscent of the Schwarzrchild singularity in general relativity. If we choose r_o to be of the order of a Planck length then the mass $M = \frac{1}{2}\sqrt{(\frac{\hbar c}{G})}$ is also of the order of the Planck mass.

VI. Spectrum of Magnetic Charges (Integral Spin Particles)

The supersymmetric transition from half-integral spin to integral spin field equations can be obtained by the transformations

$$\Phi \to i\Phi + \frac{\pi}{2}, \quad \Gamma \to i\Gamma, \quad r_o \to ir_o,$$

$$q_o^{-1} \to iq_o^{-1}, \quad \ell_o^2 \to i\ell_o^2, \quad \lambda_o^2 \to i\lambda_o^2, \tag{69}$$

from which we obtain the replacements

$$\cos\Phi \to i\sinh\Phi, \quad \sin\Phi \to \cosh\Phi, \quad \tanh\Gamma \to i\tan\Gamma. \tag{70}$$

The corresponding spherically symmetric field equation for $\hat{g}_{\mu\nu} = g_{\mu\nu} + iq^{-1}\Phi_{\mu\nu}$ are given by

$$\frac{1}{2} r_o^2 \frac{d}{d\beta}[S\exp(\rho)\frac{d\Phi}{d\beta}] = R_o^2 \sinh\Phi - (-1)^s \ell_o^2 \cosh\Phi, \tag{71}$$

$$\frac{1}{2} r_o^2 \frac{d}{d\beta}[S\exp(\rho)\frac{d\rho}{d\beta}] = R_o^2 \cosh\Phi - (-1)^s \ell_o^2 \sinh\Phi - \exp(\rho), \tag{72}$$

$$\frac{1}{2} r_o^2 \frac{d}{d\beta}[\frac{dS}{d\beta}\exp(\rho)] = \exp(\rho)[\frac{\cosh\Phi}{\cos\Gamma} - 1], \tag{73}$$

$$[\frac{d^2}{d\beta^2} - \frac{1}{4}(\frac{d\Phi}{d\beta})^2]\exp(\frac{1}{2}\rho) = 0, \tag{74}$$

where

$$R_o^2 - r_o^2 = \sqrt{[\exp(2\rho) - \lambda_o^4]}, \quad S = \frac{\exp(u)}{\cos^2\Gamma}, \quad 0 \le \Gamma \le \frac{\pi}{2}, \tag{75}$$

$$\cos\Gamma = \exp(-\rho)(R_o^2 - r_o^2) = \sqrt{[1 - \lambda_o^4\exp(-2\rho)]}, \quad -\infty < \Phi < \infty, \tag{76}$$

$$\exp(\rho)\sin\Gamma = \pm\lambda_o^2, \quad \ell_o^2 = g\, q_o^{-1}, \quad \lambda_o^2 = e\, q_o^{-1}, \quad s = 0, 1, \tag{77}$$

$$dr = f\, d\beta, \quad f = v\cos\Gamma$$

The field equations (71) - (74) are invariant under the transformation $\Phi \to -\Phi$, $\ell_o^2 \to -\ell_o^2$, which are equivalent to the symmetry operations

$$\Phi \to -\Phi, \quad g \to -g, \tag{78}$$

where the magnetic charge can assume finite arbitrary values.

The symmetry (78) implies that the total magnetic charge of the particle (boson) is zero. This boson is, of course, an integral spin particle in the form of a *magnetic charge dipole*, or simply a *magnet*. Because of the supersymmetry the magnetic charges at the two poles are redistributed in the form of the pairs of orbiton and antiorbiton or quark - antiquark pairs, without a limit to the number of pairs. In fact from (75) and (76) we see that the gravitational potential[3],[4]

$$S_B = \frac{\exp(u)}{\cos^2\Gamma} = \frac{\lambda_o^{-4}\exp(u+2\rho)}{\lambda_o^{-4}\exp(2\rho)-1} \tag{79}$$

has the form of a Bose-Einstein like ensemble of particles distribution which gives rise to *condensation* phenomenon and hence to an effective attraction between particles. For the potential

$$S_F = \frac{\exp(u)}{\cosh^2\Gamma} = \frac{\lambda_o^{-4}\exp(u+2\rho)}{\lambda_o^{-4}\exp(\rho)+1}, \tag{80}$$

The gravitational attraction is inhibited by an effective repulsion between paticles. What is due to exchange effects in quantum statistics has been replaced here by the nonlinearity of the present theory.

APPENDIX

INTEGRALS OF THE SPECTRUM OF MAGNETIC CHARGE

This appendix contains the integrals used in the discussion of the magnetic charge spectrum. We begin with

$$I_1 = \frac{1}{2}\int \frac{t^{\frac{3}{2}}\sqrt{1+m^2 t^2}}{(1+t^2)^{\frac{3}{2}}}\,dt = -\frac{1}{2}\frac{x\sqrt{1+m^2 x^4}}{\sqrt{1+x^4}} + \frac{3}{2}L_1 - L_-, \tag{A.1}$$

where

$$x^2 = t,$$

and

$$L_1 = \int \frac{\sqrt{1+m^2 x^4}}{\sqrt{1+x^4}}\,dx, \quad L_- = \int \frac{dx}{\sqrt{1+x^4}\sqrt{1+m^2 x^4}} \tag{A.2}$$

A convenient substitution is to put

$$x = m^{-\frac{1}{4}} U^{\frac{1}{2}}, \quad U = z + \sqrt{z^2-1}, \quad U^2 = 2Uz - 1,$$

$$U^{\frac{1}{2}} = \frac{1}{\sqrt{2}}[\sqrt{z+1}+\sqrt{z-1}], \quad U^{-\frac{1}{2}} = \frac{1}{\sqrt{2}}[\sqrt{z+1}-\sqrt{z-1}],$$

$$dx = \frac{m^{-\frac{1}{4}}}{2\sqrt{2}}[\frac{1}{\sqrt{z-1}}+\frac{1}{\sqrt{z+1}}]\,dz.$$

Hence we obtain

$$L_- = \frac{1}{4\sqrt{(1+m)}} [\kappa(\gamma, \tau_1) - \kappa(\omega, \tau_2)], \tag{A.3}$$

where the elliptic integrals of the first kind are defined as

$$\kappa(\gamma, \tau_1) = \int^\gamma \frac{d\gamma}{\sqrt{(1 - \tau_1^2 \sin^2\gamma)}}, \quad \cos\gamma = \frac{x^2(1+m) - (x^2\sqrt{m}-1)^2}{x^2(1+m) + (x^2\sqrt{m}-1)^2}, \tag{A.4}$$

$$\kappa(\omega, \tau_2) = \int^\omega \frac{d\omega}{\sqrt{(1 - \tau_2^2 \sin^2\omega)}}, \quad \cos\omega = \frac{x^2(1+m) - (x^2\sqrt{m}+1)^2}{x^2(1+m) + (x^2\sqrt{m}+1)^2}, \tag{A.5}$$

$$\tau_1^2 = \frac{(1-\sqrt{m})^2}{2(1+m)}, \quad \tau_2^2 = \frac{(1+\sqrt{m})^2}{2(1+m)}, \quad \tau_1^2 + \tau_2^2 = 1. \tag{A.6}$$

The behavior of the elliptic integral $\kappa(\gamma, \tau_1)$ at $x = m^{-\frac{1}{4}}$ can be seen more clearly by writing

$$\kappa(\gamma, \tau_1) = \sqrt{2}\, m^{-\frac{1}{4}} \sqrt{(1+m)} \int \frac{dz}{\sqrt{(z-1)}\sqrt{[4z^2 + \frac{(1-m)^2}{m}]}}, \tag{A.7}$$

$$\kappa(\omega, \tau_2) = \sqrt{2}\, m^{-\frac{1}{4}} \sqrt{(1+m)} \int \frac{dz}{\sqrt{(z+1)}\sqrt{[4z^2 + \frac{(1-m)^2}{m}]}}, \tag{A.8}$$

where

$$z - 1 = \frac{1}{2}\frac{1+m}{\sqrt{m}} \frac{1-\cos\gamma}{1+\cos\gamma}, \quad z + 1 = \frac{1}{2}\frac{1+m}{\sqrt{m}}\frac{1-\cos\omega}{1+\cos\omega}. \tag{A.9}$$

From the relations

$$\sqrt{(z-1)} = \frac{x^2\sqrt{m}-1}{xm^{\frac{1}{4}}\sqrt{2}}, \quad \sqrt{(z+1)} = \frac{x^2\sqrt{m}+1}{xm^{\frac{1}{4}}\sqrt{2}}, \tag{A.10}$$

and from (A.7) we see that there is a branch point at $z = 1$ or at $x = m^{-\frac{1}{4}}$ (where $x > 0$). The point $z = -1$ corresponds to an imaginary x and therefore the integral (A.8) does not have a branch point. One of the consequences of the branch point in (A.7) is the fact that the limiting process of $m \to 1$ and the integration do not commute. Thus in the forms (A.7) and (A.8) we have for $m = 1$ the results

$$\frac{1}{2}\int \frac{dz}{z\sqrt{(z-1)}} = \eta \tan^{-1}(z-1)^{\frac{1}{2}\eta}, \tag{A.11}$$

$$\frac{1}{2}\int \frac{dz}{z\sqrt{(z+1)}} = \tanh^{-1}(z+1)^{\frac{1}{2}}\Big|^{\frac{1}{2}\eta}, \quad (A.12)$$

where $\eta = \pm 1$. From (A.6) we see that for $m = 1$ we have $\tau_1 = 0$, and $\kappa(\gamma, \tau_1) = \gamma$ which is not equal to (A.11). However for $m = 1$ we have $\tau_2 = 1$ and the integral (A.5) is the same as (A.12) with $\eta = 1$. The $\eta = -1$ value in (A.12) corresponds to a complex number.

For the integral L_1 we may write

$$L_1 = L_- + m^2 L$$

where

$$L = \int \frac{x^4 dx}{\sqrt{(1+x^4)}\sqrt{(1+m^2 x^4)}}, \quad (A.13)$$

with the above substitution becomes

$$L = \frac{1}{m}L_- + \frac{1}{\sqrt{2}} m^{-\frac{5}{4}} \int [\frac{\sqrt{(z-1)}\, dz}{\sqrt{[4z^2 + \frac{(1-m)^2}{m}]}} + \frac{\sqrt{(z+1)}\, dz}{\sqrt{[4z^2 + \frac{(1-m)^2}{m}]}}], \quad (A.14)$$

where we used the easily established relation

$$U^{\frac{3}{2}}(z-1)^{-\frac{1}{2}}(z+1)^{-\frac{1}{2}} = \sqrt{2}\,[\sqrt{(z-1)}+\sqrt{(z+1)}] + \frac{1}{\sqrt{2}}[(z-1)^{-\frac{1}{2}} - (z+1)^{-\frac{1}{2}}].$$

By using the relations (A.9) we obtain the results

$$I_\gamma = \frac{1}{\sqrt{2}} m^{-\frac{5}{4}} \int \frac{\sqrt{(z-1)}\,dz}{\sqrt{[4z^2 + \frac{(1-m)^2}{m}]}} = \frac{1}{4}\frac{\sqrt{(1+m)}}{m^{\frac{3}{2}}} \int (\frac{1-\cos\gamma}{1+\cos\gamma}) \frac{d\gamma}{\sqrt{(1-\tau_1^2 \sin^2\gamma)}}, \quad (A.15)$$

$$I_\omega = \frac{1}{\sqrt{2}} m^{-\frac{5}{4}} \int \frac{\sqrt{(z+1)}\,dz}{\sqrt{[4z^2 + \frac{(1-m)^2}{m}]}} = \frac{1}{4}\frac{\sqrt{(1+m)}}{m^{\frac{3}{2}}} \int (\frac{1-\cos\omega}{1+\cos\omega}) \frac{d\omega}{\sqrt{(1-\tau_2^2 \sin^2\omega)}}. \quad (A.16)$$

Hence

$$I_\gamma = \frac{1}{4}\frac{\sqrt{(1+m)}}{m^{\frac{3}{2}}} [K(\gamma, \tau_1) - 2E(\gamma, \tau_1) + \frac{2(1-\cos\gamma)}{\sin\gamma}\Delta_1],$$

$$I_\omega = \frac{1}{4}\frac{\sqrt{(1+m)}}{m^{\frac{3}{2}}} [K(\omega, \tau_2) - 2E(\omega, \tau_2) + \frac{2(1-\cos\omega)}{\sin\omega}\Delta_2],$$

where

$$\Delta_1 = \sqrt{(1 - \tau_1^2 \sin^2\gamma)}, \quad \Delta_2 = \sqrt{(1 - \tau_2^2 \sin\omega)}$$

and

$$E_\gamma = E(\gamma, \tau_1) = \int^\gamma \Delta_1 \, d\gamma, \quad E_\omega = E(\omega, \tau_2) = \int^\omega \Delta_2 \, d\omega.$$

By combining all of these we obtain the result

$$L_1 = \tfrac{1}{4}\sqrt{(1+m)}\,[\kappa_\gamma - \kappa_\omega] + \tfrac{1}{4}\sqrt{m}\sqrt{(1+m)}\,[\kappa_\gamma + \kappa_\omega - 2E_\gamma - 2E_\omega] +$$

$$\tfrac{1}{4}\sqrt{m}\sqrt{(1+m)}\,[\tfrac{1-\cos\gamma}{\sin\gamma}\Delta_1 + \tfrac{1-\cos\omega}{\sin\omega}\Delta_2]. \quad (A.17)$$

The elliptic functions of the second kind E_γ, E_ω can also be expressed in terms of the variable z as

$$E_\gamma = E(\gamma, \tau_1) = \sqrt{2}\sqrt{(1+m)}\, m^{-\tfrac{1}{4}} \int \frac{\sqrt{[4z^2 + \tfrac{(1-m)^2}{m}]}\,dz}{\sqrt{(z-1)}\,[z + \tfrac{(1-\sqrt{m})^2}{2\sqrt{m}}]^2}, \quad (A.18)$$

$$E_\omega = E(\gamma, \tau_1) = \sqrt{2}\sqrt{(1+m)}\, m^{-\tfrac{1}{4}} \int \frac{\sqrt{[4z^2 + \tfrac{(1-m)^2}{m}]}\,dz}{\sqrt{(z+1)}\,[z + \tfrac{(1+\sqrt{m})^2}{2\sqrt{m}}]^2}. \quad (A.19)$$

The point $z = 1$, as was discussed for κ_γ, is a branch point for E_γ. In fact the limiting process $m \to 1$ and the integration, in this case also, do not commute.

We may, further, apply the above techniques to integrate

$$I_2 = \tfrac{1}{2}\int \frac{t^{\tfrac{5}{2}}\sqrt{(1+m^2t^2)}\,dt}{(1+t^2)^{\tfrac{3}{2}}} = -\tfrac{1}{2}\frac{x^3\sqrt{(1+m^2x^4)}}{\sqrt{(1+x^4)}} = +\tfrac{5}{2}L_3 - \tfrac{3}{2}L_+, (A.20)$$

where

$$L_3 = \int \frac{x^2\sqrt{(1+m^2x^4)}}{\sqrt{(1+x^4)}}\,dx, \quad (A.21)$$

$$L_+ = \int \frac{x^2\,dz}{\sqrt{(1+x^4)}\sqrt{(1+m^2x^4)}} = \frac{1}{4\sqrt{m}\sqrt{(1+m)}}\,[\kappa(\gamma,\tau_1) + \kappa(\omega,\tau_2)]. \quad (A.22)$$

The integral L_3 can also be expressed as

$$L_3 = L_+ + m^2 \int \frac{x^6 \, dx}{\sqrt{[(1+x^4)(1+m^2x^4)]}} = L_+ + m^2 L_0, \tag{A.23}$$

where

$$L_0 = \int \frac{x^6 \, dx}{\sqrt{[(1+x^4)(1+m^2x^4)]}} = \frac{1}{2} m^{-\frac{7}{4}} \int \frac{U^{\frac{5}{2}} (z^2-1)^{-\frac{1}{2}} \, dz}{\sqrt{[4z^2 + \frac{(1-m)^2}{m}]}}, \tag{A.24}$$

$$U^{\frac{5}{2}} (z^2-1)^{-\frac{1}{2}} = 2^{\frac{1}{2}} [(z-1)^{\frac{3}{2}} + (z+1)^{\frac{3}{2}}] + 3 \times 2^{\frac{3}{2}} [(z+1)^{\frac{1}{2}}] + 2^{\frac{1}{2}} [(z+1)^{\frac{1}{2}} + (z-1)^{\frac{1}{2}}].$$

Hence we have the integrals

$$\frac{1}{2} m^{-\frac{7}{4}} \int \frac{2^{\frac{3}{2}}(z-1)^{\frac{3}{2}} \, dz}{\sqrt{[4z^2 + \frac{(1-m)^2}{m}]}} = \frac{1}{4} m^{-\frac{5}{2}}(1+m)^{\frac{3}{2}} \int \left(\frac{1-\cos\gamma}{1+\cos\gamma}\right)^2 \frac{d\gamma}{\sqrt{(1-\tau_1^2\sin^2\gamma)}}, \tag{A.25}$$

$$\frac{1}{2} m^{-\frac{7}{4}} \int \frac{2^{\frac{3}{2}}(z+1)^{\frac{3}{2}} \, dz}{\sqrt{[4z^2 + \frac{(1-m)^2}{m}]}} = \frac{1}{4} m^{-\frac{5}{2}}(1+m)^{\frac{3}{2}} \int \left(\frac{1-\cos\omega}{1+\cos\omega}\right)^2 \frac{d\omega}{\sqrt{(1-\tau_2^2\sin^2\omega)}}, \tag{A.26}$$

where we used the relations

$$\sqrt{[4z^2 + \frac{(1-m)^2}{m}]} = 2 \frac{1+m}{\sqrt{m}} \frac{\Delta_1}{1+\cos\gamma} = 2 \frac{1+m}{\sqrt{m}} \frac{\Delta_2}{1+\cos\omega},$$

$$(z-1)^{\frac{3}{2}} \, dz = 2^{\frac{3}{2}} m^{-\frac{3}{4}} (1+m)^{\frac{5}{2}} \frac{\sin^4\gamma \, d\gamma}{(1+\cos\gamma)^5},$$

$$(z+1)^{\frac{3}{2}} \, dz = 2^{\frac{3}{2}} m^{-\frac{3}{4}} (1+m)^{\frac{5}{2}} \frac{\sin^4\omega \, d\omega}{(1+\cos\omega)^5}.$$

From the identity

$$\left(\frac{1-\cos\gamma}{1+\cos\gamma}\right)^2 = 1 + 4\frac{\cos\gamma}{\sin^2\gamma} + 8\left(\frac{1}{\sin^4\gamma} - \frac{1}{\sin^2\gamma} - \frac{\cos\gamma}{\sin^4\gamma}\right)$$

and the integrals

$$\int \frac{d\gamma}{\sin^4\gamma \, \Delta_1} = \frac{1}{3}[(\tau_1^2+1)\kappa_\gamma - (2\tau_1^2+1)E_\gamma] - \frac{1}{3}\cot\gamma \left[2(\tau_1^2+1) + \frac{1}{\sin^2\gamma}\right]\Delta_1, \tag{A.27}$$

$$\int \frac{\cos\gamma \, d\gamma}{\sin^2\gamma \, \Delta_1} = -\frac{1}{\sin\gamma}\Delta_1, \tag{A.28}$$

$$\int \frac{d\gamma}{\sin^2\gamma \, \Delta_1} = \kappa_\gamma - E_\gamma - \Delta_1 \cot\gamma, \tag{A.29}$$

$$\int \frac{\cos\gamma \, d\gamma}{\sin^4\gamma \, \Delta_1} = -\frac{2\tau_1^2 \sin^2\gamma + 1}{3\sin^3\gamma} \Delta_1, \tag{A.30}$$

we obtain the result

$$\int \left(\frac{1-\cos\gamma}{1+\cos\gamma}\right)^2 \frac{d\gamma}{\Delta_1} = \frac{8}{3}[(\tau_1^2 - \frac{13}{8})\kappa_\gamma + 2(1-\tau_1^2) E_\gamma] +$$

$$\frac{8}{3}\frac{1}{\sin\gamma}[-\frac{1}{2} + (1-\cos\gamma)(2\tau_1^2 + \cot^2\gamma)\,\Delta_1\,. \tag{A.31}$$

Hence

$$\frac{3}{2}(1+m)^{-\frac{3}{2}} m^{\frac{1}{2}} L_3 = (\tau_1^2 - \frac{13}{8})\kappa_\gamma + (\tau_{22}^2 - \frac{13}{8})\kappa_\omega + \frac{1}{\sin\gamma}[-\frac{1}{2} + (1+\cos\gamma)(2\tau_1^2 + \cot^2\gamma)]\Delta_1$$

$$+ \frac{1}{\sin\omega}[-\frac{1}{2} + (1-\cos\omega)(2\tau_2^2 + \cot^2\omega)]\Delta_2 + \frac{3}{8}(1+m)^{-1}(\kappa_\gamma + \kappa_\omega) + 2[(1-\tau_2^2) E_\omega]$$

$$+ \frac{9}{8}(1+m)^{-1} m^{\frac{1}{2}}[\kappa_\gamma - \kappa_\omega - 2(E_\gamma - E_\omega) + \frac{2(1-\cos\gamma)}{\sin\gamma}\Delta_1 - \frac{2(1-\cos\omega)}{\sin\omega}\Delta_2]. \tag{A.32}$$

We may express the integral

$$I_3 = \frac{1}{2}\int \frac{t^{\frac{5}{2}} \, dt}{\sqrt{(1+t^2)} \sqrt{(1+m^2 t^2)}}, \tag{A.33}$$

in the form

$$I_3 = \frac{1}{m^2}(L_3 - L_+). \tag{A.34}$$

The remaining integrals arising in the magnetic charge spectrum are listed below.

(i) $$\int \frac{t^{\frac{3}{2}} \, dt}{1+t^2} = 2\sqrt{t} - \frac{1}{\sqrt{2}}[\tan^{-1}(\frac{\sqrt{(2t)}}{1-t}) + \tanh^{-1}(\frac{\sqrt{(2t)}}{1+t})], \tag{A.35}$$

(ii) $$\int \frac{t^{\frac{5}{2}} \, dt}{1+t^2} = \frac{2}{3} t^{\frac{3}{2}} - \frac{1}{\sqrt{2}}[\tan^{-1}(\frac{\sqrt{(2t)}}{1-t}) - \tanh^{-1}(\frac{\sqrt{(2t)}}{1+t})], \tag{A.36}$$

(iii) $$\int \frac{\sqrt{t} \, dt}{\sqrt{(1-t^2)}} = \kappa(\alpha, \frac{1}{\sqrt{2}}) - 2E(\alpha, \frac{1}{\sqrt{2}}) + 2\frac{\sqrt{t} \sqrt{(1+t^2)}}{1+t}, \tag{A.37}$$

(iv) $$\int \frac{t^{\frac{3}{2}} dt}{\sqrt{(1+t^2)}} = \frac{2}{3}[\sqrt{t}\sqrt{(1+t^2)} - \frac{1}{2}\kappa(\alpha, \frac{1}{\sqrt{2}})] , \qquad (A.38)$$

(v) $$\int \frac{t^{\frac{3}{2}} dt}{(1+t^2)^{\frac{3}{2}}} = -\frac{\sqrt{t}}{\sqrt{(1+t^2)}} + \frac{1}{2}\kappa(\alpha, \frac{1}{\sqrt{2}}) , \qquad (A.39)$$

(vi) $$\int \frac{t^{\frac{5}{2}} dt}{(1+t^2)^{\frac{3}{2}}} = \frac{3}{2}[\kappa(\alpha, \frac{1}{\sqrt{2}}) - 2E(\alpha, \frac{1}{\sqrt{2}}) + 2\frac{\sqrt{t}\sqrt{(1+t^2)}}{1+t}] - \frac{t^{\frac{3}{2}}}{\sqrt{(1+t^2)}} , \qquad (A.40)$$

(vii) $$\int \frac{t^{\frac{7}{2}} dt}{(1+t^2)^{\frac{3}{2}}} = -\frac{t^{\frac{5}{2}}}{\sqrt{(1+t^2)}} + \frac{5}{3}[\sqrt{t}\sqrt{(1+t^2)} - \frac{1}{2}\kappa(\alpha, \frac{1}{\sqrt{2}})] , \qquad (A.41)$$

where

$$\kappa(\alpha, \frac{1}{\sqrt{2}}) = \int_0^\alpha \frac{d\alpha}{\sqrt{(1-\frac{1}{2}\sin^2\alpha)}} , \quad E(\alpha, \frac{1}{\sqrt{2}}) = \int_0^\alpha \sqrt{(1-\frac{1}{2}\sin^2\alpha)} d\alpha ,$$

$$\cos\alpha = \frac{1-t}{1+t}, \quad \frac{d\kappa}{dt} = \frac{1}{\sqrt{t}\sqrt{(1+t^2)}} , \quad \frac{dE}{dt} = \frac{\sqrt{(1+t^2)}}{\sqrt{t}(1+t)^2} .$$

The field equations (55)-(57) pertaining to regions of zero magnetic charge density can be integrated in the form

$$f_1(t) = \frac{t^{\frac{3}{2}} I_1(t)}{\sqrt{(1+t^2)}} + \frac{1}{4}[3m(-1)^s - 1]\frac{t^{\frac{3}{2}} \kappa_\alpha}{\sqrt{(1+t^2)}} - \frac{3}{2}m(-1)^s \frac{t^{\frac{3}{2}} E_\alpha}{\sqrt{(1+t^2)}}$$

$$+ \frac{t^2(2t^2-t+3)}{2(1+t)(1+t^2)}(-1)^s m + \frac{1}{2}\frac{t^2}{1+t^2} , \qquad (A.42)$$

$$f_2(t) = -\frac{2mG}{c^2 \ell_o}\frac{\sqrt{t}}{\sqrt{(1+t^2)}} - \frac{\sqrt{t} I_2(t)}{\sqrt{(1+t^2)}} + \frac{1}{4}[3+m(1+(-1)^s)]\frac{\sqrt{t} \kappa_\alpha}{\sqrt{(1+t^2)}}$$

$$- \frac{3}{2}\frac{\sqrt{t} E_\alpha}{\sqrt{(1+t^2)}} + \frac{1}{3}mt - \frac{1}{2}(-1)^s m \frac{t}{1+t^2} + \frac{1}{2}\frac{t(2t^2-t+3)}{(1+t)(1+t^2)} , \qquad (A.43)$$

$$\frac{df_3}{dt} = \frac{mG}{c^2 \ell_o}\frac{t^{\frac{3}{2}}}{(1+t^2)^{\frac{3}{2}}} + \frac{1}{3}\frac{t^2}{1+t^2} - \frac{1}{4}\frac{mt^{\frac{3}{2}} \kappa_\alpha}{(1+t^2)^{\frac{3}{2}}} - \frac{m^2 t^{\frac{3}{2}}}{(1+t^2)^{\frac{3}{2}}} I_3(t) , \qquad (A.44)$$

where

$$I_1(t) = \frac{3}{8}[\sqrt{(1+m)}(1+\sqrt{m}) - \frac{2}{3}(1+m)^{\frac{1}{2}}]\kappa_\gamma + \frac{3}{8}[\frac{2}{3}(1+m)^{\frac{1}{2}} - \sqrt{(1+m)}(1-\sqrt{m})]\kappa_\omega$$
$$- \frac{3}{4}\sqrt{m}\sqrt{(1+m)}(E_\gamma + E_\omega) + \frac{3}{8}\sqrt{m}\sqrt{(1+m)}[\frac{1-\cos\gamma}{\sin\gamma}\Delta_1 + \frac{1-\cos\omega}{\sin\omega}\Delta_2]$$
$$- \frac{1}{2}\frac{\sqrt{t}\sqrt{(1+m^2t^2)}}{\sqrt{(1+t^2)}}, \tag{A.45}$$

$$12m^{\frac{1}{2}}(1+m)^{-\frac{3}{2}}I_2(t) = [\pm\frac{1}{2}\tau_1\tau_2 - \frac{5}{3}\tau_1^2 - \frac{9}{2}\tau_1^2\tau_2^2 - \frac{307}{24}]\kappa_\gamma +$$
$$[\pm\frac{1}{2}\tau_1\tau_2 - \frac{5}{3}\tau_2^2 - \frac{9}{2}\tau_1^2\tau_2^2 - \frac{307}{24}]\kappa_\omega$$
$$\frac{5}{3}(2\tau_1^2+7)E_\gamma + \frac{5}{3}(2\tau_2^2+7)E_\omega +$$
$$\frac{5}{3}\frac{1-\cos\gamma}{\sin\gamma}\Delta_1[1-2\tau_1^2 + \frac{4(1-\cos\gamma)}{\sin\gamma}] + \frac{5}{3}\frac{1-\cos\omega}{\sin\omega}\Delta_2[1-2\tau_2^2 + \frac{4(1-\cos\omega)}{\sin\omega}]$$
$$- 6 m^{\frac{1}{2}}(1+m)^{-\frac{3}{2}}t^{\frac{3}{2}}\frac{\sqrt{(1+m^2t^2)}}{\sqrt{(1+t^2)}}. \tag{A.46}$$

The expressions (A.7) and (A.8) for

$$z^2 > \frac{(1-m)^2}{4m}, \tag{A.47}$$

can be expressed in the two asymptotic regions where either $t > m^{-1}$ (outer magnetic layers) or $t < m^{-1}$ (inner magnetic layers) where

$$z^2 = \frac{(1+mt^2)^2}{4mt^2}.$$

The inequality (A.47) holds in both asymptotic regions. Thus for the elliptic functions of the first and second kind we obtain

$$\kappa_\gamma \to \sqrt{2}\, m^{-\frac{1}{4}}\sqrt{(1+m)}\int\frac{dz}{2z\sqrt{(z-1)}} = \sqrt{2}\, m^{-\frac{1}{4}}\sqrt{(1+m)}\eta\, \tan^{-1}(z-1)^{\frac{1}{2}\eta},$$

$$\kappa_\omega \to \sqrt{2}\, m^{-\frac{1}{4}}\sqrt{(1+m)}\int\frac{dz}{2z\sqrt{(z+1)}} = \sqrt{2}\, m^{-\frac{1}{4}}\sqrt{(1+m)}\, \tanh^{-1}(z+1)^{\frac{1}{2}\eta},$$

$$E_\gamma \to \sqrt{2}\, m^{-\frac{1}{4}}\sqrt{(1+m)}\int\frac{2dz}{z\sqrt{(z-1)}} = 4\sqrt{2}\, m^{-\frac{1}{4}}\sqrt{(1+m)}\eta\, \tan^{-1}(z-1)^{\frac{1}{2}\eta},$$

$$E_\omega \to \sqrt{2}\ m^{-\frac{1}{4}} \sqrt{(1+m)} \int \frac{2dz}{z\sqrt{(z+1)}} = 4\sqrt{2}\ m^{-\frac{1}{4}} \sqrt{(1+m)}\ \tanh^{-1}(z+1)^{\frac{1}{2}\eta},$$

where $\eta = 1$ for $t < m^{-1}$ and $\eta = -1$ for $t > m^{-1}$.

References

(1) Behram Kursunoglu, Phys. Rev. 13, March 15, 1976.

(2) Behram Kursunoglu, *New Directions In Generall Theory of Relativity*, a 1982 Center for Theoretical Studies Report, 223 pp.

(3) Behram Kursunoglu, *Journal of Physics Essays*, December 15, Vol. 4, No. 4, pp. 437-518, 1991.

(4) Behram Kursunoglu, In the 1994 Proceedings of the Coral Gables Conference *Unified Symmetry In the Small and in the Large*. Editors, B. N. Kursunoglu, Stephan Mintz, Arnold Perlmutter, Plenum Publishing Corp., New York, 1995.

CALCULATION OF COSMOLOGICAL PARAMETERS AND THEIR APPROXIMATIONS IN THE STANDARD BIG BANG MODEL

Ralph A. Alpher
Union College and Dudley Observatory, Schenectady, New York 12308

and

Robert Herman
University of Texas, Austin, Texas 78712

" There is no excellent beauty that hath not some strangeness in the proportions."
—Francis Bacon

ABSTRACT

In this paper the properties of the standard Big Bang model are recapitulated, with emphasis on the open model. In particular the changes with time in the expansion of characteristic parameters are described. Special attention is paid to approximations for these parameters at very early times (but still in the range of validity of the standard model) and at times close to the present age of the universe (for various assumed values of the Hubble parameter). We then discuss the calculation of the present background radiation temperature using current cosmological parameters, based on both exact and approximate solutions of the governing Friedmann-Lemaître equation. Calculations carried out by

George Gamow in the late 1950s, described by some authors as displaying an unusual insight into the theoretical model, are shown to involve approximations which do not yield the functional form of the solutions required by the Friedmann-Lemaître equation, particularly at late times, and which do not give a sensible set of parameters for the conditions at crossover. Nevertheless Gamow did calculate values of the present radiation temperature which were not unreasonable, despite the fact that his formalism was incorrect. In fact, his formulas for the time history of matter and radiation densities are closer to the results of the exact solution than are the correct approximations in the vicinity of the present age of the universe. In the end, however, this work of Gamow is of only historical interest, since at the time he did the work, there was available to him a much simpler and exact approach to calculating the time histories of matter and radiation densities.

In the course of this examination of approximations to exact solutions of the Friedmann-Lemaître equation, we considered the question of determining the quantity Ω, the ratio of total density to the critical density (the density which separates solution space into open and closed models.) We have evaluated Ω numerically and find, as expected, that at early times the value for Ω is extraordinarily close to unity, and that the deviation from unity becomes increasingly small for earlier and earlier times. On the other hand deviations begin to become appreciable at times of the order of about one percent of the age of the universe, and the calculated Ω easily encompasses the low present value suggested by primeval light-element nucleosynthesis, as well as higher values which have been suggested recently, as for example, $0.1 < \Omega_o < 0.3$. We conclude that it may not be necessary to invoke that consequence of an inflationary paradigm which requires that the value of Ω be unity throughout the history of the expansion, even though it is so close to unity earlier, nor does one need to take Ω equal to one as an initial condition.

INTRODUCTION

Calculation of the present temperature of the cosmic microwave background radiation (CMBR) is a well-understood procedure with the standard Big Bang cosmological model, and is documented in many places (Alpher and Herman, 1994; Peebles, 1993; Turner, 1993). Given the impressive measurements made aboard the COBE satellite (Mather et al, 1994) the present temperature of 2.726 ± 0.01 K has become the most accurately known cosmological parameter, and consideration of its calculation is now primarily of historical interest, although the connection between this present temperature and conditions in the early prestellar state of the universal expansion and throughout the expansion are still of considerable current interest.

There persist in the literature statements about what George Gamow did in regard to the thermal history of the universe, statements which are not correct and which generally do not reveal the motivation or methodology of his approach to this calculation. To put the matter in perspective, we note that we developed a methodology and provided a theoretical calculation of the background temperature based on the standard Big Bang model in 1948, according to an algorithm which was published in papers in 1948, 1949 and 1950, as well as in a number of later review papers (Alpher and Herman, 1975, 1988, 1990, 1993, 1994) Gamow took a different approach in his calculation, using approximations for the temporal behavior of the densities of matter and radiation which were not correct in principle, or which were used at times in the expansion outside the range of their validity, all of which was in fact quite unnecessary given the results we had already published. As best we can tell, the unstated motivation for the approximations he used apparently stemmed from a desire to estimate the present CMBR temperature without invoking the early density of matter required for nucleosynthesis of the light elements. As it turns out, the numerical value of the CMBR temperature is not very sensitive to the problems with Gamow's approach; the implicit or derived value of the early matter density in Gamow's work is near the then-current as well as the present value, and thus Gamow obtained values for the present CMBR temperature of the order of several degrees to tens of degrees. It should be borne in mind that our work as well as that of Gamow on the CMBR was well in advance, about sixteen years, of the observation by Penzias and Wilson (1965), as well as of the several uninterpreted, and until recently unreported, measurements during the 1950s. It is appropriate to remind readers of the first observations of a CMBR, seen but not understood, by Adams and McKeller in the early 1940s (see Alpher and Herman, 1994).

It is surely of at least historical interest that in the three specific publications in which Gamow described his approach and results (Gamow 1952, 1954, 1956) there is no reference whatsoever to the algorithm we had developed, or to our results. Nor was there a reference to any source of information in a 1950 paper by Gamow in which he implicitly stated as given that the CMBR temperature was 3 K. All of this is the more mysterious in light of ongoing correspondence we had with Gamow during these years. For example, in response to a telegram we had sent to him in 1948 telling him of our intent to publish a 5 K estimate for the CMBR temperature in a letter to Nature (Alpher and Herman, 1948), he wrote

" . . .Then you write 'our theory uses matter density now 10^{-30} and radiation density now 10^{-32} .' Where did you get this value 10^{-32} ? From astronomical data? If so, it is quite wrong. The space temperature of about 5 K is explained by the present radiation of stars (C-cycle etc.). The only thing we can tell is that the 'residual temperature' from the original heat of the Universe is not higher than 5 K, but it could be as close to zero as one likes. Thus one cannot use 5 C (sic) and ρ_{rad} (now) = 10^{-32} as the observed boundary condition. . . ."

It seems clear to us that by 1950, and certainly by 1952, Gamow had accepted the reality of a CMBR and its temperature, but proceeded without invoking the algorithm we had already published. Gamow came around in 1967, when he suggested that the three of us publish a paper in the Proceedings of the National Academy of Sciences (Alpher, Gamow, and Herman, 1967), which we discuss later. His suggestion that we publish jointly was in the response to the fact that the exact approach by Alpher and Herman, as well as Gamow's approach for that matter, had been overlooked or ignored in the wake of the Penzias-Wilson publication in 1965.

Given what we have said above, we cannot resist quoting, with astonishment, a statement by John Maddox, Editor of Nature, (1994) in a review of cosmology:

> "The microwave background radiation, which fills even the corners of the universe, would psychologically have been more compelling evidence for the Big Bang if it been predicted before its discovery in 1965. That it was not is something of a surprise, which is nevertheless now irrelevant."

Several communications from us to Maddox concerning these remarks have gone unacknowledged by him as of this writing.

Some recent examples of a widespread problem that exists follow. Thus, in a recent article written about Gamow's contributions, and in particular in discussing his approach to estimating the present value of the CMBR temperature involving approximations and assumptions with which we take issue, Chernin (1994) wrote

> ". . . And the assumption was so simple, so natural and so implicit that nobody could have any objections, or even exact realization of it (sic)."

In a recent overview of the conflict between the Big Bang model and the Steady State model, Brush (1993) made the following comment, including a footnote (Brush's number):

> "Following the initial publication of this prediction by Alpher and Herman in 1948, Gamow published other estimates of the temperature of the radiation, including one in his popular book *The Creation of the Universe* (1952) , which was widely distributed in paperback edition in the early 1960s [4]. Yet most astrophysicists ignored these predictions before 1965."

> "[4] Gamow (1950) gave the temperature at the present epoch as approximately 3 K. In his book (see the 1961 edition, p. 40) he estimated the temperature to be 50 K. (The book was first published in 1952 and went through several editions and reprintings.) In another paper (1956) he derived values of 40 K

and 6 K from two different models. Alpher and Herman (1975, 1988, 1990) discussed these and other published estimates and the publicity given to them."

In our view Gamow did not use different models. Rather he misused the standard model by making incorrect and unnecessary approximations. Nevertheless, his papers did emphasize the possibility of simple relationships between conditions at times early and late in the expansion (albeit that his approach was formally incorrect), as well as the role of cosmological parameters in discussions of the conditions when the universe went through decoupling (see below) as this pertains to the formation of galaxies.

In retrospect it seems unfortunate that those considering the early work on Big Bang cosmology and making unwarranted inferences from Gamow's papers on the CMBR seem not to have read the already mentioned joint paper we wrote with Gamow in 1967. Gamow suggested that we publish this paper jointly in light of the fact that our early calculation of the CMBR temperature had been overlooked or ignored for some time after the Penzias-Wilson observation reported in 1965. In this joint paper Gamow in effect abandoned the relevance of his approximations and approach to calculating the CMBR temperature.

It is our plan in this paper to give the essence of the standard open Big Bang model, particularly in the context of tracing the temporal evolution of the characteristic parameters. We will describe various legitimate approximations one may make, as, for example, early in the expansion when the behavior of the expansion is controlled by the radiation content of the universe, the conditions when the behavior changed from radiation-controlled to matter-controlled, which is close to the so-called decoupling regime, and finally the behavior late in the expansion when the expansion is determined by the matter density. We shall show how all of this is used in determining the present CMBR temperature or is consistent with the accurately measured value. We discuss the ranges of validity of the several approximations and also the calculation of the density ratio Ω and its approximate values at early and late times in the expansion. We evaluate numerically the early time approximation for Ω and find, as expected, that it is extremely close to unity at times of interest in the study of the very early stages of the standard Big Bang model, and remains very close to unity up to about a million years after the big bang. We consider briefly the early seminal contributions of Gamow to the Big Bang model, and then focus on the basis of his calculations of the CMBR temperature. Finally, we compare his result with the exact value obtained from application of the algorithm we had developed earlier. We emphasize throughout the intimate connection between the major cosmological parameters. However, we shall not deal with the continuing current activity on measuring the Hubble parameter, H_o, whose value is still bouncing about in the range ~ 50 – 100 km/(secMpc) as it has for many years. It would appear that this question can only be settled by observation of a statistically valid set of objects over the universe. Where we employ H_o, we take variously as examples 50, 55 and 75 km/(secMpc).

If the measurements in the future converge on a value of the Hubble parameter greater than, say, 75, it may be necessary to resurrect a nonzero cosmological constant to obviate the model giving too short an age. We seem to be once again encountering a situation which existed in the 1940s and early 1950s, when the then-current value of the Hubble parameter gave an unacceptably short age compared to other astronomical ages, with two effects. First, a lot of work was done on models with nonzero cosmological constants. Second, and most poignant, the situation raised doubts in the minds of many about the standard Big Bang model and favored the popularity of alternatives such as steady-state models, matter-antimatter models, etc. When the observations settled down in later years and basically oscillated with values being between 50 and 100, the lower end of the range made for reasonable ages of the model. Until a few years ago, there were two schools, one favoring values of H_o near 100, and the other near 50; the former gives unacceptably short ages, the latter gives acceptable ages. We note with interest that Bartlett, Blanchard, Silk and Turner (1995) give interesting theoretical arguments for a value of $H_o \sim 30$ Km/(secMpc).

THE STANDARD BIG BANG MODEL

The basic equations describing the expansion of the standard Big Bang model are the Friedmann-Lemaître equation, which follows from Einstein's field equations, and the equation relating the change in the internal energy of the cosmos with the work done in the expansion. These equations with the cosmological constant $\Lambda = 0$, are

$$\frac{1}{L}\frac{dL}{dt} = H(t) = \left(\gamma \rho - \frac{c^2}{L^2 R_o^2}\right)^{1/2} \tag{1}$$

and

$$\frac{d}{dt}(\rho L^3) + p\frac{d}{dt}(L^3) = 0 \tag{2}$$

where $\gamma = 8\pi G/3 = 5.59 \times 10^{-7} (\text{cm sec/g})$, $L = l/l_o$ is the dimensionless proper distance in which l_o is chosen so that $L = 1$ now, $(1/L)dL/dt$ is the Hubble parameter, ρ is the sum of the densities of matter and radiation, R_o is the unit measure for the radius of curvature of the universe, taken to be equal to the radius of curvature now, c is the velocity of light and p is the total pressure. The units are c.g.s. throughout. In interpreting Eq. (2), we note that in the standard Big Bang model, energy is conserved only in a local co-moving volume L^3.

There are assumptions implicit in Eqs.(1) and (2). namely,
- the universe is homogeneous and isotropic, viewed on a sufficiently large scale;

- the content of the universe may be treated as an ideal inviscid two-component fluid, but with radiation pressure much greater than the pressure of matter;
- The content of the visible universe is a fair sample of what we can and cannot observe;
- structure is ignored, and the content of the universe may be characterized as having an average smeared-out density;
- the content of the as yet undifferentiated universe includes matter and radiation, whose characteristics are governed by baryon conservation, by radiation expanding adiabatically, and by the heat transfer between matter and radiation being small but sufficient to maintain the two constituents at the same temperature. This assumption follows from the enormous value of the ratio of the heat capacity of radiation to that of matter (Alpher, Gamow, and Herman, 1967; Alpher and Marx, 1992).

Then, baryon conservation leads to a relationship for the density of matter, as

$$\rho_m L^3 = constant \tag{3}$$

while the adiabatic expansion of radiation, which maintains a Planck distribution of energy versus frequency, and which requires that the temperature varies inversely as the proper distance, (or, $T \propto L^{-1}$ and $L = (1+z)^{-1}$ where z is the red shift) leads to a relationship for the radiation density, namely,

$$\rho_r L^4 = constant \tag{4}$$

Combining Eqs.(3) and (4) leads to a very useful power law, namely,

$$\rho_r \rho_m^{4/3} = constant \tag{5}$$

Clearly the value of this constant holds throughout the expansion, so long as the assumptions stated are valid. In particular, we note that

$$\rho_r \rho_m^{-4/3} = constant = \rho_{r'} \rho_{m'}^{-4/3} = \rho_x^{-1/3} = \rho_{ro} \rho_{mo}^{-4/3} = \zeta \tag{6}$$

where the prime in the subscripts indicates the value of the densities early in the expansion (we shall take $t = 1$ second as the canonical early time), the subscript x denotes the values when the densities of matter and radiation are equal, i.e., at the time of crossover or decoupling, and, finally, a zero in the subscript denotes values at the present epoch, when the dimensionless proper distance is taken as unity. At the risk of belaboring the obvious, let it be noted that if one knows $\rho_{r'}$, $\rho_{m'}$ and ρ_{mo} then it is a straightforward calculation to obtain a ρ_{ro} and thence, with the Stefan-Boltzmann equation, the value of the present background temperature, $T_o(t = t_o)$.

The Friedmann-Lemaître Eq. (1) can be integrated as it stands, without approximation, and the result is given by Alpher and Herman (1949) and by Heckmann (1942, 1968). We need not use these complex results here, because it is easier to deal with the solutions in parametric form. There are three sets of parametric solutions (see, for example, Alpher and Herman, 1988; Weinberg 1972), depending on whether the radius of curvature is real (a closed elliptic model), imaginary (an open hyperbolic model), or infinite (a flat or Euclidean model). We restrict our attention to the open model on the grounds that present evidence indicates that the smeared-out density of matter (taken to be baryons) is less than the critical density (Coles and Ellis, 1994). We shall in this paper use baryon densities consistent with light element nucleosynthesis. The contribution of so-called "dark matter" may bring the density closer to the critical density, but it is by no means clear how this question will be settled finally, even though a mean density equal to unity for all time appears to be demanded by some versions of an early inflationary model, or on philosophical grounds (see our discussion about Ω later). Note that the value of the critical density follows from Eq. (1) when one takes the radius of curvature to be infinite, and can be written as

$$\rho_c(t) = H^2(t)/\gamma \tag{7}$$

We can write the parametric solutions for an open model ($R_o = iR_i$) in terms of the development angle ϕ as follows:

$$L = a_o(\cosh\phi - 1) + b_o \sinh\phi \tag{8}$$

and

$$t = \frac{R_i}{c}\left[a_o(\sinh\phi - \phi) + b_o(\cosh\phi - 1)\right] \tag{9}$$

where the dimensionless coefficients are given by

$$a_o = \tfrac{1}{6} c^2 R_i^2 \kappa \rho_{mo}, \quad b_o = c R_i (\kappa \rho_{ro}/3)^{1/2} \tag{10}$$

with $\kappa = 8\pi G/c^4$. It should be noted that the coefficients contain the constant R_i, the imaginary scale factor for the radius of curvature, which can be expressed through the present value of the Hubble parameter and the present total density of the universe, by invoking Eq. (1) as

$$R_o = iR_i = c\left[\gamma(\rho_{mo} + \rho_{ro}) - H_o^2\right]^{-1/2} \tag{11}$$

It should also be noted that in order to utilize the parametric solution, Eqs. (8) through (10), we need to specify, say, the present temperature, T_o, and the derived radiation density, ρ_{ro}, of the universe, for which we will henceforth use the COBE value, 2.726 kelvin (we ignore the stated probable error of ± 0.01 here and in what follows) and the present matter density, ρ_{mo}. For the former we convert from 2.726 kelvin to radiation density by the Stefan-Boltzmann law

$$\rho_{ro} = aT_o^4/c^2 = 4.65\times10^{-34} \text{ g cm}^{-3} \tag{12}$$

where a is the radiation density constant, and c is the velocity of light. For the matter density we invoke a recent result of Olive and Steigman (1994) for Ω for baryons, as determined for light element nucleosynthesis. Recall that Ω is defined as the ratio of total density to the critical density, the latter being in turn defined as that density of the universe for which the radius of curvature R_o is infinite, with the total and critical densities functions of the epoch in general. Olive and Steigman give, for $\Omega o = \Omega(\text{now})$, $\Omega h_{50}^2 = 0.05\pm0.01$ where $h_{50} = H_o/50$ and H_o is the present value of the Hubble parameter in km/(secMpc). Then, the calculation of the present matter density would be independent of the value of the Hubble parameter, since $\rho_c = H_o^2/\gamma$ and H_o^2 cancels. The resultant baryonic mass required by BBM is given by

$$\rho_{mo} = 2.35\times10^{-31} \text{ g cm}^{-3} \tag{13}$$

Again we shall ignore the probable error of Ω here and in subsequent discussion. With these values of present matter and radiation mass densities, we can evaluate the open model parametric solution. We present in Fig. 1, an example of what Gamow called "divine curves", a plot of the principal cosmological parameters for a Hubble parameter of 55 km/(secMpc). Note that what dependence there is on the Hubble parameter enters through the calculation of $R(t)$, where the $H(t)$ dependence vanishes at early time in the expansion. Later we shall use the limits given by Olive and Steigman on η, the ratio of the number density of baryons to the number density of photons (or the reciprocal dimensionless entropy per baryon), again as found to be necessary for light element nucleosynthesis. Figure 2 shows log T as a function of log t from 10^{15} sec to beyond the present age of the universe for $H_o = 50$ and 75 km/(secMpc).

It is interesting that new observational capabilities appear to make it possible to measure the CMBR at earlier epochs. For example, Songalla et al (1994) have given recently a temperature attributed to the CMBR of 7.4 ± 0.8 K in a gas cloud, obtained by measuring absorption from the first fine-structure level of neutral carbon atoms. The red shift of the cloud is reported to be 1.776. Using $T \propto L^{-1}$, one obtains $T = 7.58$ K at this red shift. We find the age of these measurements to be 1.35×10^{17} sec and 1.88×10^{17} sec, which are approximately 0.33 of the age of the universe, for $H_o = 50$ and 75 km/(secMpc),

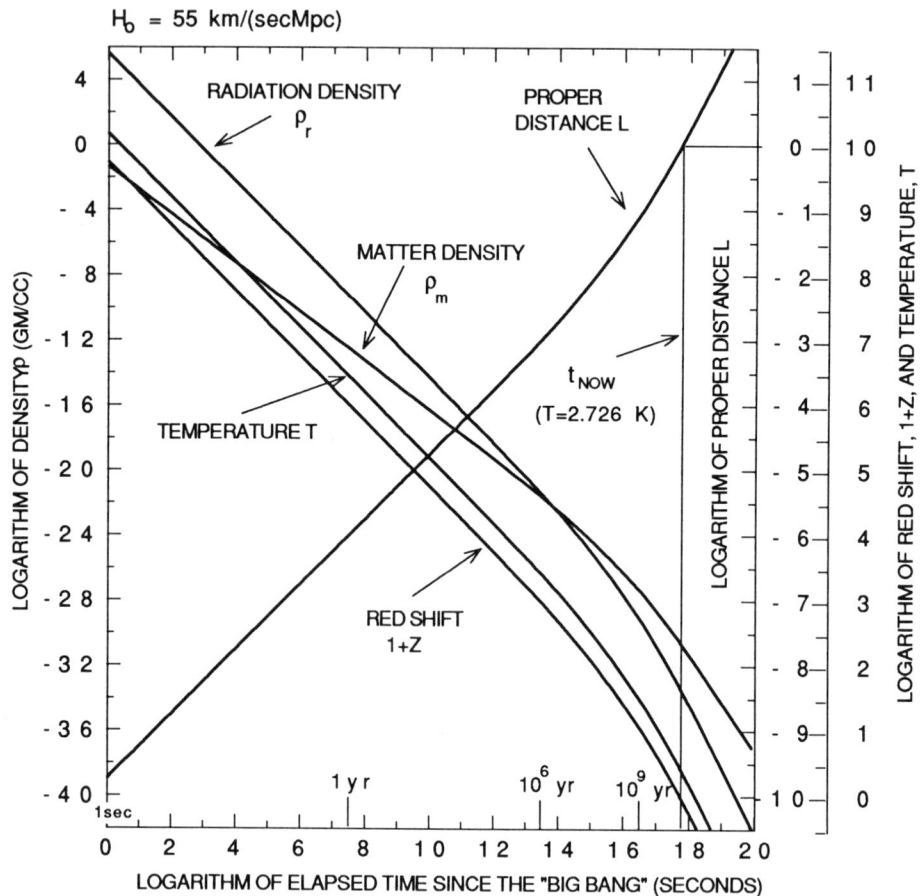

Figure 1. The temporal behavior of the major cosmological parameters as obtained from the exact parametric solution of the Friedmann-Lemaître equation, in the case of an open universe, for a sample case of H_o = 55 Km/(secMpc), a present background temperature of 2.726 K, and a present baryonic matter density of 2.34×10^{-31} gm/cc (Ω_{BBN} = 0.05). These values of the present background temperature and present matter density are used in all the Figures in this paper.

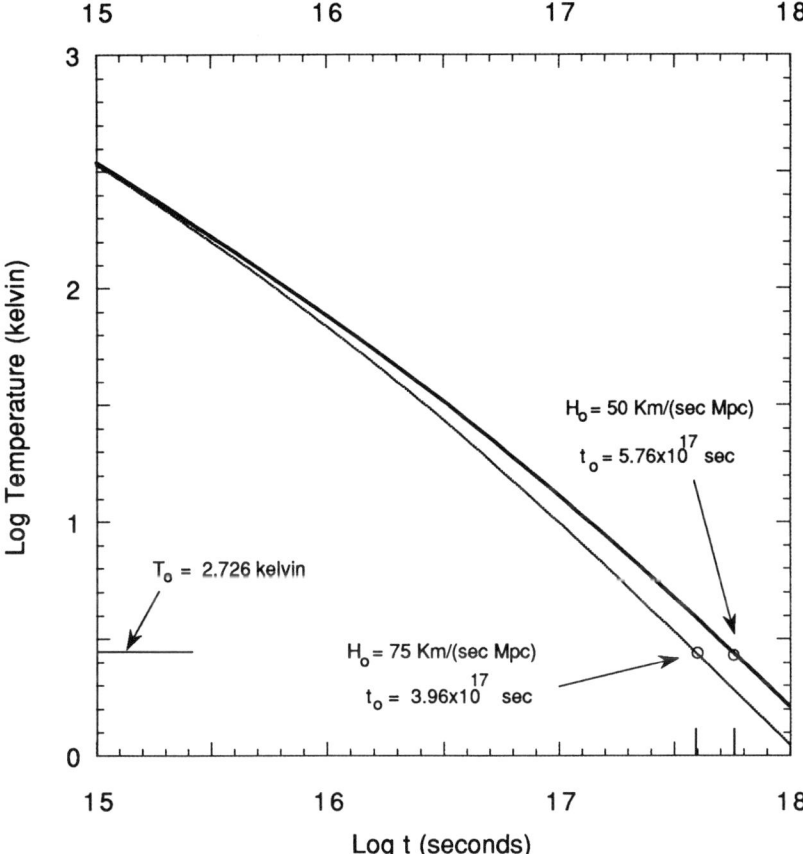

Figure 2. The temporal behavior of the temperature from the exact parametric solution of the Friedmann-Lemaître equation describing an open universe for two sample values of the Hubble parameter, as marked, for recent epochs in the expansion.

respectively. Later we shall describe how one can use late time approximations to obtain the epoch of their measurement.

Given the values of matter and radiation densities now, we note that the proper distance at the time of crossover comes directly from Eqs. (3), (4) and the power law, Eq. (6), written as $L_x = \rho_{ro}/\rho_{mo}$ with ρ_{mo} and ρ_{ro} as determined above; thus we have $L_x = 1.98 \times 10^{-3}$, so that the corresponding red shift is $z_x = L_x^{-1} - 1 = 504$. Invoking Eqs. (3) and (4), we can now calculate the matter and radiation densities at crossover. The quantity L_x is independent of the Hubble parameter, but the crossover time, which must be obtained from the exact solution of the Friedmann-Lemaître equation, has a small H_o dependence, again through the calculation of R_i as a function of ρ_{mo}, $\rho_{ro} \gg \rho_{mo}$, and the present value of the Hubble parameter. The crossover times for $H_o = 30, 50, 75$ and 100 are $9.47, 9.44, 9.39$ and 9.30×10^{13} sec. The density at cross over, $\rho_x = \rho_{mx} = \rho_{rx}$ is 3.03×10^{-23} g cm^{-3}.

Finally, given that $T \propto L^{-1}$ for the adiabatic expansion of radiation, it follows from the definition of L_x with L_o taken as unity, that $T_x/T_o = L_o/L_x = 1/L_x$ and $T_x \cong 1380$ K which is independent of the Hubble parameter.

The reader is reminded that the conditions at crossover describe the ambiance when control of the expansion switched from radiation to matter, and is frequently identified with the epoch when matter and radiation decoupled or the epoch during or near which plasma recombination occurred. These conditions then define the starting point for the free adiabatic expansion of radiation to the present day, with the temperature of matter and radiation remaining coupled even with minimal interaction because of the enormous heat capacity of radiation compared to that of matter. Certainly plasma recombination must have started at a smaller L, when the temperature was more like 3000 - 4000 K. For example, the exact parametric solution with a Hubble parameter of 75 km/(secMpc) gives the temperature as 3740 K at a z of 1370, or at a time of 4.7×10^5 years.

APPROXIMATIONS AT EARLY TIMES IN THE STANDARD BIG BANG MODEL

Using the exact solutions of the Friedmann-Lemaître equation is cumbersome for most purposes, although the parametric solutions are easily programmed for a desktop computer. Nevertheless, it is useful to consider approximations for the solutions valid for times early and late in the expansion. As will be discussed later, it was the misuse of such approximations by Gamow which has created a problem and which continues to be misinterpreted in the literature as being at least interesting and perhaps a scientific coup (Cernin 1994).

To obtain approximations at either early or late times, we rewrite Eq. (1) as follows, where we show the total density as the sum of matter and radiation densities:

$$\frac{1}{L}\frac{dL}{dt} = \left[\gamma\left(\rho_{mo} L^{-3} + \rho_{ro} L^{-4}\right) - \frac{c^2}{L^2 R_o^2}\right]^{1/2} \tag{14}$$

or

$$dL\left[\gamma\left(\rho_{mo} L^{-1} + \rho_{ro} L^{-2}\right) - \frac{c^2}{R_o^2}\right]^{-1/2} = dt$$

For sufficiently early times L becomes very small, so that in Eq. (14) we keep only the term involving $1/L^2$ and integrate, setting the constant of integration to be zero because as $t \to 0$, $L \to 0$, with the result

$$L \approx (4\gamma \rho_r)^{1/4} t^{1/2} \tag{15}$$

Using Eq. (4), we may write, since $L_o = 1$,

$$\rho_r L^4 = \rho_{ro} \tag{16}$$

Substituting this ρ_{ro} in Eq. (16), we find that L cancels, and we solve for ρ_r(early) = ρ_{re}, as

$$\rho_{re} \approx \frac{1}{(4\gamma t^2)} = 4.47 \times 10^5 \, t^{-2} \text{ g cm}^{-3} \tag{17}$$

One of the most widely used approximations at early times is that for the temperature. It follows from Eq. (17) when we use the Stefan-Boltzmann relation $\rho_r = aT^4/c^2$, that $1/(4\gamma t^2) = aT^4/c^2$, so that

$$T_e \approx \left(\frac{c^2}{4\gamma a}\right)^{1/4}\frac{1}{t^{1/2}} = \frac{1.52 \times 10^{10}}{t^{1/2}} \text{ K} = \frac{1.31}{t^{1/2}} \text{ MeV} \tag{18}$$

To continue, we use Eq. (3), written as

$$\rho_m L^3 = \rho_{mo} \tag{19}$$

to obtain an expression for ρ_m(early), written as ρ_{me}, again by eliminating L from Eq. (19) using Eq. (15), which is valid for a radiation-controlled early universe, namely,

$$\rho_{me} \approx (4\gamma \zeta)^{-3/4} t^{-3/2} \tag{20}$$

where ζ was defined in Eq. (6).

As a matter of interest if we return to Eq. (14) and consider a matter-only universe with c^2/R_i^2 neglected, then we find that $L_e = (9\gamma\rho/4)^{1/2} t^{2/3}$, and $\rho_{me} = [4/(9\gamma)]/t^2$. Thus in the matter-only universe as in the radiation-only universe, the early densities of matter and radiation depend on universal constants only. If we make use of the present temperature of 2.726 K in calculating ρ_{ro} then we can write Eq. (20) as

$$\rho_{me} \approx 1.73 \times 10^{29} \rho_{mo} \, t^{-3/2} \text{ g cm}^{-3} \tag{21}$$

In their recent analysis of light element nucleosynthesis Olive and Steigman, (1994) also give values for the range of the parameter η (which is a constant during the expansion once baryon conservation pertains because n_b and n_{ph} have the same temperature dependence), namely,

$$2.5 \times 10^{-10} < \eta < 3.9 \times 10^{-10} \tag{22}$$

Because the ratio is constant, we can rewrite the upper and lower bounds as applying at the present time. We use the integral of the Planck function, $n_{ph} \cong 20.3 \, T^3 \text{ cm}^{-3}$ to calculate the number density of photons, with ρ_{mo} in g cm^{-3}, to obtain

$$\eta = \frac{\rho_{mo}}{m_b \, n_{ph}} = \frac{\rho_{mo}}{6.88 \times 10^{-22}} \tag{23}$$

Consequently, Eq. (22) yields, for the present matter density, in g cm^{-3}

$$1.72 \times 10^{-31} < \rho_{mo} < 2.68 \times 10^{-31} \tag{24}$$

and, using the relation given by Eq. (21) for the early matter density, i.e., the matter density at one second, in g cm^{-3}, we obtain

$$2.97 \times 10^{-2} < \rho_{m'}(t = 1 \text{ sec}) < 4.63 \times 10^{-2} \tag{25}$$

This range is consistent with most prior estimates of the baryon density required for light element nucleosynthesis. Thus in 1950 we reported a value from the work of Fermi and Turkevich of 1.66×10^{-3} g cm^{-3} at $t = 1$ second. In 1975 we reported 6.02×10^{-2} g cm^{-3} at $t = 1$ second, based on the work of Wagoner, as well as Wagoner, Fowler and Hoyle, and others, prior to that time. Clearly this parameter has not changed very much over the many years of study of light-element nucleosynthesis.

APPROXIMATIONS AT LATE TIME IN THE STANDARD BIG BANG MODEL

We again return to the Friedmann-Lemaître relation, Eq. (1), rewritten in the form Eq. (14). We note that at late time L is close to unity and the term involving the density is small compared to the curvature term, and so we can rewrite Eq. (14) as

$$dL \approx (c^2/R_i^2)^{1/2}\, dt \tag{26}$$

which can be integrated. We evaluated the constant of integration by choosing l_o in $l/l_o = L$ so that $L(now) = L_o = 1$. Then for the open model with $R_o = iR_i$ and writing $L(late) = L_l$, we have

$$L_l \approx 1 + \left(\frac{c^2}{R_i^2}\right)^{1/2}(t-t_o) \tag{27}$$

We again use Eqs. (16) and (19), with the results

$$\rho_{rl} \approx \rho_{ro}\left[1+ c_1(t-t_o)\right]^{-4} \tag{28}$$

and

$$\rho_{ml} \approx \rho_{mo}\left[1+ c_1(t-t_o)\right]^{-3} \tag{29}$$

with

$$c_1 = \left(\frac{c^2}{R_i^2}\right)^{1/2} = \left[H_o^2 - \gamma(\rho_{mo}+\rho_{ro})\right]^{1/2} \tag{30}$$

where as before H_o, ρ_{mo} and ρ_{ro} are present values. We note that Eq. (30), evaluated for ρ_{mo} and ρ_{ro} as given in Eqs. (12) and (13), gives, as expected, values of an imaginary radius of curvature, namely, for $H_o = 50$ km/(secMpc), $R_i(now) = 1.90 \times 10^{28} i$ cm, and for $H_o = 75$ km/(secMpc), $R_i(now) = 1.25 \times 10^{28} i$ cm.

We can rewrite Eq. (27), which is valid in the vicinity of t_o, the age of the universe, to determine useful expressions relating L_l and red shift z_l with time, namely,

$$\frac{t}{t_o} = 1 + \frac{L_l - 1}{c_1 t_o} = 1 - \frac{1}{c_1 t_o}\left(\frac{z_l}{1+z_l}\right) \tag{31}$$

As an example of using the above we refer back to our earlier discussion of the work of Songalla et al (1994). We use their measured red shift of 1.776, with $t_o = 5.75 \times 10^{17}$ sec for $H_o = 50$ km/(secMpc) and $t_o = 3.96 \times 10^{17}$ sec, for $H_o = 75$ km/(secMpc), these approximations give = 0.30 and 0.33, respectively. The exact parametric solutions give

$\tau = 0.33$ and 0.34, respectively. Songalla et al reported $\tau = 0.25$, but did not specify the characteristics of the model they employed, so that a direct comparison is not possible.

THE DENSITY RATIO Ω AND APPROXIMATIONS THERETO

The density ratio Ω is a fundamental cosmological parameter. If the canonical Big Bang model is valid, then ultimately the observational value of Ω (now) will tell us whether the universe is open ($\Omega < 1$), closed ($\Omega > 1$), or is exactly poised in between ($\Omega = 1$). In a homogenous, isotropic universe with a smoothed-out density, Ω is defined as

$$\Omega = \rho_t / \rho_c \tag{32}$$

where ρ_t is the mean mass density of matter of all kinds and radiation and ρ_c is the critical density, i.e., the density which pertains when the radius of curvature of the universe is infinite. Both densities are functions of the epoch. It is usually assumed that in the standard Big Bang model there is no matter-radiation interconversion. This assumption leads to the relationships $\rho_m \propto L^{-3}, \rho_r \propto L^{-4}$, which must be reexamined if there is interconversion. As we shall point out later this is an unnecessary assumption, insofar as a determination of Ω is concerned, except perhaps during any nonequilibrium period associated with inflation after the Planck time. The critical density is calculated from the Friedmann-Lemaître equation with the radius of curvature taken as zero. It is then defined as before

$$\rho_c = H^2 / \gamma \tag{33}$$

where H is the Hubble parameter (epoch-dependent), γ is $8\pi G/3$, and G is the constant of gravitation. We do not consider contributions to Ω from the cosmological constant, but do not exclude contributions from nonbaryonic matter.

Conventional wisdom would aver that at early times in the standard Big Bang model the parameter Ω is very nearly equal to one, as the term involving the radius of curvature in the Friedmann-Lemaître equation is relatively negligible. However, the term is not zero, and represents the difference between two large numbers, namely, the kinetic energy and the potential energy in the expansion. Moreover, it is stated in discussions of inflationary models (we quote from an article by Guth and Steinhardt, 1989) that

> "The standard model offers no explanation of why Ω began so close to one, but merely assumes the fact as an initial condition. This shortcoming of the standard model, called the flatness problem, was first pointed out in 1979 by Robert H. Dicke and P. James E, Peebles."

We call attention to the discussions of the "flatness problem" by many of the cosmologists interviewed by Lightman and Brawer (1990). In these interviews opinions vary from basically very little of a problem to an overriding concern.

In the spirit of this paper on approximations of cosmological parameters we sought to evaluate numerically the early-time approximation for Ω.. This approximation is, for example, discussed in Kolb and Turner (1990). For the sake of completeness we examine this approximation in the following way. One rearranges the usual Friedmann-Lemaître equation to read

$$\rho_t = \rho_c - \frac{c^2}{\gamma L^2 R_i^2} \tag{34}$$

where we have replaced the radius of curvature R_o by iR_i, thereby selecting the open cosmological model for further consideration. This choice is suggested by a recent paper by Coles and Ellis (1994) which reviews the rationale for an open universe, as well as another recent paper by Bartlett, Blanchard, Silk and Turner (1995) arguing mostly on theoretical grounds for a Hubble parameter of about 30 km/(secMpc). The quantity $L = l/l_o$ is the usual dimensionless proper distance or scale factor for the expansion, where l_o is chosen so that $L = 1$ at the present epoch. Combining Equations (32) through (34) then yields

$$\Omega = 1 - c^2 / (R_i^2 L^2 H^2) \tag{35}$$

To evaluate the second term of Eq. (35), we use the exact parametric solutions of the Friedmann-Lemaître equation, again selecting for an open model with no approximation, as given in Eqs. (8) and (9), and differentiate with respect to the development angle ϕ

$$dL = (a_o \sinh \phi + b_o \cosh \phi) d\phi \tag{36}$$

and

$$dt = \frac{R_i}{c} \left[a_o (\cosh \phi - 1) + b_o \sinh \phi \right] d\phi \tag{37}$$

This enables us to write

$$\Omega = 1 - \left[\frac{A (\cosh \phi - 1) + \sinh \phi}{A \sinh \phi + \cosh \phi} \right]^2 \tag{38}$$

where

$$A = \frac{a_o}{b_o} = \frac{c R_i \kappa \rho_{mo}}{2(3\rho_{ro})^{1/2}} \tag{39}$$

The quantities a_o and b_o were defined in Eq. (10), with ρ_{mo} and ρ_{ro} as stated earlier. For $H_o = 50$ km/(secMpc), $A_{50} = 2.58$, while for $H_o = 75$ km/(secMpc) $A_{75} = 1.69$. In Fig. 3 we present a plot of $\log(1 - \Omega)$ versus $\log t$ for $H_o = 75$ km/(secMpc). Note that the relation is linear to about 10^{-4} of the present age, and that the deviations of Ω from unity are very small at early times.

Alternatively, one can replace $L^2 H^2$ in Eq. (35) with the result

$$\Omega = 1 - \left[\frac{\gamma R_i^2}{c^2}\left(\frac{\rho_{mo}}{L} + \frac{\rho_{ro}}{L^2}\right) + 1\right]^{-1} \tag{40}$$

which can be evaluated directly, again using the exact parametric solutions for L and t. We emphasize that Eqs. (38) and (40) involve no early time approximation, are therefore both valid for all epochs in the expansion for which the standard Big Bang model is valid, and can be shown to be completely equivalent.

Equation (40) is particularly convenient for reexamining the behavior of Ω at early times in the expansion, although we emphasize again that Eq. (38) can also be used as a basis. At sufficiently early time, we could in principle discard the term involving $1/L$, which requires that $\rho_{ro}/L^2 \gg \rho_{mo}/L$. This is equivalent to confining one's attention to times prior to the time of matter-radiation decoupling, where $L = L_x$, with $\rho_{mo}/\rho_{ro} = L_x$ so that L (early) $= L_e \ll L_x$. One can integrate the Friedmann-Lemaître equation at early time by neglecting the term containing the radius of curvature compared to the term containing the radiation density. First, we obtain an approximation for L^2 as a function of time, namely,

$$L_e^2 \approx (4\gamma \rho_{ro})^{1/2} t \tag{41}$$

or, for purposes of approximation in Eq. (40)

$$\frac{\rho_{ro}}{L^2} \approx \left(\frac{\rho_{ro}}{4\gamma}\right)^{1/2} \frac{1}{t} \tag{42}$$

Then, the early-time approximation for Ω follows, using Eq. (42) in Eq. (40), with the result

$$\Omega_e \approx 1 - \left[\frac{(\gamma\rho_{ro})^{1/2} R_i^2}{2c^2} + t\right]^{-1} t \approx 1 - \frac{2c^2}{(\gamma\rho_{ro})^{1/2} R_i^2} t \tag{43}$$

In Eq. (43), we ask for what value of t can we neglect the second term, t alone, with respect to the first term in the bracket in the denominator? For $H_o = 50$ km/(secMpc) the first term has the numerical value 3.22×10^{15} sec, so that t (early) $\ll 3.22 \times 10^{15}$ sec and

$$\Omega_{e,50} \approx 1 - 3.1 \times 10^{-16} t \tag{44}$$

while for $H_o = 75$ km/(secMpc), t (early) $\ll 1.39 \times 10^{15}$ sec, so that

$$\Omega_{e,75} \approx 1 - 7.2 \times 10^{-16} t \tag{45}$$

Note that these early time approximations are indistinguishable from the exact expression for Ω in the time range in which they are valid. In general we are dealing with an open matter-radiation model. Suppose we consider a model containing only radiation. Then in the full parametric solution, Eq. (38), we have $a_o = A = 0$, and $\Omega = 1 + \tanh^2 \phi$. From Eq. (9) we have $t = (R_i b_o /c)(\cosh \phi - 1)$. For early t, ϕ is small, so that $\cosh \phi - 1 \approx (1/2) \tanh^2 \phi \approx \phi^2 / 2$. We again find that Ω_e is linear in t, and we recover the second approximate form in Eq. (43).

As already mentioned, the standard big Bang model assumes that there is no matter-radiation interconversion. It has been shown [Alpher, Follin, Herman (1953)] that before the onset of nucleosynthesis reaction rates among elementary particles would have at least kept pace with the expansion, and that equilibrium would have prevailed before $T \approx 100$ Mev, which would have occurred at ≈ 170 msec. On reexamining this paper we find no obvious reason why $\Omega \sim 1$ should not have been the case at even much earlier times, with matter-radiation interconversion.

To obtain an approximation for Ω valid at late times, i.e., in the vicinity of the age of the universe, we utilize the expression for L(late), Eq.(27) in Eq.(32), to obtain

$$\left(\frac{1}{L}\frac{dL}{dt}\right)^2 = H^2 \approx \frac{c_1^2}{L^2} = \frac{c_1^2}{\left[1 + c_1 (t - t_o)\right]^2} \tag{46}$$

so that

$$\rho_c = \frac{H(t)^2}{\gamma} \approx \frac{c_1^2}{\gamma \left[1 + c_1 (t - t_o)\right]^2} \tag{47}$$

Then taking $\rho_m \gg \rho_r$ at late t, we write

$$\Omega_\ell = \frac{\rho}{\rho_c} \approx \frac{\gamma \rho_{m^*}}{c_1^2 \left[1 + c_1 (t - t_o)\right]} \tag{48}$$

We note that Ω_t so calculated agrees well with the exact calculation for Ω shown in Fig. 3, but only over a very restricted range of times near t_o.

To return to the early approximations for Ω, we note that they apply in the early radiation-dominated phase of the expansion where the relationship between temperature, radiation density and time depend only on universal constants, as given by the Friedmann-Lemaître equation specifically at early time in a radiation-controlled universe. The small correction terms multiplying t in Eqs. (44) and (45) depend on the value of the present radiation density, and the present radius of curvature, both of which are cosmological observables (again we remind the reader that this is all for an open model). The value of Ω starts arbitrarily close to unity and deviates toward smaller values as time goes on. If the present radiation density and present radius of curvature were different than what has been used here, then the small corrections multiplying t will be different. However, as can be seen from Eqs. (44) and (45), a variation in the observed present Hubble parameter within the spread of current observations is not going to alter the basic nature of these approximations. We can surely take the present radiation density as a given cosmological parameter, subject to no significant uncertainty in the present context. Thus, the deviations from unity at small t are indeed small, but are in fact based on observation, and the current uncertain value of the Hubble parameter does not alter the situation.

What we appear to have is that the value of Ω in the standard Big Bang model is extremely close to unity with no additional assumptions being required. In fact if there is any validity to the concept of a Planck time, $\sim 10^{-43}$ sec, and if the physics of the Big Bang model is still valid back to that time, then the deviation from unity is about one part in 10^{60}, and is still one part in 10^{46} at 10^{-30} sec, usually identified as the end of the inflationary era [see Kolb and Turner (1990)]. Thus one has the extremely small deviation required by various inflationary paradigms, which also purport to involve conventional physics back to the Planck time. But it appears also to be provided by the standard Big Bang model.

In addition, it is a part of some inflationary paradigms that Ω be unity throughout the expansion. In fact, the exact solution for Ω in the standard Big Bang model appears to be unity from the Planck time out to times of the order of a million years after the start of the expansion, and then deviates, finally, to encompass the value of $\Omega_o h_{50}^2 = 0.05 \pm 0.01$, where Ω_o, based on baryons, is the present value, as given for Big Bang nucleosynthesis by Olive and Steigman (1994). For example, for $H_o = 50$ km/(secMpc), the exact solution gives at $t/t_o = 6.2 \times 10^{-5}$, 0.0012, 0.011, 0.1, and 1.0, the corresponding values of Ω to be 0.99, 0.92, 0.68, 0.28 and 0.05. One can alter the time history of Ω to encompass other values of $\Omega_o < 1$ by adjusting ρ_{mo} to include nonbaryonic contributions in Eqs. (10) or (12). To have $\Omega = 1$ for all time, i.e., a critical density model with $\rho = \rho_c$, is a very special restriction. There seems to us to be no more reason to be concerned that in fact Ω is nearly equal to unity, whether it is less than or greater than one, than to be concerned that various fundamental constants have particular numerical values which cannot be calculated from first principles at the present time.

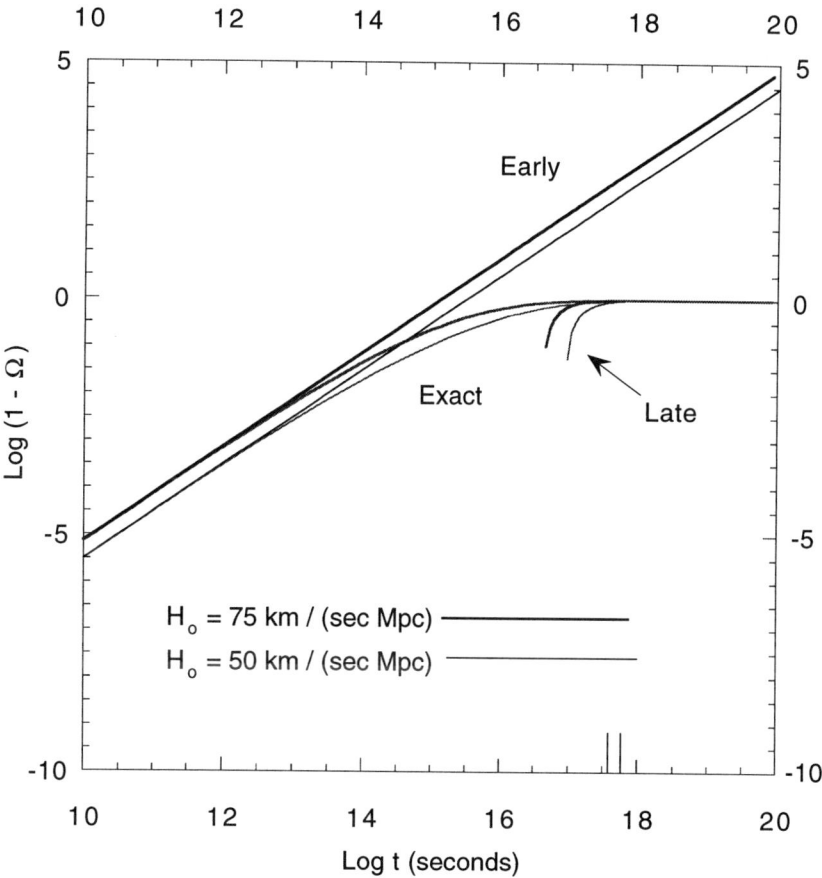

Figure 3. The temporal behavior of the deviation of Ω from unity, as calculated from the exact parametric solution of the Friedmann-Lemaître equation for an open universe, compared with a functionally correct approximation for late times in the expansion and with an extrapolation to late times of an approximation valid only for early times in the expansion, all for two sample values of the Hubble parameter. Not easily discerned in the plot is the fact that Ω is 0.05 at $\log t_o = 17.60$ and 17.76, the present epochs for $H_o = 75$ and 50, respectively, and continues to decrease as the epoch gets later.

Whatever the inflationary paradigm, we understand the need to have the early value of Ω near unity to high precision so that the inflationary model used goes smoothly into the standard model, whose parameters are based on observation. This requirement does not appear to present any particular problem, given the manner in which Ω behaves in the standard model. Some of the arguments made for Ω being exactly unity throughout the history of the expansion may be based on a failure to realize that the standard model does indeed give unity for a useful part of the age of the universe. Thus, we fail to appreciate why this so-called "fine tuning", which does depend on observational parameters, really constitutes a fundamental enigma in the standard Big Bang model. The parameters are what they are, and Ω is what it is, if observation and the standard model are correct.

Thus, while it may be that one has to have inflation early in order to deal with such problems as finding an isotropic distribution of the cosmic microwave background radiation including regions which could not have been causally connected by crossover time, as well as the apparent nonexistence of monopoles, it does not seem necessary to discard the standard Big Bang model in order to have $\Omega = 1$ with high precision during the early epochs.

GAMOW'S EARLY WORK

It is not our intention in this paper to evaluate fully the contributions of George Gamow to the development of the Big Bang model. His seminal contribution, in our view, was a 1946 paper in which he noted that various theoretical approaches toward understanding the cosmic abundance distribution of the elements had all failed, referring in particular to the so-called equilibrium theories with what he termed the "Heavy Element Catastrophe." He proposed instead that one should consider the early prestellar state of the expanding universe as the possible locale of a build up of nuclei by some agglomeration of neutrons from an initial neutron gas as the primeval state. Over the next two years Alpher, working on a dissertation with Gamow, started with this idea but developed it into a picture of thermonuclear reactions occurring in a hot dense state, in a reaction sequence starting with neutrons and with the protons arising from the decay of the neutrons, to build deuterons, and continuing on the basis of a simplified view of sequential neutron-capture reactions with intervening beta-decay to keep the sequence in the valley of stability. The extant temperature was taken to be of the order 0.1 Mev, so that deuterons would not be photodissociated. One of the consequences of this view, initially published as a letter which became known as the Alpher-Bethe-Gamow model (1948), was that the early universe was hot and dense, and expanded and cooled from that state with a short epoch of thermonuclear element building. Moreover, there was one free parameter, namely, the density of matter in the early universe to be determined as required for primordial nucleosynthesis. This preliminary model was expanded and developed further by Alpher and Herman in a series of papers (1988).

Also in 1948, Gamow, rather than trying to fit the entire distribution of abundances but nevertheless interested in obtaining the ambient condition at matter-radiation crossover as pertaining to the formation of galaxies, elected to look just at the formation of deuterons from neutrons and protons, again in a hot, dense expanding universe. In this way he derived an early density of matter, as required for obtaining the desired abundance of primordial deuterons, which he then used in an interesting and Gamowian way to relate the diameter of galaxies formed according to the Jeans criterion, with the binding energy of the deuteron as a factor in the expression. One might argue whether his motivation in this approach was to derive a "closed" form for the size of a galactic condensation, or whether he truly was trying to check the early matter density in the Alpher-Bethe-Gamow paper. As already mentioned Gamow's derived early matter density agreed with other calculations of the time.

There were a number of subsequent papers, in several of which he pursued his goal of understanding the formation of galaxies in terms of the conditions involved in early nucleosynthesis. We shall deal shortly with those of his papers in which he considered approximations for obtaining the present temperature of the universe without invoking the matter density required for early nucleosynthesis, and therefore without making use of the algorithm developed by Alpher and Herman in 1948.

The great contribution made by Gamow was to propose a prestellar state of a universe expanding from a hot dense state as the site for primordial nucleosynthesis, and as the starting point for considering the subsequent evolution of the expanding universe. One could argue that having invoked an early hot state, expansion and evolution to the present CMBR is an imbedded result, but, as discussed below, he did not consider the CMBR at all until 1950. In any event, there can be no question but that he stimulated a new starting point for cosmological considerations.

GAMOW'S APPROXIMATIONS IN CALCULATING THE CMBR TEMPERATURE

We proceed here specifically with Gamow's approach to the calculation of the CMBR temperature, which he discussed in four publications. In his 1950 paper, the first of these, he simply stated that the CMBR temperature was 3 K, with no reference or information on how he came to this number. To this day we have no insights into this strange omission. In 1952, and again in 1953 he obtained a present CMBR temperature by extrapolating the correct early time approximation for temperature, given as Eq. (18), well beyond its range of validity (in his book Gamow called this relation exact), to the then-presumed age of the present epoch, namely 10^{17} sec (3.2×10^9 yr.), which gave 50 K to be the temperature, a value which he suggested as being in "reasonable agreement" with the actual temperature of interstellar space. Had he used 17×10^9 yrs as the age, Eq. (18) would have given a temperature of 21 K. In his 1956 paper he dismissed this simple extrapolation of the early approximation, which he then reported as giving 40 K, because it led to an incorrect value of the Hubble parameter, which was certainly the case.

In later editions of his book Gamow reported in a more detailed way material he subsumed finally in his 1956 paper, the one with which we will deal primarily.

Before considering Gamow's 1956 algorithm, let us recapitulate the early and late time approximations for matter and radiation mass densities which we gave earlier in this paper. We shall at this point, again use an e in the subscript to denote early time approximations, an l to denote late time approximations, and as before a single prime or the subscript zero to denote early and present values, respectively. We again use the subscript x to denote crossover, when matter and radiation densities in the expansion were equal. Thus, with γ and ζ as defined earlier, we have

$$\rho_{re} \approx (4\gamma t^2)^{-1} \tag{49}$$

and

$$\rho_{me} \approx (4\gamma \zeta)^{-3/4} t^{-3/2} \tag{50}$$

while

$$\rho_{rl} \approx \rho_{ro} [1 + c_1(t - t_o)]^{-4} \tag{51}$$

and

$$\rho_{ml} \approx \rho_{mo} [1 + c_1(t - t_o)]^{-3} \tag{52}$$

where, as before $c_1 = c/R_i$. Note that the constant factor c_1 involves R_i, whose value depends on the present total density (which is dominated by the density of matter) and the present value of the Hubble parameter, as

$$c_1 = \gamma(\rho_{mo} + \rho_{ro}) - H_o^2 \approx \gamma \rho_{mo} - H_o^2 \tag{53}$$

Gamow chose to write his approximations as

$$\rho_{re} \approx (4\gamma t^2)^{-1} = 4.47 \times 10^5 \, t^{-2} \, g \, cm^{-3} \tag{54}$$

and

$$\rho_{me} \approx \text{constant} \times t^{-3/2} \tag{55}$$

which are correct, but then he gave

$$\rho_{rl} \approx \text{constant} \times t^{-4} \tag{56}$$

and

$$\rho_{ml} \approx \text{constant} \times t^{-3} \tag{57}$$

The late time approximations ignore the functional form which is demanded of the solutions of the Friedmann-Lemaître equation in terms of the variable $(t - t_o)$ near $L = 1$, which is by our definition the present epoch, with red shift defined as usual through $L = (1 + z)^{-1}$.

Rather than reject Gamow's approximation out of hand as not satisfying the functional form, let us first see what Gamow would have calculated using contemporary cosmological parameters. For this purpose we will use $t_o = 5.76 \times 10^{17}$ sec for $H_o = 50$ km/(secMpc) and $\rho_{mo} = 2.35 \times 10^{-31}$ g cm^{-3}, all as discussed earlier. Gamow calculated the constant in Eq. (57) to give ρ_{mo} at the present epoch, and then used the formula thus obtained to find the epoch in the expansion when ρ_{ml} so defined intersects the radiation density calculated from the early time approximation, Eq. (54). The first step gave

$$\rho_{ml} \approx 4.50 \times 10^{22} \, t^{-3} \text{ g cm}^{-3} \tag{58}$$

Gamow then calculated crossover conditions by equating Eq. (54) with Eq. (57), which yielded $t_x = 1.00 \times 10^{17}$ sec $(3.18 \times 10^9$ yr.), and hence $\rho_x = 4.44 \times 10^{-29}$ g cm^{-3}.

While Gamow chose not to use the matter density approximation early, it is of interest to see what his approximations would yield for the matter density during nucleosynthesis. Thus, we force ρ_{me} to go through the intersection of ρ_{re} and ρ_{ml}, i.e., Eqs. (54) and (57). In this way we obtain

$$\rho_{me} \approx 1.41 \times 10^{-3} \, t^{-3/2} \text{ g cm}^{-3} \tag{59}$$

The density of baryons at one second in Gamow's approximation is therefore 1.41×10^{-3} g cm^{-3}. This is to be compared with the results mentioned earlier following Eq. (25) from nucleosynthesis calculations by Fermi and Turkevich of 1.66×10^{-3} and 6.0×10^{-2} from other light element calculations.

We return to the pursuit of a value for the present CMBR temperature according to Gamow's approximations. To do this we evaluate the constant in Gamow's late-time radiation density approximation, Eq. (56), by forcing the function to go through the intersection of the early radiation density and the late-time matter density approximations, Eqs. (54) and (57), with the result

$$\rho_{rl} \approx 4.51 \times 10^{39} \, t^{-4} \, g \, cm^{-3} \tag{60}$$

When $t = t_o = 5.76 \times 10^{17}$ sec for $H_o = 50$ km/(secMpc), on obtain $\rho_{rl} = 4.10 \times 10^{-32}$ g cm^{-3}. From the Stefan-Boltzmann law, it follows that $T_o = 8.35$ K. This result is internally consistent with the use by Gamow of an implied identity $L = t/t_o$ which, for Gamow's crossover time, gives $L_x = 0.174$. It is of interest that using then-current values of the relevant parameters, Gamow in 1956 calculated ~ 6 K. Given that $\rho_m \propto L^{-3}$ and $\rho_r \propto L^{-4}$ one has, as before, that L at crossover is given by $L_x = \rho_{ro}/\rho_{mo}$ which is an exact expression; but, using the current exact present matter density $\rho_{m''}$ and Gamow's present radiation density corresponding to 8.35 K, one recovers, as expected, $L_x = 0.174$.

In the exact solution discussed earlier, we calculated $L_x = 1.98 \times 10^{-3}$ and again with $\rho_{mo} = 2.35 \times 10^{-31}$ g cm^{-3}, we obtain $\rho_{ro} = 4.65 \times 10^{-32}$ g cm^{-3} or $T(now) = 2.73$K. This is somewhat circular reasoning, since here we have used ρ_{ro} in calculating L_x. However, we state the result to show the reader the apparently small difference between Gamow's ρ_{ro} and the exact value. This does not reflect any new insight on Gamow's part but rather a fortuitous lack of sensitivity in the calculations. The difference in other cosmological parameters is not so small, as indicated in the following table in which the Gamow values have been recalculated using recent values of the parameters:

	GAMOW (recalculated)	EXACT
L_x	0.174	1.98×10^{-3}
ρ_x	4.438×10^{-29} g cm^{-3}	7.50×10^{-27} g cm^{-3}
t_x	1.004×10^{17} sec	7.98×10^{15} sec

It may be of interest to display the differences between Gamow's approximations and the late-time approximations which satisfy the functional form and the boundary conditions. We employ Eq. (34) with c_1 as before, to display $L_l(t)$

$$L_\ell(t) \approx 1 - \frac{c}{R_i}(t_o - t) \tag{61}$$

For $H_o = 50$ km/(secMpc) and $t_o = 5.76 \times 10^{17}$ sec, we have $L_1 = 0.09 + 0.91 t/t_o = 0.91 + 1.58 \times 10^{-18} t$ with $z_l = (1 - t/t_o) / (0.01 + t/t_o) = (1 - 1.74 \times 10^{-18} t)/(0.10 + 1.74 \times 10^{-18} t)$ which should be compared with Gamow's incorrect implicit assumption that $L = t/t_o$. We emphasized once more that Gamow's early radiation and matter densities have the wrong functional form but are not too bad if one is near enough to t_o.

One of the uses one might be tempted to make of Gamow's approximations is to compare his result with the temperature measured in distant clouds. For example, at $z = 2.65$ and 4.68 Gamow's CMBR temperatures are high with respect to the exact solution by about $+15$ and $+26$ percent, while the correct late time approximation is only slightly better, differing by about -10 and -19 percent, respectively. We conclude that the functionally correct late time approximations on the early side of t_o are not particularly useful. Figure 4 shows $\log L$ and $\log(1+z)$ versus $\log t$ which together with Fig. 2 is useful in making calculations of this type.

In a numerical comparison, we have prepared Fig. 5 in which we have plotted for a limited range of z, from zero to about five, the exact parametric solutions for $\log \rho_m$ and $\log \rho_r$, Gamow's approximations, and our late time approximation all versus $\log t$. The plot is for $H_o = 50$ km/(secMpc) and $t_o = 5.76 \times 10^{17}$ sec. Note that the functionally correct late time approximations are very good for times greater than the present time but become poor for times earlier and not close to t_o while Gamow's forms do not blow up far from t_o.

Thus, in a somewhat convoluted way, Gamow achieved his goal of obtaining a value for the present background temperature without invoking primordial nucleosynthesis to evaluate ρ_{me}.

A common thread in Gamow's considerations with respect to calculating the background temperature was his proposition that galaxies began to form at about the time of crossover. He employed the Jeans criterion for the diameter of the condensation, D, as follows:

$$D = \left(\frac{5\pi T_x}{3 G m_h \rho_r} \right)^{1/2} \tag{62}$$

where m_b is the mass of the baryon and T_x and ρ_x are the temperature and density at crossover, respectively. With his ρ_x from the above table we calculate T_x as $(c^2 \rho_x / a)^{1/4}$. This gives the following results: $T_x = 53.3$K, $D = 6.09 \times 10^{30}$ cm (6.43×10^{12} ly), the total mass of the condensation $M \approx \rho_x D^3 = 1.51 \times 10^{64}$ g $= 7.57 \times 10^{30}$ M$_\odot$. These results, obtained with Gamow's erroneous approximations and current cosmological parameters, are quite different from those reported in 1956 by Gamow, namely, $T_x = 170$K, $D = 10^{21}$ cm, and $M = 4 \times 10^{41}$ g (2×10^8 M$_\odot$), with $t_x = 8 \times 10^{15}$ sec and $\rho_x = 7.5 \times 10^{-27}$ gcm^{-3}.

SOME COMMENTS ON GAMOW'S CALCULATIONS

What can one say about the numbers Gamow calculated?

First, the approximations used by Gamow for matter and radiation densities late in the expansion are, strictly speaking, not correct. They do not satisfy the functional form

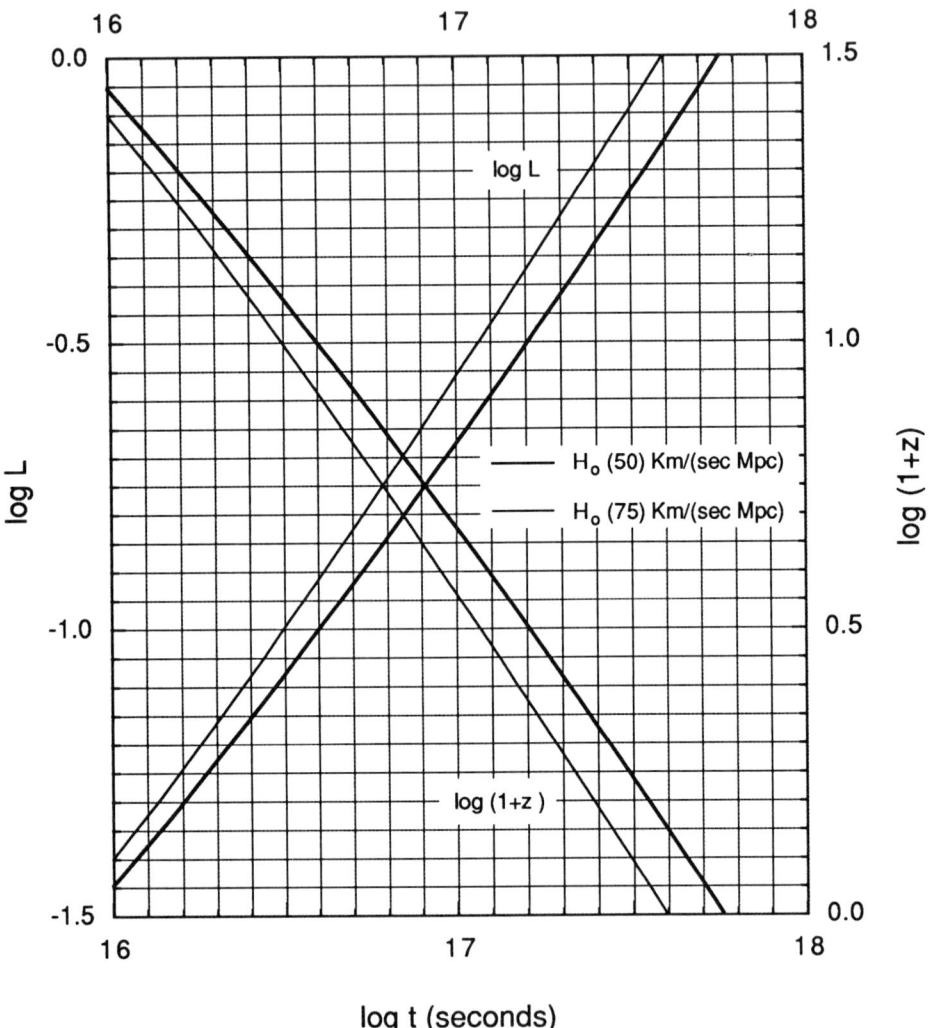

Figure 4. The temporal behavior of the dimensionless proper distance L and of the red shift z, from the exact parametric solution of the Friedmann-Lemaître equation for an open universe, for two sample values of the Hubble parameter.

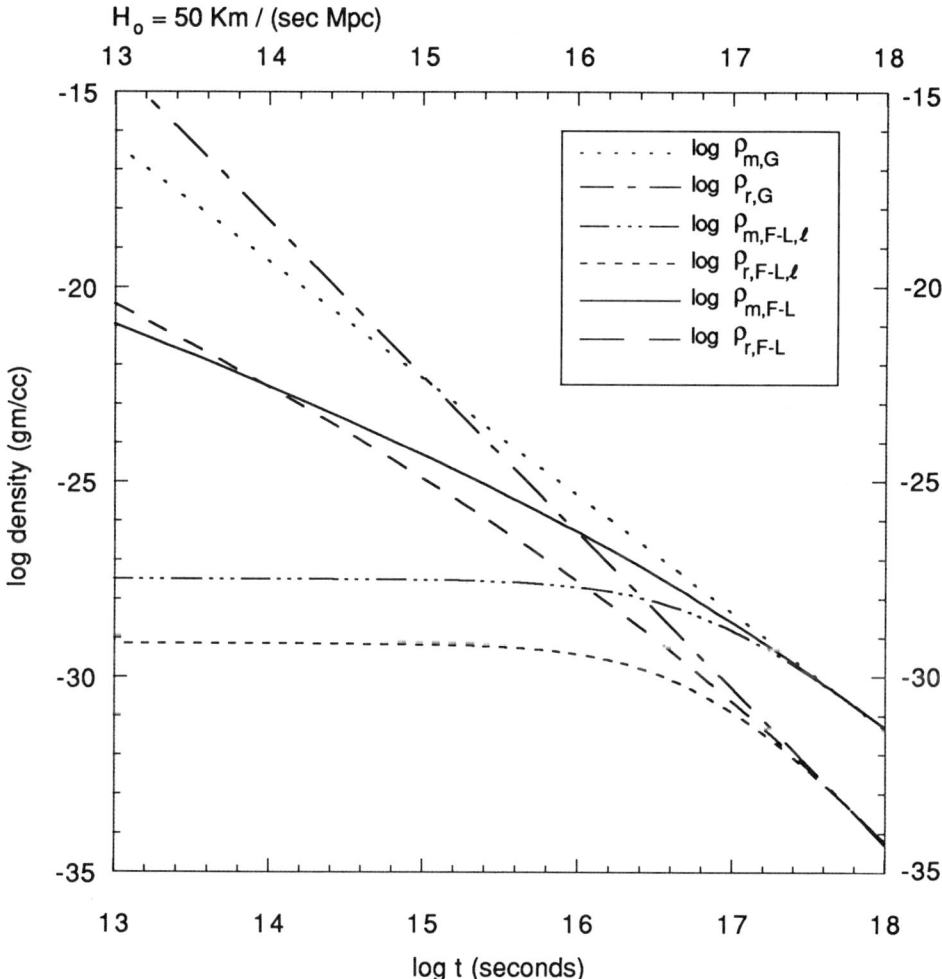

Figure 5. Comparison of calculations of the matter and radiation densities in an open model according to the exact solution of the Friedmann-Lemaître equation, to Gamow's approximation, and to a functionally-correct late-time approximation to the exact solution. Note that this last approximation fails as one goes back in time, even before crossover.

required by the Friedmann-Lemaître equation in the matter-dominated late universe. It is to be preferred that one use the simple exact power-law result, Eq. (5).

Second, the conditions at crossover, which are not that vital in discussing the properties of the standard Big Bang model, are very different with Gamow's approximations as compared to the exact solution of the Friedmann-Lemaître equation.

Third, if one has any confidence at all in the standard Big Bang model, there is no reason to ignore the determination of an early matter density as required for light element nucleosynthesis since ρ_{me}, ρ_{ml} and ρ_{mo} are related. This calculation and the agreement of calculated with observed abundances, as well as the requirement of three neutrino families in determining the initial conditions for light element nucleosynthesis, again in agreement with observation, provide one of the strongest pieces of evidence for the validity of the Big Bang model.

Fourth, the early and late mass densities of matter and radiation using the simple power law are internally consistent with current values of the cosmological parameters, with the possible sole exception being the problem of dark or missing matter. Our reading of current observations and theory suggest that the present matter density derived from the requirements of light element nucleosynthesis, and from the conversion to mass from observed luminosities of celestial objects, are lower limits. However, it is not clear to us just how far off the number may be. At the present it seems that achieving an Ω of unity for all t is not to be. Discussion of this problem is beyond the scope we have chosen for this paper. [See Cole and Ellis (1994)]

Finally, as is amply illustrated in Fig. 5, even the correct early and late approximations become very poor as one approaches crossover from either early or late times. In fact, the late matter and late radiation density approximations do not even cross. This is no problem if one uses the exact power law.

CONCLUDING REMARKS

We have included in this paper, and have already referred to, a variety of illustrations of the temporal behavior of the several cosmological parameters, all as arising from exact solutions of the Friedmann-Lemaître equation with the cosmological parameter $\Lambda = 0$, for two values of the Hubble parameter. We have also explored the region of crossover for a range of values of the Hubble parameter. Moreover, we have considered the nature of approximations to the exact solutions for times early in the expansion and for times close to the present epoch. In this context we have displayed the nature of Gamow's approximations as well as conclusions from these approximations.

Why Gamow chose to take his particular approach we believe is inexplicable and should be considered only for its historical interest. What is clear is that Gamow felt that there should be a simple connection between early times in the expansion, and conditions at the present epoch, which would increase understanding of the dynamic model of the universe. As always, he pursued his goal of understanding the conditions for the formation of galaxies in the context of the Big Bang. Moreover, he was most interested in the thermal history of the universe.

In one of his final publications, Gamow (1967) identified two important periods in the evolution of the Big Bang, the first being that of light element nucleosynthesis, and the second being that when the densities of matter and radiation were equal at crossover. He felt that during and after this regime, galaxies formed, but he saw no way to go from the conditions pertaining at crossover to galaxies forming according to the Jeans criterion, without some major modification of the Jeans formula. Using correct values of the parameters does not help, as we have seen in our earlier remarks. It is hard to dismiss Gamow's intuition in this matter, and we cannot help wondering if some modification of the Jeans formula may yet prove to be a useful approach.

Those working in cosmology today owe Gamow a great debt for his having first truly introduced some physics into considerations of the dynamic expanding universe, in the spirit of Friedmann and Lemaître, and in proposing that the early universe was initially hot and dense and expanded and cooled to its present state. All of the phenomenology of the Big Bang model emanates from this view.

In conclusion, we are moved to comment on recent attempts to validate Gamow's scientific work (as though that were necessary) as well as Gamow's persona. We have argued for some time that Gamow was one of the most creative scientists of this century, who regrettably did not receive all the formal recognition that was his due during his lifetime. However, we submit that his seminal scientific works and his popular writings were most certainly greatly appreciated during his lifetime.

The tendency now of some to attempt to raise Gamow's stature further by attributing to him achievements that have no proper basis is in our opinion doing a disservice to his memory. There are even examples of attempts to analyze Gamow's personality through his science, or to rationalize apparent flaws in his behavior. Regarding this latter matter, we both knew Gamow extremely well over many years, and can say that in all the interactions we had with him we do not recall his wandering from scientific questions and expounding on philosophical, ethical or religious questions. His comments, for example, about St. Augustine and what happened before the Big Bang were unquestionably made because it was amusing. Moreover, we continue to be distressed by various statements in the literature about Gamow such as that he suffered from crapulence and most recently by a Russian article in which it was argued that Gamow was indeed stocious because he pined for Mother Russia. We believe all this to be far fetched and highly inappropriate.

Gamow certainly had a strong personality, mainly an intense and joyful approach to science, and it is precisely this very robust enthusiasm that made some of the more conventional members of the scientific community look at him with less than affectionate approval.

There are others who seem to have the need to denigrate creative achievements when later some flaws may be perceived. But is it not true that our sustenance is drawn from the creative sea, flaws and all? This idea has been beautifully expressed by Picasso, as quoted by Gertrude Stein (1969) in her inimitable way:

> "Picasso said once that he who created a thing is forced to make it ugly. In the effort to create the intensity and the struggle to create this intensity, the result always produces a certain ugliness; those who follow can make of this thing a beautiful thing because they know what they are doing, the thing having already been invented, but the inventor because he does not know he is going to invent inevitably the thing he makes must have its ugliness."

We have no need to repair the flaws, if any, in Gamow's scientific work, particularly as it may affect his reputation. These flaws have mainly come about as a result of his passionate desire to understand everything on the back of an envelope. Nevertheless, his total creative contribution stands firmly on a series of deep and significant ideas which have shaped the course of science in a variety of disciplinary areas and have had a profound and lasting impact.

ACKNOWLEDGMENTS

This paper is dedicated to a long-time friend and colleague, the late George Gamow. Finally, we greatly appreciate the invaluable assistance of Umer Yousafzai and Faraz Syed in performing calculations and in preparing the manuscript.

REFERENCES

Alpher, R.A., Bethe, H., and Gamow, G., 1948, The origin of the elements, *Physical Review,* Vol. 73, pp. 803-804.

Alpher, R.A., Follin, Jr., J.W., and Herman, R., 1953, Physical conditions in the initial stages of the expanding universe, *Physical Review* , Vol. 92, pp. 1347-1361.

Alpher, R.A., Gamow, G., and Herman, R., 1967, Thermal cosmic radiation and the formation of protogalaxies, *Proceedings of National Academy of Science*, Vol. 58, pp. 2179-2186.

Alpher, R.A., and Herman, R., 1948, Evolution of the universe, *Nature*, Vol. 162, p. 774.

Alpher, R.A., and Herman, R., 1949, Remarks on the evolution of the expanding universe, *Physical Review*, Vol. 75, pp. 1089-1095.

Alpher, R.A., and Herman, R., 1950, Theory of the origin and relative abundance distribution of the elements, *Reviews of Modern Physics*, Vol. 22, pp. 153-212.

Alpher, R.A., and Herman, R., 1975, Big bang cosmology and the cosmic blackbody radiation, *Proceedings of American Philosophical Society*, Vol. 119, pp. 325-348.

Alpher, R.A., and Herman, R., 1988, Reflections on early work on 'Big Bang' cosmology, *Physics Today*, Vol. 41, pp. 24-34.

Alpher, R.A., and Herman, R., 1990, Early work on 'Big Bang' cosmology and the cosmic blackbody radiation", *in:* "Modern Cosmology in Retrospect," R. Bertotti, R. Balbinot, S. Bergia, and A. Messina, eds., New York: Cambridge University Press, 1990, pp. 129-157.

Alpher, R.A., and Herman, R., 1993, Origins of primordial nucleosynthesis and the prediction of the cosmic microwave background radiation, *in:* "Encyclopedia of Cosmology," N.S. Hetherington, ed., New York: Garland Pub., 1993.

Alpher, R.A., and Herman, R., 1994, Remembrances of things past: Some recollections of the development of the Big Bang model, *in:* "Unified Symmetry in the Small and in the Large," B. Kursunoglu, ed., Nova Publishing, Inc., pp. 3-54.

Alpher, R.A., and Marx, G., 1992, The creation of free energy, *in:* "Vistas in Astronomy," Vol. 35, pp. 179-214.

Bartlett, J.G., Blanchard, A., Silk, J., and Turner, M.S., 1995, The case for a Hubble constant of 30 Km(secMpc), *Science*, Vol. 267, pp. 980-983.

Brush, S.G., 1993, Prediction and theory evaluation: Cosmic microwaves and the revival of the Big Bang, *Perspectives on Science*, Vol. 1, pp. 565-601.

Chernin, A.D., 1994, How Gamow calculated the temperature of the microwave background radiation, or a few words about the fine art of theoretical physics, *Progress in Physical Science*, Russian Academy of Sciences, Vol. 164 (8), pp. 879-896.

Coles, P., and Ellis, G., 1994, The case for an open universe, *Nature*, Vol. 370, pp. 609-615.

Dicke, R.H., and Peebles, P.J.E., 1979, The Big Bang cosmology – enigmas and nostrums, *in:* "General Relativity Einstein Centenary Survey," S. Hawking and W. Israel, eds., Cambridge University Press, 1979.

Gamow, G., 1948, Expanding universe and origin of elements, *Nature*, Vol. 162, pp. 680-682.

Gamow, G., 1950, Half an hour of creation, *Physics Today*, Vol. 3, pp. 16-21.

Gamow, G., 1952, "The Creation of the Universe," Viking Press, Inc., New York, 144 pp.

Gamow, G., 1953, Expanding universe and the origin of galaxies, *Dan. Mat. Fys. Medd.*, Vol. 27, pp. 1-15.

Gamow, G., 1956, The physics of the expanding universe, Vistas *in Astronomy*, Vol. 2, pp. 1726-1752.

Gamow, G., 1967, History of the universe, *Science*, Vol. 158, pp. 766-769.

Guth, A., and Steinhardt, P.J., 1989, The Inflationary Universe, *in:* "*The* New Physics," Paul Davies, ed., Cambridge University Press, pp. 34-60.

Heckmann, O., 1942; reprinted in 1968, "Theorien der Kosmologie," Springer-Verlag, 113 pp.

Kolb, E.W., and Turner, M. S., 1990, "The Early Universe," Addison Wesley Publishing Company, 547 pp.

Lightman, A., and Brawer, R., 1990, "Origins: The Lives and Worlds of Modern Cosmologists," Harvard University Press, 561 pp.

Maddox, J., 1994, The best cosmology there is, *Nature*, Vol. 372, pp. 15-18.

Mather, J.C., et al, 1994, Measurement of the cosmic microwave background spectrum by the COBE FIRAS instrument, *Astrophysical Journal*, Vol. 429, pp. 439-444.

Olive, K.A., and Steigman, G., 1994, On the abundance of primordial helium, *Astrophysical Journal Supp. (in press)*. In a more recent preprint the authors give limits for η of 3.1 to 4.0 based on D, He^3, and Li^7, and an $\eta < 3.9$ if He^4 is included.

Peebles, P.J.E., 1988, The theoretical aspects of the nebular redshift, 33 years later, *Publ. Astron. Soc. of Pacific*, Vol. 100, pp. 670-679.

Peebles, P.J.E., 1993, "Principles of Physical Cosmology," Princeton University Press, 718 pp.

Penzias, A.A., and Wilson, R.W., 1965, A measurement of excess antenna temperature at 4080 Mc/S, *Astrophysical Journal*, Vol. 142, pp. 419-421.

Songalla, A., et al, 1994, Measurement of the microwave background temperature at a redshift of 1.776, *Nature*, Vol. 371, pp. 43-45. See also comments by D.M. Meyer, 1994, A distant space thermometer, *Nature*, Vol. 371, p. 13.

Stein, G., 1948, "Picasso," B.T. Batsford, Ltd., London, p. 9.

Turner, M.S., 1993, Why is the temperature of the universe 2.726 Kelvin?, *Science*, Vol. 262, pp. 861-866.

Weinberg, S., (1972), "Gravitation and Cosmology: Principles and Applications of the General Theory of Relativity," John Wiley & Sons, New York, 675 pp.

Estimating the Energy Spectrum of Cold Dark Matter Signal

Yun Wang

NASA/Fermilab Astrophysics Center
Fermi National Accelerator Laboratory
Batavia, IL 60510-0500

Abstract

The self-similar spherical infall model of Fillmore and Goldreich gives reasonably flat rotation curves for the initial mass density perturbation power law index $\epsilon < 2/3$. We add angular momentum to this model in a self-similar manner to obtain a more realistic model. We find that in the cold dark matter (CDM) spectrum measured on earth, the two most energetic peaks are close and have almost the same height, each containing 5% to 10% of the local halo density. The average of this density fraction is insensitive to the presence of angular momentum, if all the halo particles have the same dimensionless angular momentum j. We derive the density fractions in the two most energetic peaks as functions of ϵ and Δj (the dispersion in the dimensionless angular momentum of the halo particles) analytically. Our analytical results agree with our exact numerical calculations. In general, the spectrum of CDM signal consists of a broad thermal peak at the low energy end, and a parameter-dependent number of thin lines (with decreasing line width) at the high energy end.

The existence of dark matter has been well established by recent observations. [1] While the content of dark matter is unknown, the prime candidates are MACHOs, WIMPs and axions. WIMPs and axions are cold dark matter (CDM), sharing the property of small primordial velocity dispersion.[2] Here we study the observational signature of a CDM halo.

Before the era of galaxy formation, the phase space distribution of CDM particles is a very thin sheet near the Hubble flow $\vec{v} = H\vec{r}$, with H denoting the Hubble rate at that time. Once galaxy formation sets in, the phase space distribution of CDM particles evolves as a thin sheet wrapping around itself, because of the negligible intereaction of CDM particles with their surrounding. The phase space sheet can not tear or cross itself. For a given location in our galaxy, we expect to see a discrete set of velocities $\vec{v}^n$ ($n = 1, 2, ...$) in the CDM spectrum. The scattering by massive objects in our galaxy should smear out most of these velocity peaks. However, the velocity peaks which correspond to particles which have crossed the galaxy a small number of times should remain.[3] In particular, the first peak, which corresponds to particles falling into the galaxy for the first time, should have suffered negligible scattering. Knowledge of the first velocity peak can be important to experimental efforts to detect CDM particles. If the percentage of local halo density contained in the first velocity peak is 5% percent, the efficiency of a axion detector can be improved 9-10 times.[4]

To make quantitative calculations, a specific model must be used. Fillmore and Goldreich studied self-similar gravitational collapse for scale-free initial perturbations [5]

$$\frac{\delta M_i(r_i)}{M_i} = \left(\frac{M_0}{M_i(r_i)}\right)^\epsilon, \tag{1}$$

where M_0 is a reference mass, and $0 \leq \epsilon \leq 1$. We label a mass shell by the mass M_i it initially contains. Mass shell M_i turns around for the first time when it reaches maximum radius $r_*(M_i)$ at time $t_*(M_i)$,

$$t_*(M_i) = \frac{3\pi}{4} t_i \left(\frac{M_i}{M_0}\right)^{3\epsilon/2}, \quad r_*(M_i) = \left[\frac{8}{\pi^2} t_*^2(M_i) G M_i\right]^{1/3}, \tag{2}$$

where t_i is the initial time when Hubble law still holds. The mass shell that is turning around for the first time at time t contains mass $M(t)$ and has a radius $R(t)$ (called "current turnaround radius") given by

$$M(t) = M_0 \left(\frac{4t}{3\pi t_i}\right)^{\frac{2}{3\epsilon}} = M_i \left(\frac{t}{t_*}\right)^{\frac{2}{3\epsilon}}, \tag{3}$$

$$R(t) = \left[\frac{8}{\pi^2} t^2 G M(t)\right]^{\frac{1}{3}} = r_*(M_i) \left(\frac{t}{t_*}\right)^{q_0},$$

where $q_0 = 2(1 + 3\epsilon)/(9\epsilon)$. Let us define dimensionless variables

$$\tau \equiv \frac{t}{t_*(M_i)}, \qquad \lambda \equiv \frac{r(M_i, t)}{r_*(M_i)}. \tag{4}$$

The scaled current turnaround radius is given by $\Lambda(\tau) = R(t)/r_*(M_i) = \tau^{q_0}$.

The original version of this model [5] is not realistic because it ignores angular momentum of the halo particles. It is desirable that we add angular momentum to the equations of motion without losing the advantage of spherical symmetry and self-similarity. Let us assign the same magnitude but isotropically different directions of angular momentum to particles on the same mass shell M_i, the mass shell will remain spherical as it moves through the spherically symmetric mass distribution $M(r,t)$. To preserve self-similarity, we add angular momentum $l(M_i)$ to mass shell M_i such that

$$l(M_i) = j \frac{r_*(M_i)^2}{t_*(M_i)}. \tag{5}$$

We refer to j as the dimensionless angular momentum. To be more realistic, let us assign a distribution of magnitudes of angular momentum to each mass shell. Let each shell M_i have a fraction $f(j)dj$ of particles with dimensionless angular momentum j, where $f(j)$ is the distribution function ($\int_{j_{min}}^{j_{max}} dj\, f(j) = 1$, j_{max} and j_{min} denote the maximum and minimum dimensionless angular momentum respectively). The equation of motion for a mass shell with dimensionless angular momentum j can be written in a self-similar form:

$$\frac{d^2\lambda}{d\tau^2} = \frac{j^2}{\lambda^3} - \frac{\pi^2}{8} \frac{\tau^{\frac{2}{3\epsilon}}}{\lambda^2} \mathcal{M}\left(\frac{\lambda}{\Lambda(\tau)}\right), \tag{6}$$

where $\mathcal{M}(\lambda/\Lambda(\tau))$ is the ratio of the mass contained within $r(M_i, t)$ and the mass contained within the current turnaround radius $R(t)$ [5],

$$\mathcal{M}(\xi) = \frac{M(\xi R(t), t)}{M(t)} = \frac{2}{3\epsilon} \int_{j_{min}}^{j_{max}} dj\, f(j) \int_1^\infty \frac{d\tau}{\tau^{1+\frac{2}{3\epsilon}}} \Theta\left(\xi - \frac{\lambda_j(\tau)}{\Lambda(\tau)}\right). \tag{7}$$

For convenience, we can take $f(j) = 1/\Delta j$, with $\Delta j = j_{max} - j_{min}$.

Eq.(6) holds for all mass shells with angular momentum j. The boundary conditions at $\tau = 1$ are:

$$\lambda(1) = 1, \qquad \left.\frac{d\lambda}{d\tau}\right|_{\tau=1} = 0. \tag{8}$$

Self-similarity enables us to reduce a two-variable (r, t) system to a one-variable (τ) system. For a given shell, Eq.(6) describes the time evolution of its orbit; while at given time t, τ labels different mass shells, and Eq.(6) gives the location r and velocity dr/dt of all mass shells at time t:

$$\bar{r} \equiv \frac{\lambda(\tau)}{\Lambda(\tau)} = \frac{r}{R(t)}, \qquad \bar{v} \equiv \left[\frac{\tau}{\Lambda(\tau)}\right]\frac{d\lambda}{d\tau} = \left[\frac{t}{R(t)}\right]\frac{dr}{dt} \tag{9}$$

Fig. 1 is a typical phase space diagram ($\bar{v}$ versus $\bar{r}$) [6], showing the location and velocity of all mass shells at time t, for $\epsilon = 0.2$, $j = 0.2$. Fig. 1 shows both outer and inner caustic surfaces, where $dr/dv = 0$, which are density maxima. $dr/dv = 0$ gives $\bar{v} = q_0 \bar{r}$, i.e., velocity of particles on a caustic surface is proportional to the radius of the caustic surface. The inner caustics are the consequence of the presence of angular momentum.

The intriguing property of this model is that for $\epsilon < 2/3$, the rotation curves are somewhat flat from small radius up to the outermost caustic surface (density maximum) [5,6], which effectively defines the edge of the halo. The square of the rotation velocity is roughly given by its value at the outermost caustic ($r = r_m$):

$$v_{rot}^2(\epsilon) \equiv \frac{m(r,t)}{r} \equiv \frac{M(t)}{R(t)} c, \qquad c \equiv \frac{\mathcal{M}(\bar{r})}{\bar{r}} \simeq \frac{\mathcal{M}(\bar{r}_m)}{\bar{r}_m} \equiv c_m(\epsilon). \tag{10}$$

For our galaxy, $v_{rot} = 220$ km/s, which leads to $R(t_0)h = 1.32/\sqrt{c_m(\epsilon)}$ Mpc in our model, where t_0 is the age of the Universe today. Observational evidence seems to indicate that the current turnaround radius $R(t_0) = 1.5$ Mpc for our galaxy. The tidal disruption of the CDM phase space distribution due to the presence of nearby galaxies between the actual edge of our galaxy (~ 50 kpc) and $R(t_0) = 1.5$ Mpc does not affect the solar neighborhood ($r_\oplus \sim 8.5$ kpc) CDM distribution significantly, because the CDM particles that reach us today turned around long before the current epoch. For our purposes, $R(t_0) = 1.5$ Mpc provides a reasonable scale of length normalization. On this scale, the scale-invariant Zel'dovich spectrum of primordial density fluctuations, together with a typical CDM transfer function, gives $0.15 \lesssim \epsilon \lesssim 0.25$ [6], the ϵ range where our self-similar model gives rather flat rotation curves.

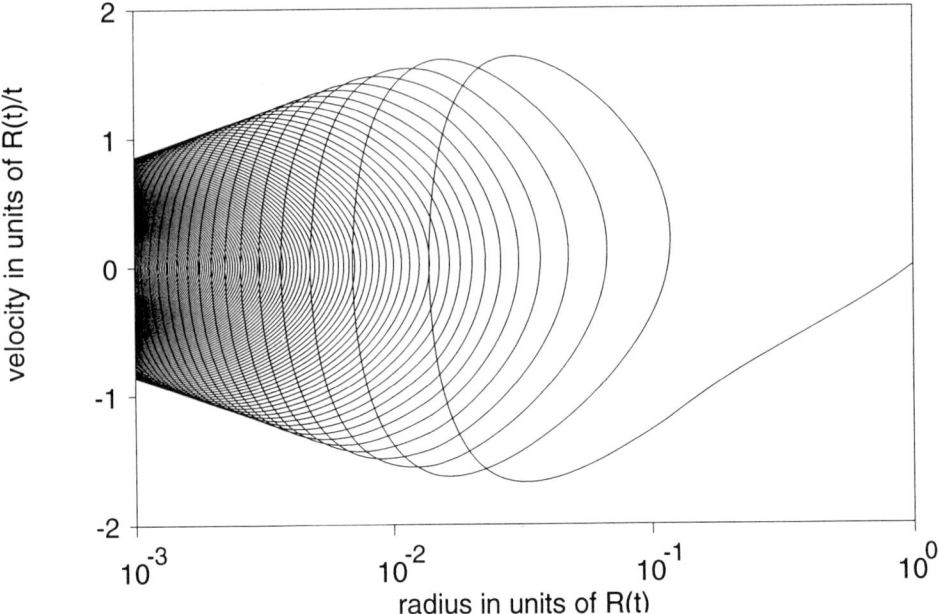

Figure 1. Phase space diagram for $\epsilon = 0.2$, $j = 0.2$.

In the absence of angular momentum, the rotation velocity in our model goes to constant at small radius [5]. With angular momentum added, the rotation velocity decreases to zero at small radius. We can define the core radius of our galaxy as the radius where the rotation velocity decreases by one half. For given ϵ, the core radius is a function of the dimensionless angular momentum j. We can find the average j needed to produce the correct core radius $r_\oplus \sim 8.5$ kpc. For $\epsilon = 0.2$, $0.1 \lesssim j_{avg} \lesssim 0.2$ [6].

In Figs. 2 and 3, we show two typical energy spectra of CDM signal [6]. The energy plotted on the x-axis is the dimensionless kinetic energy $\bar{E}_k(r_\oplus, t)$ (defined below) of the mass shells. The kinetic energy of a mass shell M_i is $E_{M_i}(r,t) = \frac{1}{2}\left(\frac{dr}{dt}\right)^2 + \frac{l^2}{2r^2}$, which can be written as

$$E_{M_i}(r_\oplus, t) = \frac{1}{2}\left[\frac{R(t)}{t}\right]^2 \left[\bar{v}^2 + \left(\frac{j}{\bar{r}_\oplus}\right)^2 \tau^{2(1-2q_0)}\right] \equiv \frac{1}{2}\left[\frac{R(t)}{t}\right]^2 \bar{E}_k(r_\oplus, t). \tag{11}$$

$E_{M_i}(r_\oplus, t) = \mathcal{O}(10^{-6})$ per unit mass. Fig. 2 shows a typical energy spectrum for negligible angular momentum ($j = 10^{-3}$), with $\epsilon = 0.2$ and $\bar{r}_\oplus = 6 \times 10^{-3}$ ($R(t_0) = 1.5$ Mpc). If $\Delta j = 0$, the energy spectrum looks quite similar for non-negligible j, but the density fraction in the first peak is very sensitive to the earth location $\bar{r}_\oplus$ relative to the first inner caustic surface for $j \gtrsim 0.1$. Fig. 3 shows a typical energy spectrum with angular momentum dispersion, the particles on a given mass shell have ten discrete values of j, $j_k = 0.4\,k/11$ ($k = 1, 2, ..., 10$), with $\epsilon = 0.2$ and $\bar{r}_\oplus = 6 \times 10^{-3}$.

The effect of scattering by massive objects in our galaxy is not shown in Figs. 2 and 3. The broadening of each velocity peak is proportional to the time the particles it contains have spent in the galactic disc. The peaks at the low energy end should have been smoothed out to form a broad thermal peak today [3]. The first two velocity peaks

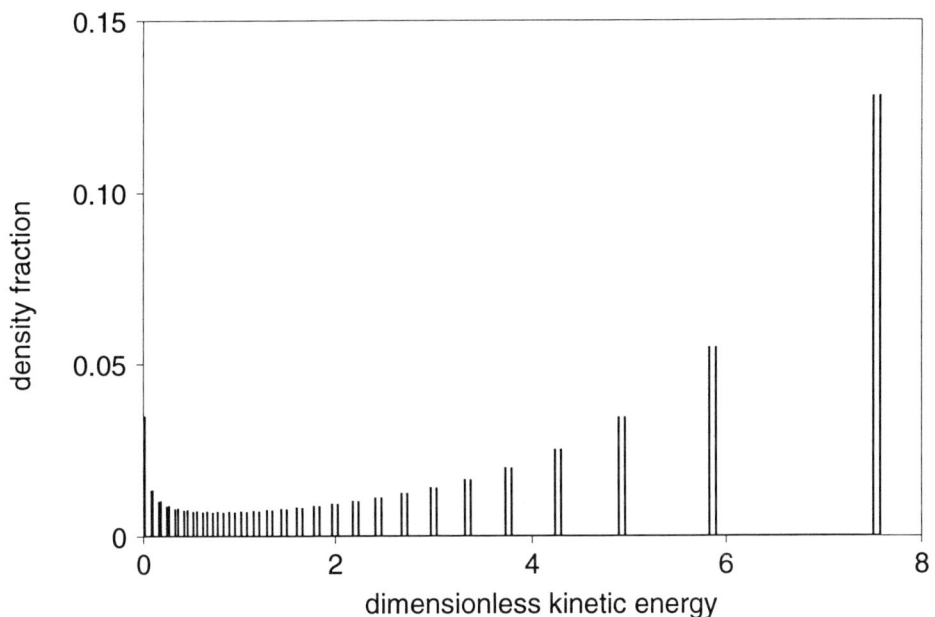

Figure 2. CDM energy spectrum at $\bar{r}_\oplus = 6 \times 10^{-3}$ ($R(t_0) = 1.5$ Mpc), for $\epsilon = 0.2$ and $j = 10^{-3}$.

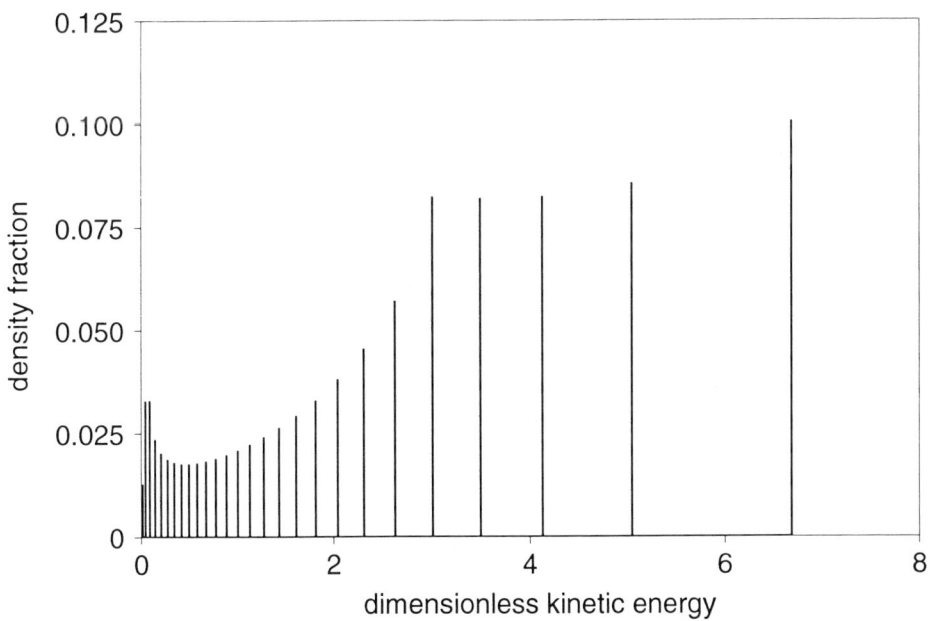

Figure 3. CDM energy spectrum at $\bar{r}_\oplus = 6 \times 10^{-3}$ ($R(t_0) = 1.5$ Mpc), for $\epsilon = 0.2$ and particles evenly divided into ten groups with dimensionless angular momentum $j_k = 0.4\,k/11$ ($k = 1, 2, ..., 10$).

on the high energy end, containing particles which are falling in or out of the galaxy for the first time, should have suffered negligible dispersion. In general, the spectrum of CDM signal consists of a broad thermal peak at the low energy end, and a parameter-dependent number of thin lines (with decreasing line width) at the high energy end.

To have a good understanding of the parameter dependence of the fraction of the local halo density in the two most energetic peaks, let us make an analytical approximation. Using Eqs.(6) and (10), and assuming that τ varies slowly compared to λ, we find

$$E_\lambda \equiv \frac{1}{2}\left[\left(\frac{d\lambda}{d\tau}\right)^2 + \frac{j^2}{\lambda^2}\right] + \frac{\pi^2 c_m}{8}\tau^{q_*}\ln\lambda \simeq E_{\lambda_m} \qquad (12)$$

where $q_* \equiv 2(2-3\epsilon)/(9\epsilon)$, $c_m(\epsilon) \equiv \mathcal{M}(\bar{r}_m)/\bar{r}_m$. Eq.(12) is the approximate conservation of energy of a given mass shell moving in a logarithmic potential which deepens as a power law with respect to time, it can be used to make useful estimates. We only use Eq.(12) between the first inner caustic (the outermost inner caustic) and the outermost caustic, when the mass shell falls out of the galaxy for the first time. Eq.(12) works well for our purpose.

Empirically, we find that the outermost caustic is given by

$$\lambda_m = \frac{1}{1.226}\left[\frac{9\epsilon}{2(1+3\epsilon)}\right]^{2/3}, \qquad \lambda_m = 1.1\,(\tau_m - 2)^{0.61}. \qquad (13)$$

Note that $\bar{r}_m = \lambda_m/\tau_m^{q_0}$. We can understand Eq.(13) as follows. For $\epsilon = 0$, the mass perturbation dominates, the mass shell M_i can only fall in (without experiencing shell-crossing) and never fall out, therefore $\lambda_m = 0$, $\tau_m = 2$ (symmetry of the initial oscillation in τ). As ϵ increases, the mass perturbation becomes smaller (see Eq.(1)), the orbit of the mass shell M_i contract less, hence λ_m and τ_m increase with ϵ.

The mass contained within the outermost caustic surface is given by

$$\mathcal{M}(\bar{r}_m) = \tau_1^{-\frac{2}{3\epsilon}}, \qquad \bar{r}(\tau_1) = \lambda(\tau_1)/\Lambda(\tau_1) = \bar{r}_m. \qquad (14)$$

For $1 \leq \tau \leq \tau_1$, shell crossing has not occured. The trajectory $\lambda(\tau)$ of a given mass shell is given exactly by

$$\lambda = \sin^2[\pi\tau/2 - \sqrt{(1-\lambda)\lambda}], \qquad 1 \leq \tau \leq \tau_1. \qquad (15)$$

For given $\bar{r}_m$, Eqs.(14) and (15) determine τ_1, thus $c_m(\epsilon)$. For $0.1 \lesssim \epsilon \lesssim 0.4$, c_m is approximately given by $c_m(\epsilon) \simeq 0.28\ln\left[1 + (11\epsilon)^4\right]$.

We are now ready to estimate analytically the density fractions of the local halo density in the two most energetic peaks if they contain particles falling in or out of the galaxy for the first time. In the absence of angular momentum, the flatness of the rotation curve (see Eq.(10)) gives the local halo density

$$\rho(r_\oplus) \simeq \frac{c_m}{4\pi r_\oplus^2}\left[\frac{M(t)}{R(t)}\right]. \qquad (16)$$

Hence the fractions of local halo density in the first two velocity peaks are given by

$$\bar{\rho}^1(\bar{r}_\oplus) \simeq \frac{1}{c_m}\left.\frac{d\mathcal{M}(\bar{r})}{d\bar{r}}\right|_{\bar{r}_\oplus}^1 = \frac{2}{3\epsilon c_m}\int_{j_{min}}^{j_{max}} dj\, \frac{f(j)}{\tau_\oplus^{2/(3\epsilon)}|\bar{v}_\oplus^1(j) - q_0\bar{r}_\oplus|}, \qquad (17)$$

where $\bar{v}_\oplus^1(j)$ denotes the dimensionless particle velocity at the dimensionless earth location $\bar{r}_\oplus$. Since particles are falling in (first peak) or out (second peak) of the galaxy for the first time, $\tau_\oplus \simeq \tau_c \simeq 2$ for $\bar{r}_\oplus \ll 1$ ($\bar{r}_\oplus = 6\times 10^{-3}$ for $R(t_0) = 1.5$Mpc). Neglecting terms of order $\alpha \equiv q_0 \bar{r}_\oplus / |\bar{v}_\oplus(j=0)| \ll 1$, the first two peaks have the same height

$$\bar{\rho}^1_{j=0}(\bar{r}_\oplus) \simeq \frac{2}{3\epsilon c_m} \frac{1}{\tau_\oplus^{2/(3\epsilon)}} \frac{1}{|\bar{v}_\oplus(j=0)|}, \tag{18}$$

$$|\bar{v}_\oplus(j=0)| \simeq \frac{\tau_\oplus}{\Lambda(\tau_\oplus)} \sqrt{\frac{\pi^2 c_m}{4}[\tau_m^{q*} \ln \lambda_m - \tau_\oplus^{q*} \ln \lambda_\oplus] + q_0^2 \left(\frac{\lambda_m}{\tau_m}\right)^2}.$$

where we have used Eqs.(9) and (12) to estimate $\bar{v}_\oplus(j=0)$.

For $0.15 \leq \epsilon \leq 0.3$ and $\bar{r}_\oplus = 6 \times 10^{-3}$, the density fractions in the first and second velocity peaks are 10–15% each. The predication of Eq.(18) differs from the exact numerical results by less than 7%, with the best agreement at $\epsilon \simeq 0.25$. This is as expected, since the rotation curve is flattest for $0.2 < \epsilon < 0.3$.

Next, let us estimate the effect of the halo particles' angular momentum on the density fractions in the two most energetic peaks. Let $\bar{r}_c(\epsilon, j)$ be the radius of the first inner caustic (the outermost inner caustic) of particles with angular momentum j. The maximum angular momentum j_c a particle can carry to reach earth in its first radial oscillation is given by $\bar{r}_c(\epsilon, j_c) = \lambda_c / \Lambda(\tau_c) = \bar{r}_\oplus$, when the first inner caustic occurs at the earth location. $j_c = j_c(\epsilon, \bar{r}_\oplus)$ should decrease with increasing ϵ (shallower potential well) for given $\bar{r}_\oplus$, and increase with increasing $\bar{r}_\oplus$ for given ϵ. Eq.(12) gives

$$j_c^2 \simeq \frac{(\pi^2 c_m/4)[\tau_m^{q*} \ln \lambda_m - \tau_\oplus^{q*} \ln \lambda_\oplus] + q_0^2 [(\lambda_m/\tau_m)^2 - (\lambda_\oplus/\tau_\oplus)^2]}{(1/\lambda_\oplus^2 - 1/\lambda_m^2)}. \tag{19}$$

For $j \lesssim 0.5$, we can use $\tau_\oplus \simeq \tau_c \simeq 2$, $\lambda_\oplus \simeq \bar{r}_\oplus 2^{q_0}$.

If all halo particles carry the same dimensionless angular momentum j, the mass shells which turned around earlier have smaller angular momentum $l(M_i)$ (see Eqs.(5) and (2)). Hence for a given j, the mass shells which turned around at a sufficiently early time (i.e. the particles which have experienced sufficient number of radial oscillations) can reach us today. If $j \leq j_c(\epsilon, \bar{r}_\oplus)$, the most energetic halo particles that reach us today are falling in or out of the galaxy for the first time. If $j > j_c(\epsilon, \bar{r}_\oplus)$, the most energetic particles that reach us today have undergone a number of radial oscillations. The density fractions in the two most energetic peaks oscillate as the observer's distance to the galactic center changes; the sum over the two peaks goes to 1 at the distances corresponding to inner caustics, when the two peaks overlap. Our numerical calculations show that the mean density fraction in the two most energetic peaks (of equal height) is roughly independent of j, and given by the value at $j = 0$ as predicted by Eq.(18).

In Fig. 3, $\epsilon = 0.2$ and $\bar{r}_\oplus = 6 \times 10^{-3}$, which gives $j_c \simeq 0.1$ (see Eq.(19)); since particles on a mass shell are evenly divided into ten groups with dimensionless angular momentum $j_k = 0.4 k/11$ ($k = 1, 2, ..., 10$), not all particles in a mass shell contribute to the same peak in the energy spectrum on earth.

Let us consider the case when the halo particles have dimensionless angular momentum evenly and continuously distributed between j_{min} and j_{max}, i.e., $f(j) = 1/\Delta j$. If $j_{min} > j_c(\epsilon, \bar{r}_\oplus)$, the case is similar to that of halo particles with one dimensionless angular momentum $\frac{1}{2}\Delta j$. If $j_{min} < j_c(\epsilon, \bar{r}_\oplus)$, the two most energetic peaks contain particles which are falling in or out of the galaxy for the first time. We can estimate the density fractions in these peaks by integrating over j in Eq.(17) with $n = 1$. The particle velocity at earth location, $\bar{v}_\oplus^1(j)$, can be estimated using Eqs.(9) and (12). Again neglecting terms of order $\alpha \equiv q_0 \bar{r}_\oplus / |\bar{v}_\oplus(j=0)| \ll 1$, we find that both peaks have the same height

$$\frac{\bar{\rho}^1_{\Delta j}(\bar{r}_\oplus)}{\bar{\rho}^1_{j=0}(\bar{r}_\oplus)} = \left(\frac{j_c}{\Delta j}\right) \left[\arccos\left(\frac{j_{min}}{j_c}\right) - \arccos\left(\frac{\min(j_{max}, j_c)}{j_c}\right)\right]. \tag{20}$$

For $j_{max} > j_c$, the first and second velocity peaks overlap.

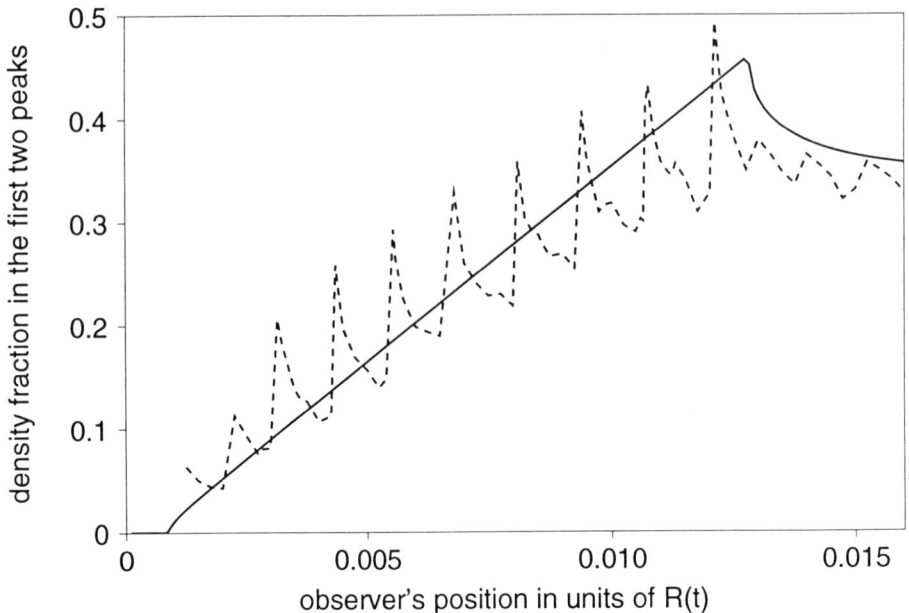

Figure 4. The density fraction of the local halo density in the two most energetic peaks as a function of the earth location, for $\epsilon = 0.2$, $j_{min} = 0$, $\Delta j = 0.2$. The solid line is analytical. The jagged dashed line is numerical, with particles evenly divided into ten groups with dimensionless angular momentum $j_k = 0.2\,k/11$ ($k = 1, 2, ..., 10$).

Note that for $j_{min} = 0$, $j_{max} > j_c$, there is a factor of $\pi/2$ enhencement to the height of the most energetic peak, over the naive reduction factor of j_c/j_{max} (particles with $j_{max} \geq j > j_c$ can not reach earth in their first radial oscillation). This enhencement is due to the earth location overlapping with the first inner caustic of mass shells with dimensionless angular momentum j_c. Fig. 4 shows the sum of the density fractions of the local halo density in the first and second velocity peaks as a function of the earth location, for $\epsilon = 0.2$, $j_{min} = 0$, $\Delta j = 0.2$. There is good agreement between the analytical formula and the numerical results. The jagged structure in the numerical curve is the consequence of using a set of 10 discrete angular momentum components to approximate the continuous angular momentum dispersion. The turn over in the density fraction occurs when $j_{max} = j_c(\bar{r}_\oplus)$; for earth location beyond this radius, $j_{max} < j_c(\bar{r}_\oplus)$, the first and second velocity peaks separate.

In summary, we have studied a simple self-similar spherical infall model with angular momentum dispersion, it produces fairly flat rotation curves for the range of the initial density fluctuation power law index ϵ given by the standard CDM theory of structure formation, and should provide a reasonable first order approximation to reality. We find the density fractions of the local halo density in the two most energetic peaks in a cold dark matter spectrum to be at least 5-10% each, depending on the angular momentum dispersion. Our results should help optimize the design of cold dark mattter detectors. Once a cold dark matter signal is detected, we should be able to extract a wealth of information on the history of our Galaxy from the energy spectrum of the signal.

Acknowledgments

This paper is based on work done in collaboration with Pierre Sikivie and Igor Tkachev. This work has been supported by the DOE and NASA under Grant NAGW-2381.

References

1. For reviews, see *e.g.*, V. Trimble, Ann. Rev. Astron. Astrophys. **25** (1987) 425.

2. J. Ellis, J. S. Hagelin, D. V. Nanopoulos, K. A. Olive, and M. Srednicki, Nucl. Phys. B **238**, 453 (1984); R. D. Peccei and H. Quinn, Phys. Rev. Lett. **38**, 1440 (1977); Phys. Rev. D **16**, 1791 (1977); S. Weinberg, Phys. Rev. Lett. **40**, 223 (1978); F. Wilczek, *ibid.* **40**, 279 (1978).

3. J. R. Ipser and P. Sikivie, Phys. Lett. B **291**, 288 (1992); J. R. Ipser and P. Sikivie, Phys. Rev. D **35**, 3695 (1987).

4. P. Sikivie, Phys. Rev. Lett. **51**, 1415 (1983); K. Van Bibber et al., private communication.

5. J. A. Filmore and P. Goldreich, Ap. J. **281**, 1 (1984); E. Bertschinger, Ap. J. S. **58**, 39 (1985).

6. P. Sikivie, I.I. Tkachev, and Y. Wang, in preparation (1994).

Moment and Wavelet Analysis of Correlations in Multihadron and Galaxy Distributions

P. Carruthers
Department of Physics
University of Arizona Tucson, AZ 85721, USA

Introduction

Here we consider recent developments in the analysis of textures of systems composed of many "points". For clarity we focus on two important examples, for which experimental progress has been decisive. The first is that of multihadron production[1], in which a large collection of final states in the momentum phase space can be prepared having identical initial conditions. The second is that of galaxy distributions, in which the points live in ordinary space-time. In this case there is no ensemble since there is only one specimen, presumably created by the big bang. Although the system is very large, it possesses correlations of very long range, making the usual procedure of creating a fake ensemble by partitioning the system into arbitrarily chosen subsystems suspect as a method of deriving correlation functions.

For simplicity we consider systems composed of one species of particle. Typically one considers charged hadrons without regard to sign. However recent experiments have found a strong and important effect for like-sign charges for high resolution, which must be taken into account. In the case of galaxies we do not discriminate among the different types, spiral, elliptic, etc. At the expense of writing more complicated equations, it is possible to handle such details.

The classic approach to texture analysis is by means of correlation functions[2]. Unfortunately these are hard to measure beyond second order. However it is now possible to measure factorial moments of the count distributions to fifth order and high resolution in the hadronic case. These moments are found in principle by integrating the density correlation functions $\rho_p(x_1...x_p)$ over some patch of phase space called Ω. Defining a sequence of densities of which the first two are

$$\rho_1(x;s) = \sum_i \delta(x - s_i) \quad (1)$$

$$\rho_2(x,x;s) = \sum_{i+j} \delta(x - s_i)\delta(x - s_j) \quad (2)$$

and performing suitable averages leads to

$$\langle n \rangle_\Omega = \int_\Omega \rho_1(x) d^3x \quad (3)$$

$$\langle n(n-1) \rangle_\Omega = \int_\Omega d^3x \int_\Omega d^3x' \rho_2(x,x') \quad (4)$$

Another formulation for densities in particle physics uses the sequence of inclusive differential cross sections. These formulations are equivalent; for uniformity of presentation we shall use the delta function definition of densities.

The dependence of the moments on the location, shape and size of Ω tells us much about the correlation function. Of particular interest has been the search for domains (and appropriate variables) in which scaling might occur. This development has been much influenced by the emergence of fractal thinking in many fields.

By now there is a detailed technology on this topic. This allows the model- independent description of data in a form of considerable utility. We shall refer to earlier papers for technical details.

In the past decade a powerful variant of the Fourier transform, "wavelet analysis" has shown great promise for the analysis of textures and also great data compression capability. Unlike the Fourier method, wavelets allow simultaneous localization in conjugate variables. For example, when listening to music we make somehow a frequency analysis at each moment in time. The vast possibilities of this approach have already led to new insights into many subjects, such as signal analysis, pattern recognition, etc. In physical systems we expect that advances both in phenomenology and the reformulation of dynamical problems. We briefly examine the behavior of correlations in simple cascade models using the simplest wavelet, the Haar basis.

The Situation for Multihadron Production

Ten years ago the field of multiparticle production was rejuvenated by the suggestion[3] of Białas and Peschanski, to study the dependence of "bin-averaged" factorial moments as a function of rapidity bin-width δy. Recall that the rapidity $y = (\frac{1}{2} \ln \frac{E+p_z}{E-p_z})$ is additive under change of inertial frames and is a natural variable for longitudinal kinematics. When true momentum is not measured it is well approximated by pseudorapidity $\eta = -\ln \tan \theta/2$, with θ the angle of the final particle with respect to the collision axis. From Eqs. 3-4 and their generalization to higher order we write the factorial moment $F_p(\delta y)$ as

$$F_p(\delta y) = \frac{1}{M} \sum_{i=1}^{M} \frac{\langle n_i(n_i - 1)......(n_i + p + 1)\rangle}{\langle n_i \rangle^p} \tag{5}$$

We have chosen the Ω to be M adjacent bins of width δy. Note that for Poissonian count statistics the F_p are unity. Typically one expects dynamical correlations due to resonance decays and statistical correlations among like sign particles, (such as $\pi^-\pi^-$, etc.) usually called Bose-Einstein correlations.

In the case of the second moment we can measure $\rho_2(y, y')$ and directly integrate Eq.4 to get $F_2(\delta y)$, as shown in Fig.1. The bending of the curve is best attributed to the existence of a correlation length[4]. Much attention has been given to the possibility of scaling for very small y. To the accuracy of the data given in Fig. 1 there is no conclusive evidence for scaling. However recent data show good scaling for like sign pions which can dominate for small phase space cells (in this case the best variable seems to be the invariant momentum transfer (Q^2) between particle pairs.)

The density correlations and the corresponding moments inevitably contain contributions of lower order. The systematic way to remove these is by going over to cumulants C_p, and the factorial cumulant moments, as we have discussed elsewhere.[5]

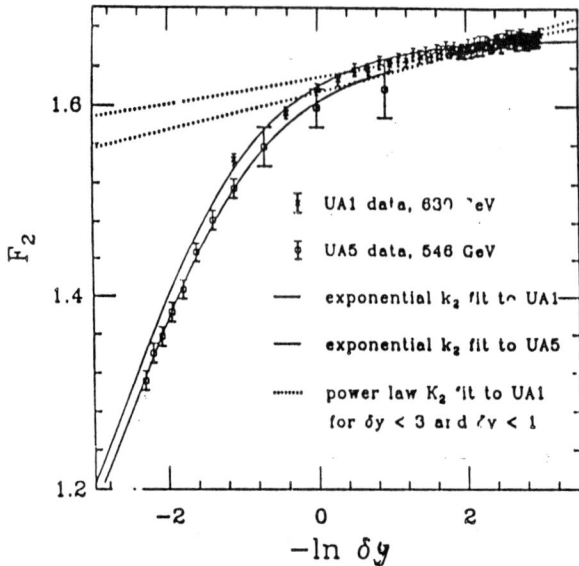

Fig.1 The dependence of the second order factorial moment F_2 on the bin resolution δy shown. The solid curve results by integrating an exponential fit of the two particle correlation. The dotted lines are fit to agree with the high resolution data.

In summary we can mention some salient results

1. The moments F_p (or K_p) typically increase with order p and with collision energy. They also increase (that is fluctuations increase) with shrinking bin size.

2. For complex targets (involving nuclei, typically) there are essentially no true (cumulant) correlations beyond second order.[6]

3. For most multiparticle production processes the pth order cumulant can be built[4] from (symmetrized) linked products of two particle cumulants

$$Cp(1,2,...p) = \frac{Ap}{(p/2)} \sum_{perm} C_2(1,2)C_2(2,3)....C_2(p-1,p) \qquad (6)$$

with the C's normalized to the single particle densities $\rho(1),...\rho_1(p)$.

4. If A_p is properly chosen[7] as $(p-1)!$ Eq. 6 integrates to give the factorial cumulant moments characteristic of the negative binomial distribution

(NBD), which provides a good description of charged particle counts in much of phase space. For later reference we write the NBD distribution as

$$P_n^k = \frac{(n+k-1)!}{(k-1)!n!} \frac{(\bar{n}/k)^n}{(1+\bar{n}/k)^{n+k}} \tag{7}$$

The parameter k can be any positive real number. It depends on the reaction, on the energy, and the size of the phase space volume in which the counting takes place. In terms of correlations, $1/k$ is given by the integral of the two particle cumulant correlation C_2 over the domain Ω.

Although a variety of dynamical models (typically cascade processes) suggest the NBD, more work needs to be done. As in the case of the Maxwell distribution for molecular velocities, the form of Eq. 7 seems to be independent of dynamical details, a situation that has both good and bad features.

5. An interesting "statistic" is $P_o(\Omega)$, the probability of finding nothing[8] in Ω. This is usually called the "rapidity gap" probability. In the case of galaxies it is the void probability, discussed in the next section. In each case the data are well described by negative binomial count statistics.

6. Some important technical modifications of the bin-averaging approach have been developed by the Tucson group. To appreciate these, first note that for the one-dimensional case F_2 is given by the integral over adjacent boxes[9] of side δy, centered on the diagonal $y_1 = y_2$ in the $y_1 - y_2$ plane. (For higher order one has hypercubes.) Two points close by in this plane but not in the same bin do not contribute to the moment. For highly clumped events the problem is serious because the moments jump around when the bin size is reduced. As a consequence these rare but important events give rise to spurious statistical fluctuations in the domain of greatest interest.

The situation is much improved if the integration over boxes is replaced by a strip. Originally[4] the strip domain was used as a convenient approximation, but later we realized[5,8] that the strip domain gives a general approach, close numerically to the box method but without the spurious fluctuations mentioned above.[10] The strip width ϵ now provides the resolution scale; all pairs of points closer than ϵ contribute to the strip moment on an equal footing[8,9]. This approach is now widely used in multihadron data analysis. Note that should scaling occur, the method is numerically close to that used to define correlation dimensions, in which case <u>all</u> pairs are less than a shrinking value ϵ. In addition to the Grassberger-Hentschel-Procaccia algorithm[11] one can define[10] a "star integral", which has the additional merit of reducing computational needs, allowing the analysis of events having very high multiplicity, as expected in the planned accelerators LHC and RHIC.

For details about these developments consult the references.

The Situation for Galaxy Counts and Correlations

Although we reviewed this topic fairly recently[8], new results and techniques are constantly emerging in this popular field, often known as "large-scale structure". The classic reference is the book by Peebles[12]. The method of correlation functions is developed in detail, and a conjecture about the structure of correlations known as the "hierarchical model" is put forward. In fact it is basically the same as the "linked pair" structure of Eq.6 except that the magnitude of the coefficients was left as a free parameter.

Historically galaxy counts were studied by Hubble over sixty years ago[12]. By counting galaxies in his photographic plates (therefore a projection onto the sphere) he got a good fit using the log-normal distribution. The conclusion is that a random placement such as multinomial or Poisson is ruled out. Except for some important theoretical work by Neyman[14] and collaboration in the 50's and 60's, the problem of count statistics of galaxies seems to have attracted little attention. About this time systematic galaxy catalogues began to appear. In 1983 C.C. Shih and I showed[15] that the new UA5 multiplicity data were well described by the negative binomial distribution. Soon another former student, Ming Duong-Van noticed[16] that the Zwicky data[17] fit the same function, even though one could not imagine the dynamics to be more different. We mention that the log-normal is a good approximation to the negative binomial.

As recently as ten years ago I was told by a well-known observer that I was not allowed to consider the problem until the data improved (!). Besides the NBD, or its close approximation, the gamma distributions, other similar count distributions were proposed[18]. Meanwhile considerable new information emerged about counts and curious structures showing the existence of long range correlations.

From the point of view of count probabilities the void probability is particularly interesting. For a volume v there is a general formula in terms

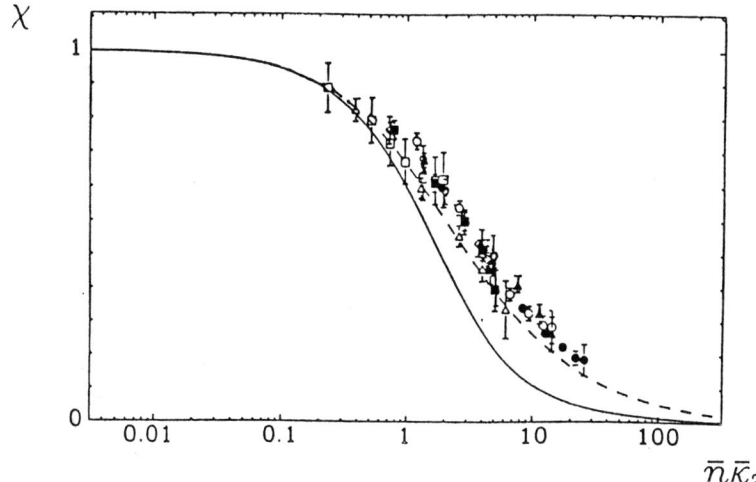

Fig.2 The void probability scalaing function χ (Eq. 9) is shown as a function of the average galaxy number $\bar{n}$ times the normalized second factorial cumulant moment, the latter being the same as $1/k$ in the case of the negative binomial distribution. The dashed curve is the prediction of the negative binomial distribution.

of normalized factorial cumulant moments K_p (e.g. K_2 is $[\langle n(n-1)\rangle - \langle n\rangle^2]/\langle n\rangle^2$):

$$P_o(v) = \exp\left[\sum_{n=1}^{\infty} \frac{(-\bar{n})^n}{n!} K_n(v)\right] \quad (8)$$

Since $\bar{n}(v)$ can be large, as often happens also for the K_n, the convergence of (8) is problematic. In the case of linked pairs, or hierarchical structure $K_n \infty [K_2(v)]^{n-1}$ which leads to a scaling law

$$\frac{-\ln P_o(v)}{\bar{n}(v)} \equiv \chi(v) = f(\bar{n}(v) K_2(v)) \quad (9)$$

In the case of the NBD the coefficients lead directly to

$$-\frac{\ln P_o(v)}{\bar{n}} = \frac{\ln(1+\bar{n}/k)}{(\bar{n}/k)} \quad (10)$$

in agreement with Eq. 7, as must be the case. Here $1/k = K_2(v)$ relates the K parameter to the 2-particle correlation function.

Fig. 2 shows good agreement with (10). Curiously, hadronic (UA5) data follow[8] the same scaling law (Hegyi), where the void is a symmetric gap in pseudorapidity. Note that Eq. (8) is sensitive to all orders of correlation. This is of interest because it is so difficult to measure correlation functions beyond second order.

Because of the existence of large structures in the distribution of luminous matter, and the known existence of only one event, the usual ensemble formulation of correlation functions is suspect in principle. In typical condensed matter systems one can assume homogeneity and create a false ensemble by chopping a large system into equivalent subsystems. One sits on each particle in turn, computes distances to the others and averages over all chosen particles. We mention that not only is the universe not very homogeneous, as measured by luminous matter, but it is also not isotropic, at least for the Zwicky data, as shown[16] by the count statistics for different directions. Although we are not ready to discard correlation functions the posed problem merits close consideration.

Consider the (normalized, cumulant) two particle galaxy correlation function derived by the Princeton school. Their notation for this is $\xi(r)$ the latter obeying a scaling law

$$\xi(r) \cong (r_o/r)^\delta \qquad (11)$$

with $\delta \cong 1.7$ and $r_o \cong 5$ Mpc. This formula is said to be valid for r up to 10-20 Mpc. Eq. 11 allows calculation of the conditional probability of finding a galaxy at $\vec{r}$ if one is known to be at the origin. The number N(R) contained in a sphere of radius R is (ρ is the galaxy number density)

$$N(R) = \rho \int_0^R 4\pi r^2 dr (1 + \xi(r)) = 4\pi \frac{R^3}{3}\rho[1 + \frac{3}{3-\delta}(\frac{r_o}{R})^\delta] \qquad (12)$$

For $R < r_o$ this suggests a simple fractal with dimension $3 - \delta \approx 1.3$. For large $R, N \propto R^3$ and the standard uniform homogeneous universe is recovered. Many improvements to this "standard model" can be considered. Besides the trouble defining correlation functions, which might not affect 2nd order, one should consider the possibility of multiple scales. In fact the multifractal formalism has been applied to the galaxy distribution analysis. Is the galaxy fractal on all scales? Despite the simplicity of Eq. 12 and its interpretation the debate continues[19].

What about higher orders? Here the moment method rescues the situation somewhat from the difficulties associated with correlation functions. Chmaj et al[20] have applied the bin moments (Eq.5) for various geometries and sample depth to data from the Zwicky catalogue. They find good scaling for moments F_2 through F_5. In addition the magnitudes appear to agree with NBD, although the size of the plots (on log-log scales) makes it hard to read the precise magnitudes. We note that since Eq. 5 is defined solely by counting, the connection to Eqs. 3-4 is not really necessary.

Another interesting discovery is that the cluster-cluster correlation $\xi_{cc}(r)$ seems to have the same scaling law in Eq. 11, with a larger r_0.

Since negative binomial counts, as well as the related linking structure of correlations, apparently occur both in gravity-driven dynamics and in strong interaction physics (probably QCD) it would seem that a general statistical phenomenon is at work. We mention the following: 1) each system has a scaling regime (large Q^2 for QCD) 2) cascade mechanisms akin to turbulence are likely 3) dissipation and instabilities are likely to lead to a reduction of the number of operative degrees of freedom and fractal structures. However these remarks have to be regarded as wishful speculations at the moment.

Analysis of Textures by Wavelet Analysis

In the past ten years a new mathematical technique known as wavelet analysis has shown great promise for texture analysis and data compression. A key physical and mathematical aspect of this method is the ease of multi-resolution analysis, whereby the behavior of the examined system at different scales can be studied, and in some cases decoupled. In a short time an enormous literature has appeared. These methods are now prominent in many fields, with physics lagging behind.

The method of Fourier analysis is based on the set of functions e^{inx} created from one master function e^{ix}. Despite the power of this technique it is not the best for some problems. The wavelet approach builds all functions from localized functions, which act as a microscope, using translations to localize and dilations to magnify the inspected function. The associated group is known as the affine group. A standard reference is the book by Daubechie[21].

A given distribution of points can be represented by its Fourier or wavelet transform. It turns out that the latter has certain advantages. Both methods fit into the general topic of the representation of functions by orthogonal basis sets. Although non-orthogonal bases (such as oscillator coherent states) provide useful wavelets we shall here use only orthogonal wavelets of compact support, since non-orthogonal wavelets can introduce unphysical correlations.

An attractive feature of wavelets is their ability to provide simultaneous information about conjugate variables such as frequency and time. Such information is cumbersome in the Fourier description of the same data.

A few definitions and examples will give some orientation about the general approach. Here we will only state some very basic aspects of wavelets and the related concept of a multiresolution analysis.

Given an arbitrary and for simplicity one-dimensional function $\epsilon(x)$, we seek to approximate it in terms of a set of box "scaling" functions

$$\Phi_{Jk}^{H}(x) = \Phi(2^J x - k) = \begin{cases} 1 & \text{for } k2^{-J} \leq x \leq (k+1)2^{-J} \\ 0 & \text{otherwise} \end{cases} \quad (13)$$

which all belong to the same 'resolution' scale J and are constructed from $\Phi^J(x) = \Phi(2^J x)$ by a simple translation governed by an integer k. We could also say, that the $\Phi_{Jk}(x)$ evolve from the scaling function $\Phi_{00}(x)$ by discrete dilations and translations only. Within a given scale the box functions $\Phi_{Jk}(x)$ are orthogonal with respect to the shift index k. For the approximation of the function $\epsilon(x)$ at the finest resolution 2^{-J} we get:

$$\epsilon(x) \to \epsilon_J(x) = \sum_k \epsilon_{Jk} \Phi_{Jk}(x) \quad (14)$$

This is a fancy way to describe a histogram with resolution 2^{-J}. If we were to approximate $\epsilon(x)$ with the box functions belonging to the rougher resolution scale J - 1, which are again orthogonal with respect to the shift index k, but are not orthogonal to the box functions Φ_{Jk} of the finer 'resolution scale' J, evidently some detail is lost compared to the approximation (14). This detail is the difference between approximations (14) with resolution scales J and J -1; it can be fully expressed in terms of the difference functions $\Psi_{(J-1)k}(x) = \Psi(2^{J-1}x - k)$ with

$$\Psi_{00}(x) = \Psi(x) = \begin{cases} 1 & \text{for } 0 \leq x < 1/2 \\ -1 & \text{for } 1/2 \leq x < 1 \\ 0 & \text{otherwise} \end{cases} \quad (15)$$

This is the basic Haar wavelet.

The histogram ϵ_J can be decomposed into Haar wavelets with expansion coefficients $\tilde{\epsilon}_{kj}$:

$$\epsilon_J(x) = \epsilon_{00}\phi_{00} + \sum_{j=0}^{J-1}\sum_{k=0}^{2^J-1} \tilde{\epsilon}_{jk}\psi_{jk}(x) \quad (16)$$

The $\tilde{\epsilon}$ depend only on adjacent levels and Eq. 16 can be regarded as a multi-resolution scale analysis. We[22] have applied these ideas to the cascade process known[23] as the p-model, which is analytically soluble and several other toy cascade models.

In the p-model an initial energy defined on the unit line splits into two unequal parts, defining two parallel cascades by the same rule as the first split. At level j ($j = 0$ the initial state and $j = J$ the final stage) the ϵ_{jk} of Eq. are interpreted as the energy density in bin k at the jth level of the cascade. Give the probability of the number of combinations that can occur at level j one can calculate the sequence of correlation functions $\langle \epsilon_{j_1 k_1} \epsilon_{j_2 k_2} \rangle$, $\langle \epsilon_{j_1 k_1} \epsilon_{j_2 k_2} \epsilon_{j_2 k_2} \rangle$, etc.

However the most interesting result is found by calculating correlations of the wavelet coefficients: $\langle \tilde{\epsilon}_{j_1 k_1} \tilde{\epsilon}_{j_2 k_2} \rangle$, etc. This correlation is exactly diagonal. Hence the Haar wavelets are eigen-functions of the correlation matrix.

This result seems natural since the self-similar scaling cascade is being analyzed by scaling wavelets with decoupled levels of the cascade according to the multiscale resolution character of the wavelet.

It will be interesting to apply these ideas to more realistic QCD cascades and to cascades of matter density fluctuations. Each of these theories has a large domain of scale-invariant dynamics.

The Haar basis is discrete, compact, orthogonal and complete. Many other discrete bases are known and proving their merit. Recently Pando and Fang have made progress on large scale structure problems[24]. Because of the data compression abilities of wavelets it should be possible to simplify dynamical calculations in field theory. Likewise the pattern recognition data compression might allow one to make particle detectors more efficient, reducing their cost.

Acknowledgment.

This work was supported in part by the U. S. Department of Energy, Divisions of High Energy and Nuclear Physics.

References

1. E.A. DeWolf, I.M. Dremin and W. Kittel, Usp. Fiz. Nauk 163 (1993) 3.

2. P. Carruthers, Phys. Rev. A43 (1991) 2632; Int. J. Mod. Phys. A4 (1989) 5587.

3. A. Białas and R. Peshchanski, Nucl. Phys. B273 (1986) 703; ibid, B308, (1988), 857.

4. P. Carruthers and I. Sarcevic, Phys. Rev. Lett. 63 (1989) 1562.

5. H.C. Eggers, P. Lipa, P. Carruthers and B. Buschbeck, Phys. Lett. B301 (1993) 298.

6. P. Carruthers, H.C. Eggers and Ina Sarcevic, Phys. Rev. C44 (1991) 1629.

7. E. A. deWolf Acta Physica Polonica B21 (1990) 611.

8. S.D.M. White, MNRAS 186 (1979) 145 P. Carruthers, Astrophysical Journal 380 (1991) 24 S. Hegyi, Phys. Lett. B309 (1993) 443 J. N. Fry *et al* Astrophysical Journal301 (1984) 1.

9. P. Carruthers, H.C. Eggers, Q. Gao and I. Sarcevic, Int. J. Mod. Phys. A6 (1991) 3031; P. Carruthers, H.C. Eggers and Ina Sarcevic, Phys. Lett B254 (1991) 258.

10. H.C. Eggers, P. Lipa, P. Carruthers and B. Buschbeck, Phys. Rev. D48 (1993) 2040.

11. I. Grassberger, Phys. Lett. 97A (1983) 227; H.G.E. Hentschel and I. Procaccia, Physica 8D (1983) 435.

12. P.J.E. Peebles, "The Large-Scale Structure of the Universe" (Princeton, NJ, 1980).

13. E. Hubble, Astrophysical Journal 79 (1934) 8.

14. J. Neyman, p. 135 in "Mathematical Models in Physical Sciences", ed. S. Drobot (Prentice-Hall, Inc., Englewood Cliffs, N.J., 1963).

15. P. Carruthers and C.C. Shih, Phys. 127B (1983) 242.

16. P. Carruthers and Minh Duong-Van, Lett. B 131B (1983) 116.

17. F. Zwicky, E. Herzog, P. Wild, M. Karpowicz and G.T. Kuwal "Catalogue of Galaxies and Clusters of Galaxies" (Cal Tech, Pasadenaa, 1961-68).

18. P. Crane and W.C. Saslaw Astrophysical Journal 301 (1986) 1; F. Lucchin and S. Matarrese, Astrophysical Journal 330 (1988) 535; R. Balian and R. Schaeffer Astronomy and Astrophysics 220 (1989) 1; 226 (1989) 373.

19. P.H. Coleman and L. Pietronero, Phys. Rev. 213 (1992) 311.

20. T. Chimaj, W. Doroba and W. Słominski, Z. Phys. C50 (1991) 333.

21. I. Daubechie, "Ten Lectures on Wavelets" SIAM (1992).

22. M. Greiner, P. Lipa and P. Carruthers, Phys. Rev. E51 (1995) 1948; M. Greiner, J. Giesemann, P. Lipa and P. Carruthers, "Wavelet Correlations in Hierarchical Branching Processes", sumbitted to Z. Physik C.

23. C. Meneveau and K. R. Sreenivasan, Phys. Rev. Letters 59 (1987) 1424.

24. Jesus Pando and Li-zhi Fang, University of Arizona preprint (1995).

DILATON-DRIVEN INFLATION IN STRING COSMOLOGY

Ram Brustein[1]
Theory Division, CERN
CH-1211, Geneva 23
Switzerland

INTRODUCTION

I present an outline of cosmological evolution in the framework of string theory. The main emphasis is on a phase of dilaton-driven kinetic inflation and its possible observable consequences, in particular, a background of stochastic gravitational radiation. The results concerning the produced spectrum of gravitational radiation were obtained in [1, 2]. More details on various aspects of the suggested outline and additional references may be found in [3-11].

POTENTIAL-DRIVEN INFLATION

Inflationary evolution of the universe requires a source of energy to drive the expansion. The conventional expectation is that the energy source is dominated by potential energy of scalar fields, called inflatons [12]. The inflatons are expected to posses non-vanishing potential energy during some phase in their evolution in which inflationary expansion takes place. Eventually, the inflatons settle down to the true minimum of their potential where the potential energy vanishes, thus depriving the universe of the necessary source to drive its accelerated expansion. The inflationary phase ends and the universe continues to expand sub-luminally until today. If one tries to implement similar ideas in the framework of string theory, an apparent problem is immediately encountered [13, 14]. String theory does indeed contain many scalar fields, called moduli, which seem particularly suitable for the job of inflatons [15, 16]. Among the moduli the dilaton ϕ is an important and universal field whose expectation value determines the string coupling parameter $g_s^2 = \langle exp(\phi) \rangle$. It couples to all other fields with gravitational strength. If some scalar field, for example, one of the moduli fields acquires a non-vanishing potential so does the dilaton. The type of generated dilaton potential depends on the details of the model. Two types are distinguished, perturbative $V(\phi) \sim exp(-\alpha\phi/M_{pl})$, and non perturbative $V(\phi) \sim exp(-exp(-\beta\phi/M_{pl}))$, with par-

---

[1]Present address: Department of Physics, Ben-Gurion Universty, Beer-Sheva 84105, Israel.

ticular numerical parameters α, β. The equations of motion for the resulting string dilaton-gravity, assuming isotropic and homogeneous universe

$$ds^2 = -dt^2 + a^2(t)dx_i dx^i$$
$$\phi = \phi(t), \qquad (1)$$

are the following

$$H^2 = \frac{8\pi}{3M_{pl}^2}\left(\frac{1}{2}\dot\phi^2 + V(\phi)\right)$$
$$\ddot\phi + 3H\dot\phi = -\frac{dV}{d\phi} \qquad (2)$$

The Hubble parameter, H, is related to the scale factor, a in the usual way, $H \equiv \frac{\dot a}{a}$ and V is the potential. Consider, for example, the (unrealistic) case of exponetial potential $V = V_0 \exp(-\alpha\phi/M_{pl})$. We can solve eqs. (2) explicitly

$$a(t) = a_0 \, t^{16\pi/\alpha^2}. \qquad (3)$$

If the potential is steeper than the critical steepness $\alpha = 4\sqrt\pi$, the dilaton kinetic energy becomes dominant over potential energy and the expansion is subluminal. The generic situation in string theory is that the potentials in several models are steeper than critical and therefore potential-driven inflation requires special situations and is generally speaking hard to obtain. Recently, some progress has been made towards characterizing requirements from models in which potential-driven inflation could be supported [16, 17].

DILATON-DRIVEN KINETIC INFLATION

The outline for cosmological evolution that I present here relies heavily on the fact that the kinetic energy of the dilaton tends to dominate the energy density. Instead of trying to fight this tendency, one accepts it and turns this feature into a virtue, using it to drive kinetic energy dominated inflationary evolution. Kinetic inflation was also discussed in [18]. The evolution starts when the dilaton is deep in the weak-coupling region ($\phi \ll -1$) and Hubble parameter, H, is small. The evolution in this epoch is shown below to be accelerated expansion dominated by the dilaton kinetic energy and determined by the vacuum solution of the string dilaton-gravity equations of motion [3]. To describe the first phase in more detail, look for solutions of the effective string equations of motion in which the metric is of the isotropic, FRW type with vanishing spatial curvature and the dilaton depends only on time. One finds three independent first order equations for the dilaton and H

$$\dot H = \pm H\sqrt{3H^2 + U + e^\phi \rho} - \frac{1}{2}U' + \frac{1}{2}e^\phi p \qquad (4a)$$

$$\dot\phi = 3H \pm \sqrt{3H^2 + U + e^\phi \rho} \qquad (4b)$$

$$\dot\rho + 3H(\rho + p) = 0 \qquad (4c)$$

where $U = e^\phi V$. Some sources in the form of an ideal fluid were included [3] as well. The ($\pm$) signifies that either a (+) or (−) is chosen for both equations simultaneously.

The solutions of equations (4a-4c) belong to two branches, according to which sign is chosen. In the absence of any potential or sources the (+) branch solution for $\{H, \phi\}$ is given by

$$H^{(+)} = \pm\frac{1}{\sqrt{3}}\frac{1}{t - t_0}$$
$$\phi^{(+)} = \phi_0 + (\pm\sqrt{3} - 1)\ln(t_0 - t) , \quad t < t_0 \tag{5}$$

This solution describes either accelerated contraction and evolution towards weak coupling or accelerated inflationary expansion and evolution from a cold, flat and weakly coupled universe towards a hot, curved and strongly coupled one. I assume that the initial conditions are such that the latter is chosen. In general, the effects of a potential and sources on this branch are quite mild. After a period of time, of length determined by the initial conditions, a "Branch Change" event from the dilaton-driven accelerated expansion era into what will eventually become a phase of decelerated expansion has to occur. It occurs either when curvatures and kinetic energies reach the string curvature or when quantum effects become strong enough. The correct dynamical description of this phase should, therefore, be stringy in nature. If the value of the dilaton is small throughout this stage of evolution, dynamics can be described by classical string theory in terms of a two-dimensional conformal field theory. This stage is not yet well understood. At the moment, the only existing examples are not quite realistic [19, 20]. More ideas about this stage may be found in [21, 22]. The value which the dilaton takes at the end of this epoch ϕ_{end} is an important parameter. After the "Branch Change" event, the universe cools down and may be described accurately, again, by means of string dilaton-gravity effective theory. Now, however, radiation and matter are important factors. The dilaton remains approximately at the value ϕ_{end}. The universe evolves as a regular Friedman-Robertson-Walker (FRW) radiation-dominated universe.

TENSOR PERTURBATIONS AND RELIC GRAVITATIONAL WAVES

The phase of accelerated evolution, described in the previous section, produces a typical and unique spectrum of gravitationl radiation. The basic mechanism is by now well known [23] (see [24, 25] for recent reviews). Quantum mechanical perturbations exist as tiny wrinkles on top of the classically homogeneous and isotropic background. These wrinkles are then magnified by the accelerated evolution and become classical stochastic inhomogeneities. Below I sketch the derivation of the spectrum of perturbations. Many technical elements are omitted here and can be found in gory details in [1]. The classical solution (5) in conformal time η, where $dt \equiv a d\eta$, is given by

$$g_{\mu\nu} = diag(a^2(\eta), -a^2(\eta)\delta_{ij}) \quad i,j = 1,2,3 \tag{6}$$

where

$$a^2(\eta) \sim |\eta|^{1/2}, \quad \phi(\eta) \sim -\sqrt{3}\ln|\eta| + \phi_0 \tag{7}$$

for $\eta \to 0_-$. One expands the metric around the classical solution $g = g_{cs} + \delta g$ where g_{cs} is given in the previous equation and $\delta g_{ij} = -a^2(\eta)h_{ij}(\eta, \vec{x})$. The resulting equation of motion for each of the two independent perturbation components is given in Fourier space by

$$h_k'' + 2\frac{a'}{a}h_k' + k^2 h_k = 0 \tag{8}$$

and has the general solution $h_k = A_k + B_k \ln|k\eta|$. Initial conditions corresponding to quantum fluctuations at short scales $h_k \sim 1/(a\sqrt{k})\, exp[i(\vec{k}\cdot\vec{x}-k\eta)]$, determine h_k

$$|h_k| \simeq \frac{\ln|k\eta|}{\sqrt{k}a_{HC}} \simeq \ln|k\eta|. \qquad (9)$$

The amplitude of stochastic tensor perturbations in x space is characterized by $|\delta h_k| \sim k^{3/2}|h_k|$. From eq.(9) we obtain

$$|\delta h_k|^2(\eta) \sim \left(\frac{H_{max}}{M_{Pl}}\right)^2 |k\eta_{max}|^3 (\ln|k\eta|)^2. \qquad (10)$$

The end of the dilaton-driven epoch is assumed [5] to take place when the curvature scale H reaches the string scale M_s. In the Einstein frame, in which M_p is constant, the string scale depends upon the dilaton as $M_s = \exp(\phi/2)M_p$. Thus we assume the dilaton driven era to end at conformal time $|\eta| = \eta_1$ where: $H_1 \simeq (\eta_1 a(\eta_1))^{-1} = M_s(\eta_1) = \exp(\phi(\eta_1)/2)M_p$. At the end of the dilaton driven era we thus have:

$$|\delta h_k(\eta_1)| \sim \frac{H_1}{M_p}(k\eta_1)^{3/2}\log(k\eta_1) \qquad (11)$$

This is the final result for the primordial spectrum of tensor perturbations. From the primordial spectrum one wishes to compute the observable spectrum today. A nice feature of gravitational waves is that gravitons are affected practically only by the evolution of the background curvature since right after the "Branch Change" era. Thus the spectrum that should be seen today should mainly reflect what happened in the very early universe processed through presumed known background evolution. While frequencies shift according to the evolution of the background scale factor throughout the evolution, amplitudes of tensor perturbations freeze while outside the horizon and evolve only when inside the horizon. If the dilaton-driven era is followed by a stringy phase characterized by an almost constant value of H, we expect scales which went out of the horizon during the dilaton-driven era to keep moving further outside and to reenter only much later, during the radiation, or possibly even matter dominated era. If we assume this to be the case for all (comoving) scales larger than η_1^{-1}, we must also assume that h_k remains frozen, for all these scales, at the value given in eq.(11) until reentry.

The result is [2] that the part of the processed spectrum which lies below a certain maximal frequency ω_{max}, the highest frequency amplified during the dilaton-driven era, is presently given by

$$|\delta h_\omega| = \sqrt{H_0/M_s}\, z_{eq}^{-1/4} z_{out}^{1/2} \exp(\frac{1}{2}\phi_{end})\, (\frac{\omega}{\omega_{max}})^{\frac{1}{2}} \log(\frac{\omega}{\omega_{max}}) \qquad (12)$$

where $z_{out}(k) = a_{re}(k)/a_{ex}(k)$ is the red-shift while the scale k^{-1} was outside the horizon, z_{eq} is the red-shift from the matter-radiation equality epoch until today, M_s is the present value of string scale (usually estimated to be about $2-5\cdot 10^{17} GeV$) and

$$\omega_{max} = \sqrt{H_0 M_s}\, z_{eq}^{-1/4} z_{out}^{-1/2}. \qquad (13)$$

The fraction of energy in gravitational waves in units of the critical density is given by

$$\frac{d\Omega}{d\log\omega} = z_{eq}^{-1}\exp(\phi_{end})\,(\frac{\omega}{\omega_{max}})^3 \log^2(\frac{\omega}{\omega_{max}}). \qquad (14)$$

Equations (12-14) should be taken as good estimates and not as numerically accurate expressions.

The processed spectrum of gravitational radiation is presented graphically in Figure 1,

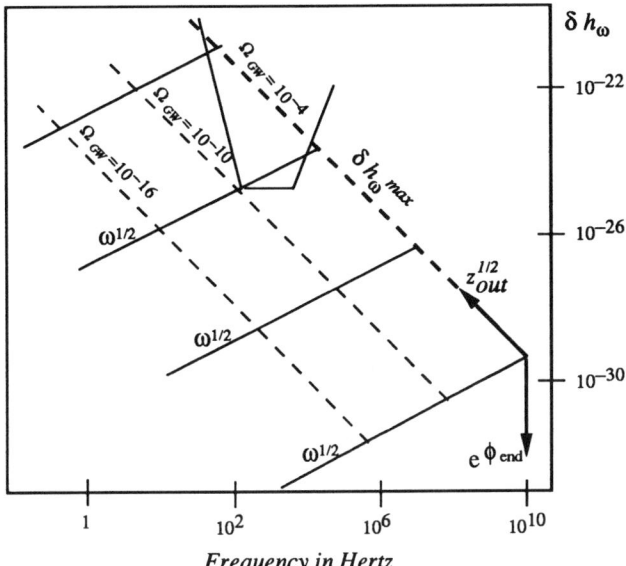

Figure 1. The characteristic spectral amplitude of gravitational waves $|\delta h_\omega|$. The solid lines show several individual spectra for different values of z_{out} and $\phi_{end} = 0$. The thick dashed line shows the maximum amplitude $|\delta h_\omega^{max}|$ as a function of z_{out} for $\phi_{end} = 0$. The dashed lines are lines of fixed ϕ_{end} and therefore lines of constant energy density. Ω_{GW} is the maximal amount of gravitational energy density at a given ϕ_{end}. Also shown in the figure is a triangular shape marking the sensitivity goals for detection of stochastic background $h_{3/yr}$, of the "Advanced LIGO".

Two possible devices may be able to detect the predicted stochastic gravitational wave background, in the lower frequency region $1 - 10^4$ Hz, large interferometers, such as the planned LIGO[26] and VIRGO[27] and in the higher range of frequencies $10^6 - 10^9$ Hz, room-size microwave cavities. For a given set of parameters the amplitude grows as $|\delta h_\omega| \sim \omega^{1/2}$ and therefore it may seem that the best sensitivity for detection is at the high end of the spectrum $\omega = \omega_{max}$. However, the noise in a given interferometer grows as $h_n \sim \omega^{5/4}$ [28]. Therefore for a given interferometer the best sensitivity actually is in the lowest frequency range available. Microwave cavities may be operated as gravity wave detectors [29] for the high frequency range $10^6 - 10^9$ Hz. For the MHz range specific suggestions [30, 31] have been implemented [32], but not operated as gravitational radiation detector. As can be seen from Figure 1, the required sensitivity for detection at the MHz region is $h_c \sim 10^{-26}$ corresponding to $h_{3/yr}$ of the same order and therefore to a noise level of $h_n \sim 10^{-23}$ [28], assuming a bandwidth of MHz. With attainable Q factors of the order of 10^{11}, this sensitivity goal does not seem out of reach. For the GHz region the required sensitivity is $h_c \sim 10^{-28}$ corresponding to $h_n \sim 10^{-24}$.

ACKNOWLEDGMENT

Research supported in part by an Alon Grant. I would like to thank M. Gasperini, M. Giovannini, V. Mukhanov and G. Veneziano for enjoyable and fruitful collaboration and S. Finn, P. Michelson P. Saulson and N. Robertson for discussions about gravity wave detectors.

REFERENCES

[1] R. Brustein, M. Gasperini, M. Giovannini, V. Mukhanov and G. Veneziano, "Metric perturbations in dilaton-driven inflation", CERN-TH. 7544/94, Phys. Rev. D (1995), in Press.

[2] R. Brustein, M. Gasperini, M. Giovannini and G. Veneziano, CERN preprint.

[3] G. Veneziano, Phys. Lett. B265 (1991) 287.

[4] M. Gasperini and G. Veneziano, M. Gasperini and G. Veneziano, Astropart. Phys. 1 (1993) 317;
M. Gasperini and G. Veneziano, Mod. Phys. Lett. A8 (1993) 3701;
M. Gasperini and G. Veneziano, Phys. Rev. D50 (1994) 2519.

[5] R. Brustein and G. Veneziano, Phys. Lett. B329 (1994) 429.

[6] M. Gasperini, "Phenomenological aspects of the pre-big-bang scenario in string cosmology", in Proc. of the 2nd Journee Cosmologie, Paris, June 1994 (World Scientific P.C., Singapore), Torino University Preprint DFTT-24/94;
M. Gasperini, "The inflationary role of the dilaton in string cosmology", in Proc. of the First Int. Workshop on Birth of the Universe and Fundamental Physics, Rome, May 1994 (Springer-Verlag, Berlin), Torino University Preprint DFTT-29/94.

[7] G. Veneziano, "Strings, Cosmology, ...and a Particle" talk given at PASCOS '94, preprint CERN-TH.7502/94 (November 1994).

[8] R. Brustein, "The role of the superstring dilaton in cosmology and particle physics", in Proc. of the XXIX Rencontres de Moriond, Meribel, March 1994;
R. Brustein, "Cosmology and models of supersymmetry breaking in string theory", Proc. of the SUSY94 Workshop, Ann Arbor, May 1994.

[9] M. Gasperini and M. Giovannini, Phys. Rev. D47 (1993) 1529.

[10] M. Gasperini and M. Giovannini and G. Veneziano, "Primordial magnetic fileds from string cosmology", preprint hep-th/9504083.

[11] K. Behrndt and S. Forste, Nucl. Phys. B430 (1994) 441;
E. J. Copeland, A. Lahiri and D. Wands, Phys. Rev. D50 (1994) 4880;
C. Angelantonj, L. Amendola, M. Litterio and F. Occhionero, Phys. Rev. D51 (1995) 1607.

[12] A. Guth, Phys. Rev. D23 (1981) 347;
A. Linde, Phys. Lett 108B (1982) 389;
A. Albrecht and P.J. Steinhardt, Phys. Rev. Lett. 48 (1982) 122;
A. Linde, Phys. Lett 129B (1983) 177.

[13] R. Brustein and P. J. Steinhardt, Phys. Lett. B302 (1993) 196.

[14] B. Campbell, A. Linde and K. Olive, Nucl. Phys. B335 (1991) 146.

[15] P. Binetruy and M.K. Gaillard, Phys. Rev. D34 (1986) 3069;
M. Cvetic and R. L. Davis, Phys. Lett. B296 (1992) 316;

[16] T. Banks, M. Berkooz, S.H. Shenker, G. Moore and P.J. Steinhardt, "Modular cosmology", preprint hep-th/9503114.

[17] T. Damour and A. Vilenkin, "String theory and inflation", preprint hep-th/9503149.

[18] J. Levin, Phys. Rev. D51 (1995) 1536.

[19] E. Kiritsis and K. Kounnas, Phys. Lett. B331 (1994) 51;
E. Kiritsis and K. Kounnas, in Proc. of the 2nd Journ99 Cosmologie, Paris, June 1994 (World Scientific P.C., Singapore).

[20] A. A. Tseytlin, Phys. Lett. B334 (1994) 315.

[21] E. Martinec, Class.Quant.Grav.12 (1995) 941.

[22] N. Deruelle and V.F. Mukhanov, "On matching conditions for cosmological perturbations", preprint gr-qc/9503050.

[23] V. Mukhanov and G. V. Chibisov, JETP Lett. 33, (1981) 532;
A. Guth and S. Y. Pi. Phys. Rev. Lett. 49 (1982) 1110;
A.A. Starobinski, Phys. Lett. B117 (1982) 175;
S.W. Hawking, Phys. Lett. B115 (1982) 295;
J.M. Bardeen, P.S. Steinhardt and M.S. Turner, Phys. Rev. D28 (1983) 679.

[24] V. Mukhanov, H.A. Feldman and R. Brandenberger, Phys. Rep. 215 (1992) 203.

[25] A. R. Liddle and D. H. Lyth, Phys. Rep. 231, (1993) 1.

[26] A. Abramovici et al , Science 256 (1992) 325.

[27] B. Caron et. al, "Status of the VIRGO experiment", preprint Lapp-Exp-94-15.

[28] K. S. Thorne, in 300 Years of Gravitation, S. W. Hawking and W. Israel, Eds. (Cambridge Univ. Press, Cambridge, 1987.

[29] L. P. Grishchuk, Proc. 9th Int. Conf. on General Relativity and Gravitation, ed. E. Schmutzer, Cambridge Univ. Press, 1983.

[30] F. Pegaro, E. Picasso, L. Radicati, J. Phys. A11 (1978) 1949.

[31] C. M. Caves, Phys. Lett B80 (1979) 323.

[32] C. E. Reece et al., Phys. Lett A104 (1984) 341.

GRAVITY DRIVEN INFLATION

Janna J. Levin

Canadian Institute for Theoretical Astrophysics
McLennan Labs
60 St. George Street
Toronto, ON M5S 1A7

The theory of inflation[1] has become the new paradigm for the early universe. The inflation, or accelerated growth of the universe, provides a dynamical explanation for the observed smoothness on large-scales, the apparent local flatness, and the lack of undesirable monopole relics. Further, seeds to catalyze the formation of galactic structures on large-scales are predicted. Though inflation occurs in the first fraction of a second, the industry of designing inflation models aspires to explain the universe we live in today; that is, to explain the specific features in the cosmic background radiation and the formation of structure.

For all of its successes, there remain some imposing questions. In particular, there is no model of inflation predicted from a fundamental theory. It might be hoped that quantum gravity would provide such a prediction. However, the most successful attempt at unifying gravity with particle theory to date, namely string theory, is inconsistent with the standard inflationary paradigm[2]. While one might be willing to abandon strings or inflation, the point is we are no closer to a consistent union of the evolution of our universe and a fundamental field theory. Though superstrings may not survive as the fundamental theory, some of the salient features must. By investigating elements common to strings and other particle theories in a cosmological context, our larger view of physics can be tested.

A common element of many high-energy theories is a dynamical Planck mass. In the low-energy string action the dilaton acts as a dynamical Planck field and thus supplants the fundamental constant of the Einstein theory. Outside of superstrings, dynamical Planck fields are often generated in particle theories. Even simple quantum corrections to a field theory in a curved spacetime will contribute to a variable Planck field. In a higher dimensional or Kaluza-Klein approach[3], the variable Planck mass has a geometric interpretation. It is related to the radius of the compact internal dimensions.

A completely new source of inflation is predicted in such extensions of Einstein gravity, as we showed in[4,5]. The phenomenon is manifest in string cosmology but is not unique to string theory. The nonminimal coupling of the Planck field to gravity allows for an unusual elasticity associated with the kinetic energy of the field. The kinetic energy density can grow with time, fueling a more rapid expansion of the universe, that is, an inflation. It is worth stressing that there is no potential and no cosmological constant. Independently, Gasperini and Veneziano considered the specific case of superstrings[6,7].

In standard inflation, a potential energy density drives an era of accelerated expansion. The characteristic feature is the potential. Previously, string theories were shown to interrupt potential dominated inflation[2]. The kinetic energy in the dilaton field overwhelmed the potential energy. As a result, standard inflation could not proceed unhindered.

Since the Planck field can actually drive inflation, in lieu of the potential, string theory may not only be compatible with, but actually predicts inflation. I found an analogous type of behavior in Kaluza-Klein models of additional spacetime dimensions[8]. In vacuum, the shear from contracting dimensions is able to drive an inflation of a three-volume. When the extra dimensions are integrated over, the scenario is equivalent to a four-dimensional model of a dynamical Planck field.

The task at hand is to uncover a successful end to the scenario. Currently, both string cosmology[9] and Kaluza-Klein cosmologies[8] are unable to exit the inflationary phase. Instead the universe is ushered toward a future singularity. The graceful exit problem is more serious than the usual obstacle which plagues potential-driven inflation. In string cosmology, the inflationary branch of solutions is totally distinct from the branch of solutions which describes our universe today. There is no overlap. A branch change is needed to move from inflation to a more temperate evolution. The nature of the graceful exit will be elaborated on here.

Fortunately, the graceful exit problem does not plague all models of a Planck-driven inflation. As will be described, there are entire families which are able to both inflate and match onto an expanding universe today. A means by which to heat the universe, thereby completing the model, will also be described.

The gravitational action can be written in generality as

$$A_G = \frac{1}{16\pi} d^4x \sqrt{-g} \left[-\Phi \mathcal{R} + \omega (\partial \Phi)^2 \right] . \tag{1.1}$$

The fundamental Planck scale of the Einstein theory is replaced by the field $\Phi = m_P^2$ and $\mathcal{R}$ is the Ricci scalar. The kinetic coupling constant which determines the theory, $\omega(\Phi)$, is left general. To condense the notation it is worthwhile to introduce the parameter

$$f(\Phi) \equiv (1 + 2\omega(\Phi)/3)^{1/2} . \tag{1.2}$$

In string theory, the dilaton is described by the action (1.1) with $\omega = -1$. In a Kaluza-Klein model with $n > 1$ contracting dimensions, the radius of the internal dimensions obeys the action (1.1) with $\omega = -1 + 1/n$. Notice that in both cases it happens that $f < 1$.

In a flat, Friedman-Robertson-Walker universe, the Einstein equation which determines the expansion rate of the universe is simply

$$H^2 = \frac{8\pi}{3\Phi} (\rho_\Phi + \rho) , \tag{1.3}$$

where the undecorated ρ represents the energy density in everything but the Planck field. The kinetic energy density in Φ is given by

$$\frac{8\pi}{3\Phi} \rho_\Phi = \frac{2\omega}{3} \left(\frac{\dot\Phi}{\Phi} \right)^2 - H\dot\Phi . \tag{1.4}$$

As a consequence of the direct coupling of the Planck field to gravity, ρ_Φ involves H directly. The pressure associated with this kinetic energy is roughly a measure of the change in energy with unit volue, $p_\Phi \sim -dE_\Phi/dV$. As a result of the direct coupling to the Ricci scalar, the kinetic energy acquires a unique elasticity. In fact, for certain couplings f, the kinetic energy can actually grow leading to a negative pressure. In full glory, the pressure associated with the kinetic energy can be written

$$\frac{8\pi}{3\Phi}p_\Phi = \frac{2}{3}\left(\frac{\dot\Phi}{2\Phi}\right)^2 \left(1 + \omega \pm f - 2\frac{d\ln f}{d\ln \Phi}\right) . \tag{1.5}$$

The origin of the two branches is discussed below. As in standard Einstein gravity, a negative pressure can lead to inflation.

Inflation refers to an accelerated growth of the scale factor. The scale factor is accelerated if the following condition[5] is satisfied:

$$f \pm 1 - \frac{\Phi}{f^2}\frac{df}{d\Phi} < 0 . \tag{1.6}$$

If ω is a negative constant so that $f < 1$, as is the case for strings and Kaluza-Klein, then condition (1.6) will only be satisfied for the $-$ sign branch. The condition becomes $f - 1 < 0$, which is automatically satisfied. If $\omega(\Phi)$ is variable, then it is possible to satisfy condition (1.6) for the $+$ sign branch. The branch taken turns out to be important.

The physical relevance of these two branches can be seen by solving the quadratic Einstein equation (1.3) for H

$$H = -\left(\frac{\dot\Phi}{2\Phi}\right) \pm \sqrt{\frac{f^2}{4}\left(\frac{\dot\Phi}{\Phi}\right)^2 + \frac{8\pi}{3\Phi}\rho} . \tag{1.7}$$

The two branches in condition (1.6) and expression (1.5) reflect these two solutions of Einstein's equations. For comparison, the standard model Hubble expansion is given by

$$H_{\text{stand}} = \pm\sqrt{\frac{8\pi}{3M_o^2}\rho} . \tag{1.8}$$

The standard Einstein equation (1.8) also allows two branches, one expanding (+) and one contracting (−). The expanding branch is singled out as the physically relevant one. In the case of a dynamical Planck mass on the other hand, both branches can expand if $\dot\Phi < 0$ and $f(\Phi) < 1$. In fact, for both Kaluza-Klein and strings, the + branch expands without inflation while the − branch inflates.

The pathology of the − branch can now be seen. Even if a mechanism exists to stabilize the Planck mass, the universe would ultimately contract. Today the universe is described by + branch solutions of the form (1.8). A branch change is needed to connect smooly onto our expanding cosmology. To induce such a branch change requires negative energies so that the total energy density drops to zero. Obviously this is no mean feat.

If instead we allow for a variable $\omega(\Phi)$, then this hardship can be circumvented. The branch which inflates can also be the branch which smoothly connects onto an expanding universe today. Entire families of couplings can satisfy the inflationary condition (1.6) on the healthier, + branch. Some examples of toy models include

$$f_1(\Phi) = \ln(M_o^2/\Phi)$$
$$f_2(\Phi) = \left(\frac{1}{2\ln(M_o^2/\Phi)}\right)^{1/2}$$
$$f_3(\Phi) = \frac{\Phi}{M_o^2}$$
$$f_4(\Phi) = \frac{\Phi}{M_o^2 - \Phi} \quad . \tag{1.9}$$

The question remains if any such coupling can be generated from quantum corrections to a low-energy effective action or non-perturbatively from a high-energy theory.

An accelerated expansion alone does not an inflationary model make. The universe must expand enough to envelop our entire observable unviverse. Furthermore, the universe must then heat up to restore the standard hot big bang picture. Gravitational particle production due to the gravity-driven inflation may be able to heat the universe[10].

During inflation quantum fluctuations in any underlying field theory can be amplified. For wavelengths well within the event horizon, the amplified fluctuations can propogate as particles. The gravitational field transfers energy into the virtual quantum field thereby creating a hot bath. On wavelengths which exceed an event horizon, the mode cannot propogate as a particle but instead contributes to the fluctuation in an inhomogeneous classical background. The long wavelength fluctuations are the usual density perturbations generated during inflation. The short wavelength fluctutations are the analog of Hawking-Unruh radiation.

In de Sitter inflation, the short wavelength fluctuations are redshifted away as rapidly as they are generated. The equilibrium of classical de Sitter is therefore left undisturbed. The main contribution of these high-frequency quantum fluctuations is to build up the long wavelength $\delta\rho/\rho$ as the modes cross outside the event horizon. In kinetic inflation however, particle production can run away when the expansion is fiercest. The back-reaction of the spacetime in turn drains energy from the Planck field. As the Planck field slows, inflation would in principle be exited.

A quick sketch of a gravity-driven, kinetic inflation unfolds as follows. The elastic nature of the kinetic energy in a variable Planck mass can drive an epoch of inflationary expansion. The kinetic energy in the Planck field can grow as inflation proceeds. Consequently, high-energy physics becomes increasingly important. The final stage of a gravity-driven inflation will thus be marked by the influence of quantum mechanics through gravitational particle production. It must still be shown that conversion of the classical kinetic energy into particles is efficient enough to appease the demands of successful inflation[11]. If a hot universe can be created from this cold beginning, a cohesive model of gravity-driven inflation is within reach.

1. A. H. Guth, *Phys. Rev.* D **23**, 347 (1981).
2. R. Brustein and P.J.Steinhardt, *Phys. Lett.* B**302**, 196 (1993).
3. Th. Kaluza, *Sitzungsber, Preuss. Akad. Wiss. Phys. Math. Kl.* 966 (1921); O. Klein, *Z. Phys.* 37 (1929) 895.
4. K. Freese and J.J. Levin, unpublished report UMAC-93-23.
5. J.J. Levin, *Phys. Rev.* D **51**, 462 (1995).
6. G. Veneziano, *Phys. Lett.* B**265**, 287 (1991).
7. M. Gasperini and G. Veneziano, *Astropart. Phys.* I, 317 (1993).
8. J.J. Levin, *Phys. Lett.* B **343**, 69 (1995).
9. R. Brustein and G. Veneziano, *Phys. Lett.* B**329**, 429 (1994).
10. J. R. Bond and J. J. Levin, in preparaion.
11. J.J. Levin, *Phys. Rev.* D **51**, 1536 (1995).

SECTION III

CURRENT STATUS OF GRAVITY WAVE DETECTION

THE SEARCH FOR GRAVITATION WAVES

Barry C. Barish
(for the LIGO Project)

California Institute of Technology 256-48
Pasadena, California 91125

INTRODUCTION

The Laser Interferometer Gravitational-Wave Observatory (LIGO)[1] is a large project supported by the National Science Foundation and is part of an International effort to search for gravitational waves emitted by Astrophysical sources. The approach is to measure the small change in distance that occurs when a gravitational wave is incident on the earth by use of long baseline laser interferometers. Construction of the vacuum system and conventional facilities for this device has recently begun, while research continues on the development of improved techniques for improving the sensitivity of the interferometers. The detailed design effort for the initial detector has also recently gotten underway. Completion of the facility and initiation of the detector commissioning and early running is expected to begin approximately in the year 2000.

The initial goal LIGO is the direct detection of gravitational waves from cosmic sources[1]. In the longer run, the facility is aimed at the exploration of the Universe with this entirely new probe (e.g. gravitational waves). The U.S. facility will consist of facilities at two widely separated sites, Hanford, Washington and Livingston, Louisiana, where laser interferometers that will measure the displacements of suspended test masses as they are perturbed by gravitational waves crossing the arms of the interferometer. The initial LIGO detector has three interferometers (two in Washington and one in Louisiana) which will operate in coincidence. LIGO will be part of an international network of gravitational-wave detectors (another major facility, VIRGO, is being developed by a French-Italian collaboration near Pisa, Italy).

The LIGO facility is being designed and built by a collaboration[*] from the California Institute of Technology and the Massachusetts Institute of Technology. The construction of the facilities has recently begun with completion scheduled for 1999, followed by turn-on and commissioning period through 2001, by whic time the design sensitivity should be reached. Research and development activities on the advanced techniques that will enable improved sensitivity either through enhancements or development of second generation interferometers are being pursued at both Caltech and MIT, and within a number of collaborating groups.

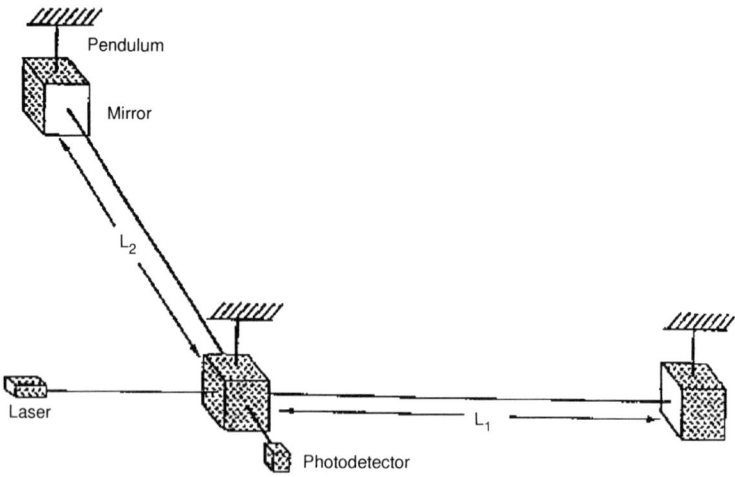

Figure 1. A simplified schematic diagram illustrating the basic principle of a laser interferometer for sensing gravitational waves.

THE DETECTION METHOD

The sensors for gravitational waves are long-baseline laser interferometers which measure the distances between suspended test masses, that are arranged in an L-shaped pattern. The basic principle of detection is illustrated by the simplified schematic diagram of a Michelson interferometer, shown in Figure 1. A gravitational wave crosses the detector and produces a distortion of the local metric that causes distances, as measured by a beam of light, to vary in the plane transverse to the direction of propagation of the wave. The pattern of distortion is quadrupolar: distances appear to stretch along one axis and to shrink along an orthogonal axis at any given time. If such a wave were incident on the interferometer in Figure 1 from above and optimally polarized, then the arm lengths L_1 and L_2 would be changed by ΔL_1 and ΔL_2, respectively. Since the metric defines how

---

[*]This project is supported by the National Science Foundation under cooperative agreement PHY-9210038. The current LIGO collaboration members are: A. Abramovici, W. Althouse, F. Asiri, B. Barish, B. Bochner, R. Bork, J. Camp, J. Carri, J. Chapsky, L. Chu, P. Csatorday, D. Durance, S. Elieson, T. Evans, F. Fernandez, P. Fritschd, M. Gamble, J. Heefner, Y. Hefetz, G. Hiscott, J. Giaime, A. Gillespie, L. Jones, D. Jungwirth, S. Kawamura, Y. Kommemi, J. Kovalik, E. Kruzel, A. Kuhnert, B. Lantz, A. Lazzarini, P. Lindquist, T. Lyons, J. Mason, T. Mast, N. Mavalvala, O. McCullough, S. Meshkov, M. Patlan, F. Raab, M. Rakhmanov, M. Regehr, K. Reithmaier, P. Saha, G. Sanders, R. Savage, V. Schmidt, D. Shoemaker, L. Sievers, R. Spero, G. Stapfer, K. Thorne, S. Vass, R. Vogt, R. Weiss, S. Whitcomb, J. Worden, H. Yamamoto, M. Zucker.

distances are measured, the apparent displacements of the mirrored test masses, ΔL_1 and ΔL_2, are simply related to the original arm lengths; for arm lengths much smaller than the wavelength of the gravitational wave we have

$$\Delta L = \Delta L_1 - \Delta L_2 = h \times L \qquad (1)$$

where h is the amplitude of the gravitational wave (sometimes referred to as the strain) and where it is assumed that $L \approx L_1 \approx L_2$.

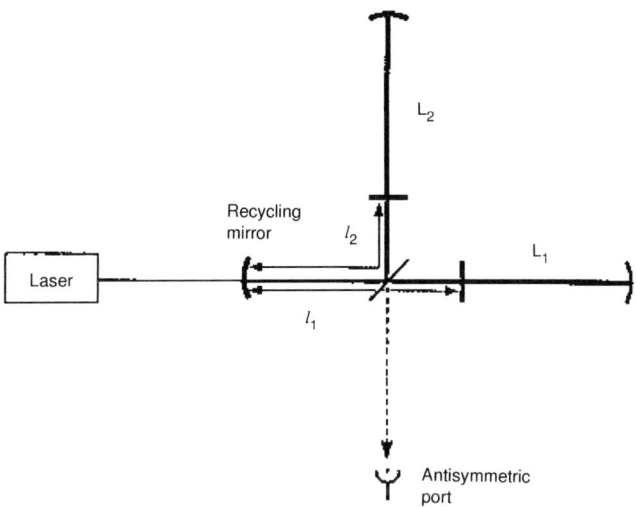

Figure 2. Schematic diagram of a laser interferometer incorporating Fabry-Perot arm cavities and power recycling.

The small motions of the test masses are detected by measuring the phase shifts in light beams that propagate along the two arms. A beam splitter at the vertex of the interferometer splits the light into two beams which propagate along the arms, are reflected from the end masses and interfere again at the beam splitter. In the balanced configuration, the phase of the two interfering beams is equal and the light is returned toward the laser. A photo detector, monitoring light at the other beam-splitter port, measures only a small amount of light due to imperfections in the optics. If a gravitational wave (or some other perturbation) were to unbalance the two arm lengths, then the destructive interference condition for light propagating toward the photodetector is disrupted and a photocurrent appears in the photodetector.

The initial LIGO interferometers will use the more sophisticated interferometer configuration shown in Figure 2, which employs Fabry-Perot cavities in the arms and power recycling[2]. By resonating the light in each arm, the Fabry-Perot cavities (of length L_1 and L_2) produce larger phase shifts on the light returning to the beam splitter for a given displacement of the test masses. The power recycling mirror allows the light incident on the Fabry-Perot cavities to be increased above the laser power level by resonance in the cavity labelled by l_1 and l_2.

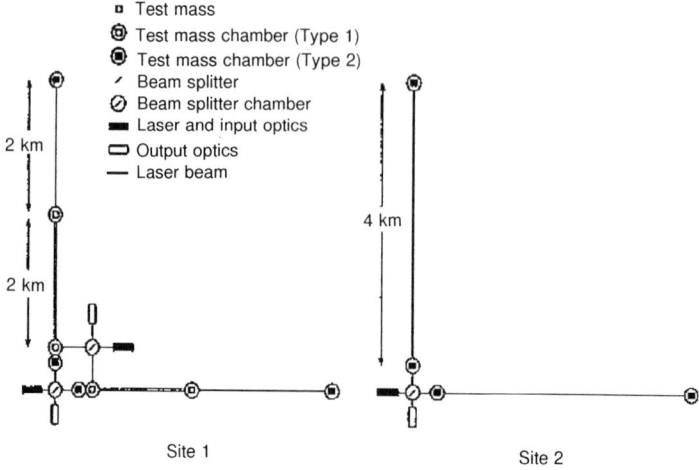

Figure 3. Schematic layout of the LIGO facilities at the Washington (left) and Louisiana (right) sites.

THE LIGO FACILITIES AND INTERFEROMETERS

The LIGO facilities at each site will consist of 4 feet diameter L-shaped vacuum pipes for interferometers with arm lengths of 4 km. Buildings at the vertex and ends of the "L" will house test masses, lasers and electronics. The Washington facility will also have test-mass chambers in midstation buildings, associated with an additional interferometer with 2-km-long arms, allowing a fast ($\Delta t \sim$ 1msec) local coincidence. A slower ($\Delta t \sim$ 30msec) coincidence corresponding to the time difference for a gravitational wave to strike the two sites provides a triple coincidence. Schematic layouts of the two sites are shown in Figure 3. The overall vacuum system will be capable of achieving pressures of 10^{-9} Torr in each of three long beam tube modules, to reduce noise associated with fluctuations in the number of residual gas molecules in the arm cavities.

The LIGO facilities are now under construction. The rough grading of the arms has been completed at the Washington site and the clearing and grading has commenced in Louisiana. The detailed design concept for the beam tube modules has been successfully tested in a full scale qualification test and the architectural and engineering design for the buildings and enclosures is underway. The vacuum equipment design activities are also underway.

The LIGO facilities will support both multiple interferometers and an evolution of the initial interferometers with increasing sophistication and sensitivity. The anticipated sensitivity of the detectors to burst sources and the strengths of possible sources are compared in Figure 4. In general, the sensitivity to a burst depends on details of both the detector noise and the source waveform. A useful quantity to characterize the detector[3] is h_{rms}, the rms strain of a gravitational-wave burst (of optimal source position and polarization) equivalent to detector noise, assuming that the waveform is centered at frequency f with a bandwidth comparable to f. For purposes of estimating detected event rates, Figure 4 shows detector sensitivity plotted in terms of the quantity $h_{3/yr} = 11 h_{rms}$, the rms strength of a burst source that could be confidently distinguished from detector noise if it occurred three times per year with random polarization and source direction. When sources are plotted in terms of the rms strain produced at earth in a bandwidth Δf about frequency f, the plot has a simple meaning: if the dashed source line is tangent to the detector line, we can estimate an average of three such detected sources per year. (Such a source would be detected with a signal to noise ratio of 11 if it had optimal polarization and source direction.) If the source line is a factor of two above the detector curve, we can expect a factor of two increase in signal to noise ratio and a factor of eight increase in the number of detected events. The curve labelled LIGO Early Detector represents an estimate of the sensitivity of a detector using available technology, expected to be reached by the end of 2001. The curve labelled LIGO Advanced Detector represents an estimate of the cumulative effect of a number of changes to the early detector design as technical improvements are made in various areas. For both detector curves, the sensitivity is determined at the lowest frequencies by the transmission of seismic noise to the test masses. At intermediate frequencies, detector sensitivity is limited by thermal noise. Photon shot noise limits the sensitivity at the highest frequencies.

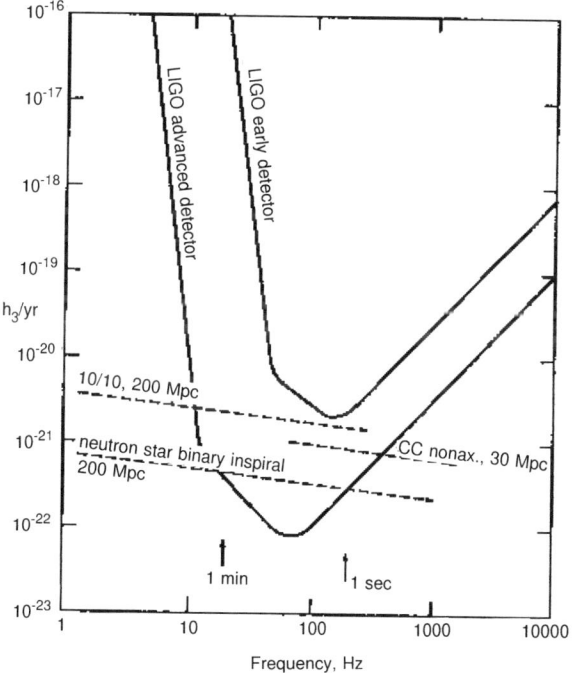

Figure 4. Estimated sensitivity of LIGO detectors (solid lines) and estimated strengths of possible sources (dashed lines). The dashed one labelled 10/10 refers to the inspiral of two 10-solar-mass black holes. The dashed curve labelled CC Nonax refers to a nonaxisymmetric stellar core collapse. The labelled arrows indicate approximate times before final coalescence for the neutron star binary inspiral come.

The best understood source of gravitational waves is the inspiral of binary neutron star systems. In Figure 4, the detector sensitivity curves are compared to the signal strength for a neutron star inspiral at 200 Mpc, the best guess for the distance within which three such events per year can be expected[4]. More speculative source estimates are shown for the inspiral of two 10-solar-mass black holes and for the nonaxisymmetric collapse of a stellar core. Sensitivity curves such as those shown in Figure 4 are based on the assumption that the detector sensitivity is determined by stationary gaussian noise sources. The LIGO detector will rely on a system of local vetoes, based on environmental monitors at each of the sites, and the operation of multiple interferometers in coincidence to achieve immunity to nonstationary, nongaussian noise.

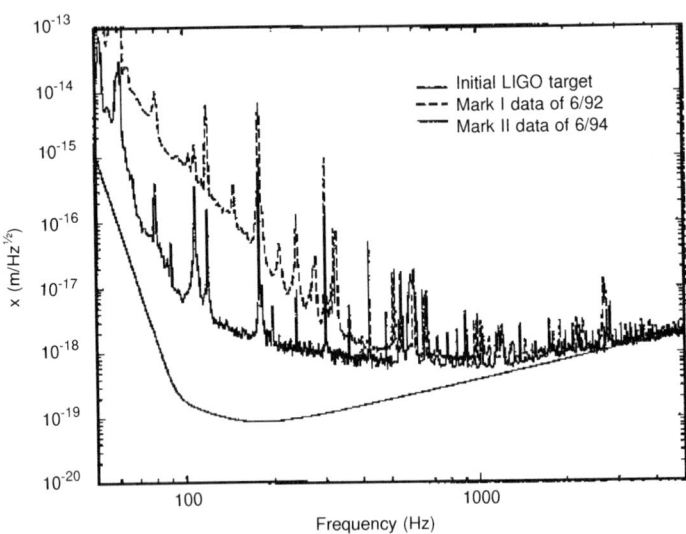

Figure 5. Comparison of the displacement sensitivity of the Mark II and the Mark I interferometers.

The design of LIGO detectors is supported by ongoing research and development on ultrasensitive laser interferometers. A 40-meter interferometer at Caltech using Fabry-Perot cavities in the arms is being used to experimentally characterize noise sources and to develop and demonstrate technology for LIGO. A 5 meter prototype is being built at MIT to study phase noise. In addition, specialized optical tests are underway, and the interferometer performance is tested through modeling.

The improvement of the noise-equivalent displacement sensitivity of the Caltech 40-meter interferometer prototype is shown over a two year period in Figure 5[*], along with the initial LIGO goal for comparison. The largest improvement has been at frequencies below a few hundred Hertz and is attributed primarily to the improved seismic isolation stacks. Careful tuning of the servo electronics and careful alignment of the optical beams is required to reduce noise contributions from the damping systems[7] below the levels shown in Figure 6.

The engineering design of the initial LIGO interferometers will be finalized in stages over the next two years. Research on laser interferometers during this time will focus on achieving, as closely as possible in laboratory-scale interferometers, the displacement sensitivity and the optical phase sensitivity goals of the initial LIGO

---

[*] The quantity x, given in the following figures, is the calibrated displacement (rms) of the East End test mass that is equivalent to the rms amplitude of differential arm length fluctuations ΔL in a 1-Hz bandwidth at frequency f.

interferometers. The 40-meter interferometer at Caltech and a new suspended-mass interferometer at MIT will be used for this work.

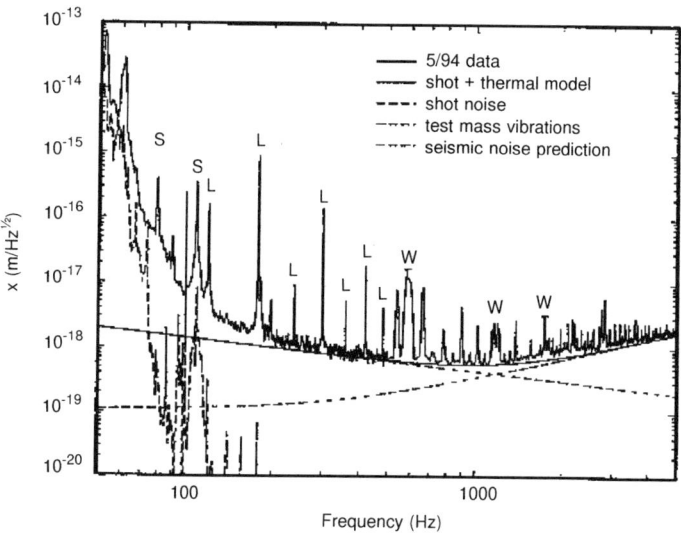

Figure 6. Comparison of Mark II data of 5/94 with a model of the interferometer noise. Some of the known peaks in the spectrum have been labelled: L refers to power line harmonics, W corresponds to violin resonances in the test mass suspension wires, and S corresponds to a suspension resonance arising from the rocking motion of the test-mass-control-block system.

STATUS AND PROSPECTS

LIGO is presently entering the construction phase, which is scheduled for completion in 1999. The interferometers will be installed and commissioned at that time leading to a physics program in this exciting new area with increasing sensitivity over the next several years. The initial goal is to provide convincing detections of gravitational wave signals and the longer range goals are to use gravitational waves as a new probe of the Universe.

REFERENCES

1. A. Abramovici, W.E. Althouse, R.W.P. Drever, Y. Gursel, S. Kawamura, F.J. Raab, D. Shoemaker, L. Sievers, R.E. Spero, K.S.Thorne, R.E. Vogt, R. Weiss, S.E. Whitcomb and M.E. Zucker, *Science*, **256**, 281, (1992).

2. See R.W.P. Drever, in **Gravitational Radiation**, eds. N. Deruelle and T. Piran, 321, North Holland, (1983), B. J. Meers, *Phys. Rev. D*, **38**, 2317, (1988).

3. K.S. Thorne, in *Three Hundred Years of Gravitation*, eds. Hawking, S. and Israel, W., (Cambridge University Press, Cambridge, 1987), p 330.

4. R. Narayan, T. Piran and A. Shemi, *Astrophys. J.*, **379**, L17, (1991), E. S. Phinney, *Astrophys. J.*, **380**, L17, (1991).

5. Y. Gursel and M. Tinto, *Phys. Rev. D*, **40**, 3884, (1987).

6. M.E. Zucker in *Proceedings of the Sixth Marsel Grossmann Meeting on General Relativity*, eds. Sato, H.and Nakamura, T. (World Scientific, Singapore, 1991), p. 224.

7. S. Kawamura and M.E. Zucker, *Appl. Opt.*, **33**, 3912, (1994).

REDUCING THERMAL NOISE IN INTERFEROMETRIC DETECTORS OF GRAVITATIONAL WAVES

Peter R. Saulson
Department of Physics
Syracuse University
Syracuse, New York 13244

INTRODUCTION

The initial sensitivity of the new interferometric gravitational wave detectors (exemplified by the Laser Interferometer Gravitational-Wave Observatory, or LIGO[1]) is unlikely to satisfy their users for long. The expected strengths of frequent gravitational wave signals are such that the first long baseline instruments are not likely to detect them at large signal-to-noise ratios. Initial detections would need to be followed up by observations with more sensitive detectors, to learn more details of the signal waveforms and of the distribution of their sources on the sky. It is possible that the first instruments might not register any statistically significant signals at all, thus making sensitivity improvement all the more urgent.

As a benchmark for high performance in the crucial range of signal frequencies from 10 Hz to 100 Hz, we can take the so-called *standard quantum limit*, given by

$$h_{QL} = \frac{1}{\pi f L}\sqrt{\frac{\hbar}{m}},$$

where L is the length of the interferometer arms, m is mass of one of the test masses, and f is the signal frequency of interest. The quantity h_{QL} represents the limit to the precision of the read-out of the difference in length of the interferometer arms, as imposed by the quantum nature of light. It can be thought of as the result of the sum of the shot noise (which improves as the laser power in the interferometer grows) and the fluctuating radiation pressure (which gets worse as the power grows). Although it does not represent a "hard" limit to interferometer performance, it will almost certainly represent substantial improvement over the sensitivity of the first generation instruments. The LIGO interferometers will have $L = 4$ km, and (at least in the early generations of detectors) $m \approx 10$ kg.

Reaching quantum-limited sensitivity in this frequency range will pose a challenge to the designers of the interferometers' optics. There are well-conceived plans for attaining optical performance at this level, but it is beyond the scope of this contribution to describe them.[2] Even if the interferometers' design would allow optical read-out noise at this level, such good performance can only be realized if, in addition, the actual motion of the test masses is sufficiently low. This is generally described as the *displacement noise* problem. One of the most important of the sources of displacement noise is *thermal noise*.

THERMAL NOISE

Thermal noise is a generalization of the phenomenon of Brownian motion. The thermal noise displacement power spectrum for a mode of resonant frequency f_0, mass m, and quality factor Q is given by[3]

$$x^2(f) = \frac{k_B T}{2\pi^3 m Q f_0^2 f} \frac{1}{[(1 - f^2/f_0^2)^2 + 1/Q^2]}.$$

In the initial LIGO interferometer, one of the most important degrees of freedom will be the simple pendulum mode of the test mass on its wire suspension. With a mass of 10 kg, a resonant frequency of 1 Hz, and a Q of order 10^6, it is expected that the resulting strain noise above the resonance will be about

$$h_{\text{th,pend}}(f) = 1.3 \times 10^{-18}/\sqrt{\text{Hz}} \left(\frac{1\text{ Hz}}{f}\right)^{5/2}.$$

A second class of modes that will also make an important contribution to the noise budget is the internal modes of vibration of the test masses. Even though the lowest of these modes will have a resonant frequency of about 15 kHz, the gently-sloping $1/f$ spectrum below resonance will nevertheless be one of the strongest noise sources near 100 Hz. Quite a few modes actually contribute to the sum.[4] Assuming the modes are characterized by quality factors of order 10^6, then the strain noise will be given by

$$h_{\text{th,int}}(f) = 3.4 \times 10^{-22}/\sqrt{\text{Hz}} \left(\frac{1\text{ Hz}}{f}\right)^{1/2}.$$

THERMAL NOISE VS. THE QUANTUM LIMIT BENCHMARK

Substantial advances are required before thermal noise in the pendulum mode can be reduced to the level of the standard quantum limit benchmark. A quality factor of a few times 10^7 would required to reach the goal at 100 Hz, and but $Q = 10^{10}$ is necessary at 10 Hz, assuming that the test masses are 10 kg. Fused silica and niobium flexures, among others, offer the possibility of achieving such quality factors, especially when one imagines using fine ribbons instead of round wires for flexures. Experiments to be carried out in the near future[5,6] should be able to verify whether such specifications can be attained.

Reducing the thermal noise from the test mass internal modes may prove to be the most challenging aspect of reaching the quantum limit benchmark. While Qs of only 10^7 are required to meet this goal at 10 Hz, a quality factor of 2×10^8 is necessary to make thermal noise negligible at 100 Hz.

Trying to reduce the internal mode thermal noise by scaling current designs up in mass is a discouraging prospect; although there is an explicit dependence of $h_{\text{th}}(f) \propto m^{-1/2}$, the reduction in frequency of modes as the mass is scaled up transforms the scaling of the noise from a particular mode to $h_{\text{th}}(f) \propto m^{-1/6}$. An important additional effect is that modes of higher order will contribute to the total noise in a larger mass; all modes with wavelength longer than the beam diameter need to be considered.[4] This will nullify any improvement in noise by increasing the mass.

It is uncertain whether grades of fused silica can be found that would exhibit 2×10^8 at 100 Hz. The purest grades of fused silica can probably achieve[7] the 10 Hz requirement of $Q \approx 3 \times 10^7$. Better understanding of what level of internal friction is achievable in fused silica is urgently needed.

COOLING TO REDUCE THERMAL NOISE?

Substantial variations on present designs may be required. One possibility might be to lower the temperature of the test masses, since $h_{th}(f) \propto T^{-1/2}$. Going from room temperature to that of liquid helium offers a possible improvement of more than a factor of 8. One difficulty with this idea that has yet to be overcome is to design a way to conduct the substantial heat load away from the test masses – dissipated laser power may be measured in hundreds of mW, and the heat load from the warm walls of the vacuum system will also be significant. (The latter problem can be overcome by building a cold shield around the test mass.[8]) What makes this difficult is that thermal conductivity must be supplied while meeting the need to keep the mechanical quality factors of the various modes high – this need is what dictates the use of very fine wires or other flexures as the only attachments to the masses.

Even if this tricky bit of cryogenic engineering could be accomplished, there is another difficulty with the naive application of cooling to present designs. The losses in fused silica have a temperature dependence that would tend to negate the benefits of cooling. Fused silica losses rise steeply as the temperature is lowered below room temperature,[9] reaching a peak (with $Q \approx 10^3$) at about 50 K. Below that temperature losses decline somewhat, but have not been seen to return anywhere near the room temperature values; at 4 K, fused silica has $Q \approx 10^4$. So to achieve the benefits of cooling, one would almost certainly have to abandon the use of fused silica as the test mass material.

The prospects for finding a simple substitute for fused silica are uncertain at best. Not only does fused silica have a high mechanical quality factor, but its very low optical absorption and birefringence and excellent homogeneity in its index of refraction all play important roles in ensuring that present optical designs can work well. Sapphire has excellent transparency and high mechanical Q. However, it has substantial birefringence, and without better control of the orientation of crystal axes would cause large wavefront distortions that would degrade the quality of the dark fringe and thus the performance of power recycling.

NOVEL INTERFEROMETER CONFIGURATIONS

Transmission through optical components has been built into the initial interferometer designs in several ways. One is the transmission through the input coupling mirror of the Fabry-Perot cavities adopted by most of the large projects. (The delay line design of GEO 600, as well as more ambitious delay lines that can do substantial amounts of beam folding, make no use of transmission through any of their cavity mirrors.) All interferometer designs adopted so far make crucial use of transmission through the beam splitter, as a way of generating coherent beams for the two arms and of then recombining the beams to generate the interference fringe. Without the need for light to pass through the bulk of an optical component, one would have a much wider choice of materials. This might include crystalline silicon, whose mechanical Q has been shown to reach 3×10^7 at 300 K. Its Q improves as temperature is reduced, so that if it could be cooled successfully one would actually reap the benefits that one would naively expect. Sapphire has exhibited even higher Q, 2×10^8 at room temperature. For non-transmissive applications, its birefringence is not a problem. Sapphire also exhibits reduced losses at low temperature; $Q \approx 4 \times 10^9$ has been reported at 4 K.[10]

One way to obtain substantial benefits from the use of these alternative mirror materials might be to abandon the use of Fabry-Perot cavities as beam folding elements. A Michelson delay line could be made of multiple individual mirrors, each of the size of one of the mirrors in the first generation interferometers, but made of a high Q material. (This optical geometry has been dubbed the *schellenbaum* by the MPQ group.) In this design, light would still need to be transmitted through the beam splitter. The latter component does contribute some thermal noise, but at a level down by a factor of the number of bounces in the arms. So if the arms had of order ten round trips, then arm mirrors with Q up to 10^9 would dominate the thermal noise in an interferometer whose beam splitter had $Q \approx 10^7$. Such a design could reach thermal noise levels required for quantum-limited performance.

The possibility of constructing a non-transmitting beam splitter has begun to receive some attention. One intriguing idea is to form a diffraction grating on the surface of a mirror.[11] With proper choice of grating parameters, diffraction orders higher than the first could be eliminated. Whether such beam splitters can be made that will meet the necessary specifications concerning, for example, optical losses, remains to be seen.

ACKNOWLEDGMENTS

This work was supported in part by the National Science Foundation, under grant PHY-9113902.

1. A. Abramovici, W.E. Althouse, R.W.P. Drever, Y. Gürsel, S. Kawamura, F.J. Raab, D. Shoemaker, L. Sievers, R.E. Spero, K.S. Thorne, R.E. Vogt, R. Weiss, S.E. Whitcomb, and M.E. Zucker, *Science* 256:325 (1992).
2. See the discussion and references in Chapter 12 of P.R. Saulson. "Fundamentals of Interferometric Gravitational Wave Detectors," World Scientific, Singapore (1994).
3. P.R. Saulson, *PhysRevD* 42:2437 (1990).
4. A. Gillespie and F.J. Raab, *PhysicsLettersA* (1995), in press.
5. D.G. Blair, L. Ju, and M. Notcutt, *RevSciInstrum* 64:1899 (1993).
6. J.E. Logan, talk at 7th Marcel Grossman Meeting (1994).
7. V.B. Braginsky, private communication.
8. Work of the GEO Collaboration, reported at Snowmass 94.
9. D. Tielbürger, R. Merz, R. Ehrenfels, and S. Hunklinger, *PhysRevB* 45:2750 (1992).
10. See the references cited in V.B. Braginsky, V.P. Mitrofanov, and V.I. Panov. "Systems with Small Dissipation", University of Chicago Press, Chicago (1977).
11. R.W.P. Drever, talk at Snowmass 94.

SECTION IV

NEUTRINOS AND MUONS

A Search for $\bar{\nu}_\mu \to \bar{\nu}_e$ Oscillation at LAMPF

Jeremy Margulies

Physics Department, Temple University

February 3, 1995

Abstract

I describe the methods, and preliminary results, of a search for $\bar{\nu}_\mu \to \bar{\nu}_e$ oscillation, carried out with a scintillator-doped imaging Cerenkov detector (the Liquid Scintillator Neutrino Detector [1]) behind a proton beam dump at the Los Alamos Meson Physics Facility. The production of $\bar{\nu}_e$ in the beam dump is less than 8 parts in 10^4 of $\bar{\nu}_\mu$ production. This rarity of $\bar{\nu}_e$ produced by conventional decays in the neutrino source allows us to carry out a sensitive search for $\bar{\nu}_e$ appearance from nonstandard processes. In time with the LAMPF proton beam, we observe 10 $e^+\gamma$ coincidences which are candidates for $\bar{\nu}_e p \to e^+ n$ followed by $n+p \to D_2\gamma(2.2 \text{ MeV})$. We predict a background from beam-unrelated sources of 0.7 events, and from conventional neutrino processes of 0.67 events.

The Liquid Scintillator Neutrino Detector has now collected data through two running periods at the Los Alamos Meson Physics Facility (September-October 1993 and August-November 1994). A central purpose of the instrument is to search for $\bar{\nu}_\mu \to \bar{\nu}_e$ oscillation; I review a preliminary analysis of this search here.

The primary beam at LAMPF consists of 800 MeV protons from the LAMPF linac. The protons are produced in macropulses of ~ 600 μsec in duration, at a repetition rate of 120 Hz. (There is a 201 MHz microstructure within the macropulses, but it plays no role in the $\bar{\nu}_e$ appearance search.) The largest source of neutrinos for the LSND experiment arises in the final target area of the linac beam. Here the protons strike a 30-cm-deep water target, producing secondary π^+ and π^-. The production target is followed by a 1 meter drift space, in which 3% of the pions decay in flight. The balance of the pions are stopped in a water-cooled copper beam stop, where the surviving π^+ decay at rest, predominantly to $\mu^+\nu_\mu$, while essentially all of the π^- (except a part in $\sim 10^5$) undergo nuclear capture. The LAMPF proton beam also intercepts two upstream targets, which yield π^+ and π^- in about the same ratio as the water target. The integrated flux of protons on the water target during data-taking in 1993 was 1787 Coulombs, and in 1994, 5904 Coulombs.

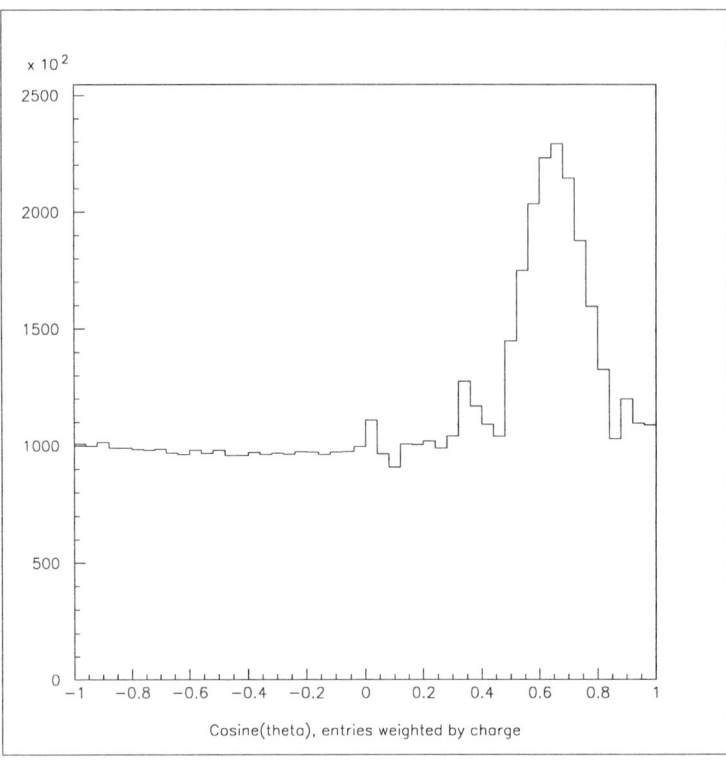

Figure 1. Accumulated phototube charge vs. the cosine of the angle between reconstructed electron direction and the vertex-phototube line of sight, histogrammed over all hit phototubes in a sample of ~5 × 10³ Michel electrons. The Cerenkov cone is at 47°.

The dominant decay chain contributing to the LAMPF neutrino source is π^+ decay at rest to $\mu^+\nu_\mu$, followed by μ^+ decay at rest to $e^+\nu_e\bar{\nu}_\mu$. This decay chain is free of $\bar{\nu}_e$; the largest conventional source of $\bar{\nu}_e$ at LAMPF is the decay chain $\pi^- \rightarrow \mu^-\bar{\nu}_\mu$ followed by $\mu^- \rightarrow e^-\bar{\nu}_e\nu_\mu$. In this decay chain, only the 3% of the π^- which decay in flight contribute significantly, and only ~10% of their μ^- daughters evade nuclear capture to decay at rest to $e^-\bar{\nu}_e\nu_\mu$. Simulation of neutrino production from the LAMPF proton targets [2, 3] yields, for the 93 and 94 runs combined, a $\bar{\nu}_\mu$ flux (at the center of the detector) of 3.8±0.3 × 10¹³ cm⁻² from μ^+ decay at rest to $e^+\nu_e\bar{\nu}_\mu$, and a $\bar{\nu}_e$ flux of 3.0±0.2 × 10¹⁰ cm⁻² from μ^- decay at rest. (The calculated flux of ν_μ from π^+ decay in flight was 3.4±0.25 × 10¹² cm⁻²; the decay in flight fluxes are important in the calculation of backgrounds to $\bar{\nu}_e$ appearance. The ν_μ spectrum from π^+ decay in flight has a broad peak at 40-60 MeV, and falls approximately quadratically to nil by 260 MeV.)

The main detector for the experiment is an approximately cylindrical (horizontal axis) tank, 5.7 meters in diameter by 8.3 meters in length, holding 167 metric tons of mineral oil (CH_2), doped with 0.031 gm/l of the scintillator butyl-PBD. The oil is viewed by 1220 8" Hamamatsu phototubes which line the tank inner surface, providing a 25% photocathode coverage at the inscribed locus of the phototube faces. The line of sight from the beam dump to the tank center lies 12° from the direction of the

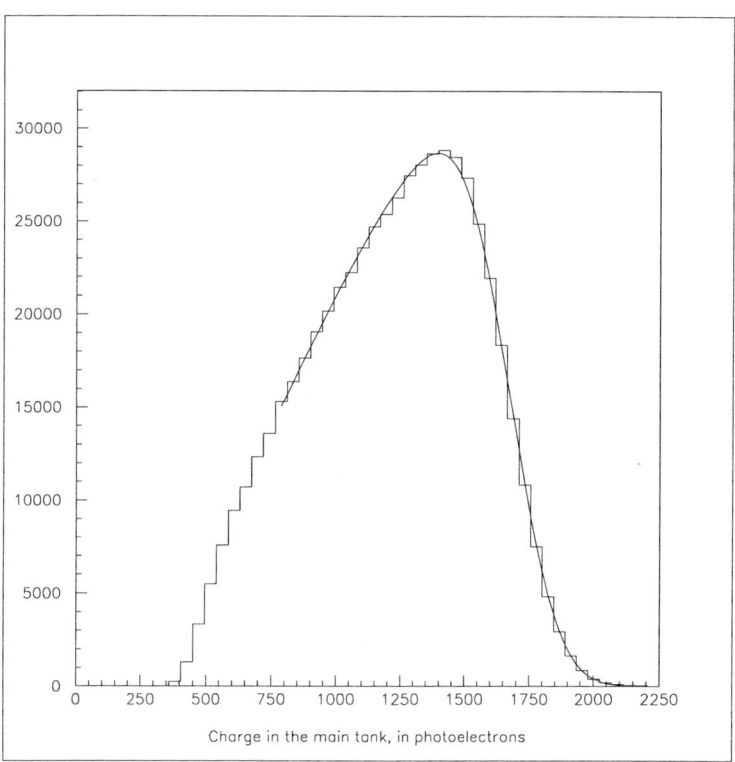

Figure 2. The spectrum of collected charge for a sample of ~5 × 10^5 Michel electrons. The fit yields a scale of 31 photoelectrons/MeV and an 8% resolution at the Michel endpoint.

LAMPF proton beam; tank center is 29.5 meters from the beam stop. The main tank is shielded, except on the bottom, by a 292-phototube active liquid scintillator veto shield [4] and several arrays of plastic scintillator sheets. The entire instrument is housed in a tunnel, under 2 kg/cm² of overburden.

We collect, from an average 45 MeV electron in the main tank, 1400 photoelectrons, of which 260 are from prompt Cerenkov light [5]. The average 45 MeV electron reconstructs with a position resolution of 25cm and an angular resolution on the Cerenkov cone of 12°. Figure 1 illustrates, for a sample of Michel daughters of cosmic muons which have stopped in the tank, the angular distribution of the collected charge with respect to the reconstructed direction of the electron. The peak at the Cerenkov angle of 47° is evident. Michel electrons are also the basis of the electron energy calibration; Figure 2 shows the spectrum of collected charge for the same subsample of Michels. The superimposed curve shown in the figure is fit with the hypothesis of a flat energy scale and a fractional energy resolution that varies as $1/\sqrt{E}$. The fit yields a scale of 31 photoelectrons/MeV and a resolution at the Michel endpoint of 8%. (Taking into account correlations of the scale with position and direction of the electron, and the time during the run, brings the endpoint resolution below 7%.)

It is useful to consider the raw processes in the LSND detector which trigger data acquisition. These originate in cosmic rays and in ambient radioactivity. The

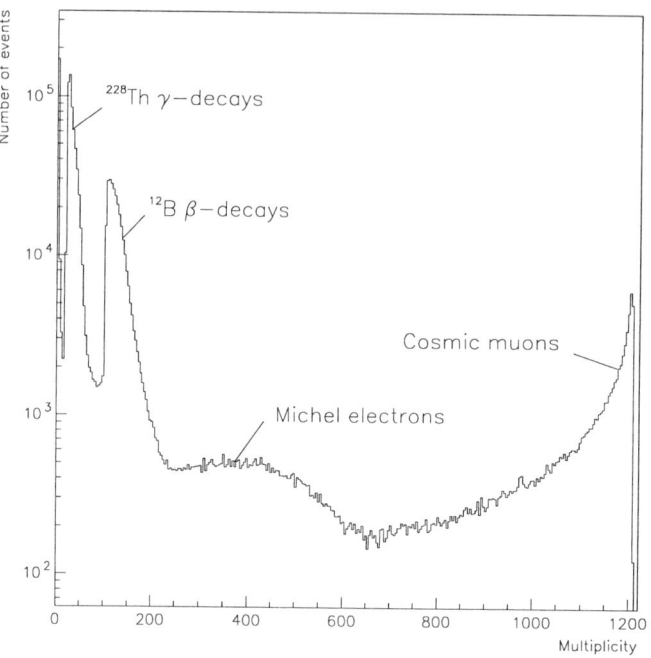

Figure 3. The hit multiplicity of LSND triggers. Approximately 10 hours of data are shown.

triggering protocol is as follows. If within a 200 nsec interval 100 or more phototubes in the main tank fired their discriminators, then we wrote from the front-end memory buffers to tape a 500 nsec history of the full detector (main tank and veto system) approximately centered in time on the burst of hit tubes. If, within the millisecond following a trigger of 125 or more tubes in the tank (in 1993, 300 or more) 21 or more phototubes in the main tank fired in a 200 nsec interval, we acquired the data for this low-threshold event as well, as a candidate 2.2 MeV gamma from neutron capture. The 100-tube trigger corresponds to ~4 MeV in electron energy, the 21-tube trigger to ~0.7 MeV in gamma energy. If 6 or more tubes fired in the active veto system during a summing interval, neither primary (i.e., positron candidate) nor secondary (i.e., gamma candidate) events occurring in the following 15.2 microseconds were written to tape. This measure suppressed the writing of Michel daughters of stopped cosmic muons. To suppress the acquisition of cosmic rays for which the 6-tube threshold in the active veto was inefficient, we did not acquire primary events if they occurred in coincidence with 4 or more hit tubes in the veto system. Events which hit 6 or more tubes in the veto system or 18 or more in the main tank and which occurred in the 51.2 microsecond preceeding a primary were also written to tape (no more than four such 'activity events' – the four closest in time to the primary – were written to tape with any primary). The trigger operates independently of the state of the LAMPF beam, tagging individual events as in-time or out-of-time with the LAMPF proton macropulse. The beam-unassociated background to any neutrino signature can be

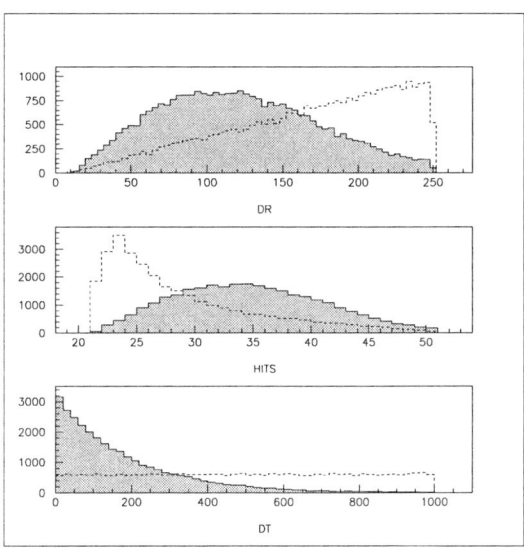

Figure 4. The distribution in primary-gamma spatial separation (DR, in cm), number of hit phototubes in the main tank, and primary-gamma time separation (DT, in μsec) for 2.2 MeV capture gammas (shaded histograms) and accidental gammas (dashed lines).

determined by counting its occurrence in the beam-off sample, and scaling into the beam-on sample by the beam-on/beam-off ratio of 0.075.

Figure 3 histograms the number of hit phototubes in the main tank over a sample of consecutive LSND triggered events. From the figure it is clear that primary triggers are dominated by ^{12}B decays (arising from the capture of stopped cosmic μ^- on ^{12}C) and Michel electrons, originating in stopped cosmic muons that outlive the 15.2 microsecond veto interval. The cosmic muon events are accumulated by the 51.2 microsecond event histories, and as primaries for which fewer than 4 phototubes fired in the veto system. Ambient radioactivity dominates the secondary triggers, owing primarily, we believe, to the decay of Th^{228} in the phototube glass. Events in the tank with multiplicity below the secondary trigger threshold enter the data stream through the 51.2 microsecond histories, owing predominantly to veto events which activate few tubes in the main tank. We suspect that many of these veto events which accompany a quiet tank originate in radioactivity in the active veto. We estimate a cosmic ray rate of $\sim$4 kHz, with an inefficiency of the 6-tube veto condition of a few 10^{-5}. The ambient gamma rate, integrated over the detector tank, is 1.2 kHz.

We have observed in LSND the neutrino-nucleus interactions $\nu_\mu C \to \mu^- X$, $\nu_e{}^{12}C \to e^{12}N_{groundstate}$, and $\nu_e C \to eX$. The cross section for $\nu_\mu{}^{12}C \to \mu^- X$, averaged against the LAMPF ν_μ flux from π^+ decay in flight, has been measured to be $(8.3\pm0.7_{stat.}\pm1.6_{syst.}) \times 10^{-40}$ cm^2 [6, 7]. The observed rate of $\nu_e{}^{12}C \to e^{12}N_{groundstate}$ is consistent with a recent measurement [8] and with theoretical prediction [9]; this provides a check on our calculation of the neutrino flux from μ^+ decay at rest. A report on the $\nu_e C$ cross sections will appear elsewhere [10, 11].

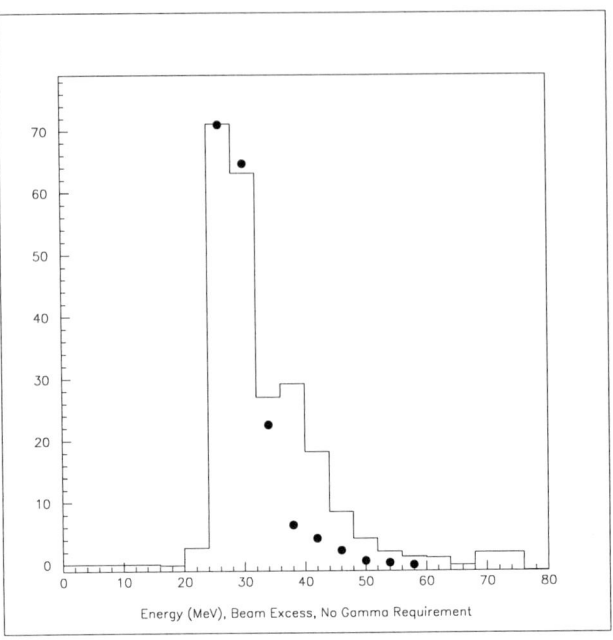

Figure 5. Beam excess spectrum of e^+ candidates in the $\bar{\nu}_e p \to e^+ n$ search, 93 and 94 data, with no requirement on a coincident gamma. The solid dots are the calculated number of events from conventional neutrino sources.

We detect $\bar{\nu}_e$ by the reaction $\bar{\nu}_e p \to e^+ n$, followed by neutron capture on a free proton, $n+p \to D_2\gamma(2.2 \text{ MeV})$. Candidate positrons are required to exhibit a well-fit Cerenkov cone and track location, and an adequate fraction of prompt light, consistent with a single-track event above Cerenkov threshold. The 2.2 MeV gamma ray from neutron capture is identified by a likelihood technique which discriminates against gamma rays (owing to ambient radioactivity in the tank) which occur in accidental space-time coincidence with the candidate positron. Figure 4 shows the basis of this discrimination. By identifying cosmic ray neutrons, we can determine the distribution in number of hit phototubes, neutron capture time, and neutron-gamma vertex separation of 2.2 MeV capture gammas. The mean number of hits is 34, the neutron capture has a time constant consistent with the calculated value of 186 microseconds, and the neutron-gamma separation has a mean of 1.25 meters. Accidental gammas can be identified by looking in the last 200 microseconds of the one millisecond secondary windows. The accidental hit distribution is notably softer than for capture gammas; the accidentals are flat in time, and have a distribution in primary-gamma spatial separation that is distinctly different from that of the capture gammas. Both for accidentals and for capture gammas, the distributions in spatial separation, time separation, and number of hits are only weakly correlated. From the parameters N_γ (the number of hits), Δt, and Δr (relative to the primary trigger time and vertex) of any secondary trigger we compute the likelihood $L_{correlated} = P_c(N_\gamma)P_c(\Delta r)P_c(\Delta t)$ that it arose from capture of a neutron created at the primary vertex, and the likelihood $L_{accidental} = P_a(N_\gamma)P_a(\Delta r)P_a(\Delta t)$ that it is accidental. These quantities are only approximate likelihoods, in part because of the neglect of small

Source	Events, 36-60 MeV
μ^- Decay at Rest	1.70
$\bar{\nu}_\mu p \to \mu^+ n$	0.73
μ^- & $\pi^- e_2$ Decay in Flight	0.02
$\nu_\mu C \to \mu^- X$	1.12
$\nu e \to \nu e$	5.09
$\nu_e C^{13} \to e^- X$	3.92
$\nu_e C^{12} \to e^- X$	1.42
μ^+ & $\pi^+ e_2$ Decay in Flight	0.16
$\pi^+ e_2$ Decay at Rest	0.51
$\nu C \to \nu C \pi^0$	0.12
Total Background = 14.8 events	

Table 1. Conventional neutrino sources which contribute to backgrounds plotted in Figures 5 and 6. The numbers on the right are the backgrounds on the restricted interval (of e^+ energy) 36-60 MeV, with no requirement on an associated gamma.

correlations between the N_γ, Δr, and Δt distributions, and in part because of the neglect of small spatial variations in the shape of the distributions. However, the ratio $R_\gamma = (L_{correlated}/L_{accidental})$ is a useful variable by which to separate correlated from accidental gammas. For example, a cut at $R_\gamma > 30$, for which data will be shown below, admits 23.5% of correlated gammas but only 0.7% of accidentals.

For the analysis discussed here of data from the 1993 and 1994 run cycles, candidate positrons for the $\bar{\nu}_e p \to e^+ n$ search were selected by these criteria (numbers in parentheses are the corresponding cut efficiencies):

1. the fit to the Cerenkov cone and to the vertex, and the fraction of late light, were consistent with an electron (0.775 ± 0.030);

2. no significant early charge was present in the 500nsec event data window (for rejection of multiple vertices; 0.898 ± 0.020);

3. <2 hits in the active veto system occurred in the 500nsec data window of the event (0.814 ± 0.020);

4. no activity events occurred in the 40μsec prior to the event (0.516 ± 0.020);

5. the positron vertex (i.e., the track midpoint) reconstructed more than 35 cm from the phototube faces (events occurring near the boundary of the fiducial volume defined by this cut can be pushed out of the volume or pulled into it by reconstruction; we compute that on net these effects reduce the effective volume of target liquid by 0.880 ± 0.040).

The efficiencies of cuts 1 and 2 above are measured from a sample of $\sim 5 \times 10^5$ Michel daughters of cosmic muons; the efficiencies of cuts 3 and 4 are measured from data collected with random laser strobes of the main detector; the last efficiency is derived from Monte Carlo simulation of the track reconstruction. Cut 4, which relies on the 51.2 microsecond event histories, is intended to reject Michel daughters of long-lived muons (either low-energy cosmics or muons produced in ν_μ or $\bar{\nu}_\mu$ charged current reactions). Deadtime in the data acquisition system and the occasional overwriting of data in the front end buffers (caused by tardiness in the real-time trigger decision) yield an inefficiency of 0.03 ± 0.01; the overall positron efficiency is therefore 0.250 ± 0.017.

Figure 5 shows the beam-on - beam-off energy spectrum of events selected by the positron criteria. Shown also in the Figure is the spectrum of events expected from conventional processes. A list of the processes taken into account in computing the superimposed backgrounds is given in Table 1. Calculation of the (anti)neutrino backgrounds proceeds from fluxes derived from the beam-stop Monte Carlo [2], conventional weak cross sections, a Fermi Gas model of $\nu_e{}^{12}C$ scattering [12], and a cross section estimate of Fukugita et. al. [13] for $\nu_e{}^{13}C$ scattering. The $\nu_\mu C \to \mu^- X$ background is computed using the measured rate of this process in LSND [6]. Both the $\nu_\mu C \to \mu^- X$ background, and the $\bar{\nu}_\mu p \to \mu^+ n$, require the muon to be missed, either because it is too low in energy to reach the 18-tube activity threshold or because it is not resolved from its electron daughter by reconstruction. The calculation of the rate of missed muons requires Monte Carlo simulation [14] of the detector muon response, especially for muons of a few MeV kinetic energy. The small background from coherent π^0 production is computed by assuming a cross section that rises as the cube of the neutrino energy above threshold, and which matches, at 200 MeV, to a linear extrapolation down in energy from measured cross sections [15] in the GeV region. Over the entire energy interval shown in Figure 5, it is $\nu_e C \to e^- X$ that is the largest source of electrons. The upper limit of the electron spectrum for this reaction is 36 MeV. Above this limit, in the region 36-60 MeV (the upper energy limit making a 2.5 standard deviation allowance for energy resolution above the $\bar{\nu}_e p \to e^+ n$ endpoint) we observe 83 positron candidates beam-on and a scaled beam-off of 30.8, for a beam-on excess of 52.2 events. The summed conventional sources over 36-60 MeV (tabulated by process in Table 1) yield 14.8 events.

We have looked for beam neutron backgrounds to $\bar{\nu}_e p \to e^+ n$, by looking for primary-gamma pairs in which the primary reconstruction quality indicates a particle below Cerenkov threshold, but the pair otherwise is consistent with an $e^+\gamma$ coincidence with the positron in the 36-60 MeV range. The data are consistent with the absence of a beam excess of such pairs; at the 90% confidence level the excess is <3% of the total rate of such pairs when the beam is on. The beam-associated neutron background which passes electron particle identification is therefore <3% of the total beam-unassociated (i.e., scaled beam off) neutron background. Examination of the beam off backgrounds which meet the positron criteria in the 36-60 MeV window show <10% to be neutrons; from the scaled beam off background of 30.8 events in the positron sample, we bound the beam neutron background in the 83 beam-on candidates to be less than $0.03 \times 0.1 \times 30.8 \sim 0.1$ event.

Requiring an associated gamma to appear in coincidence with the candidate positron, with a gamma likelihood ratio $R_\gamma > 30$, results in the excess spectrum shown in Figure 6; the calculated conventional backgrounds are again superimposed. Whereas in the 36-60 MeV sample with no gamma requirement the conventional backgrounds without an associated neutron (the largest being neutrino-electron elastic scattering) dominate, after the application of a tight cut on associated gammas conventional processes with an associated neutron dominate the background (the largest now arising from μ^- decay at rest in the beam stop). This accounts for the change in shape of the plotted background: processes without an associated neutron scale down by 0.007, the probability of an accidental gamma with $R_\gamma > 30$, while processes with an associated neutron scale down by 0.235, the efficiency for correlated gammas to pass the $R_\gamma >$

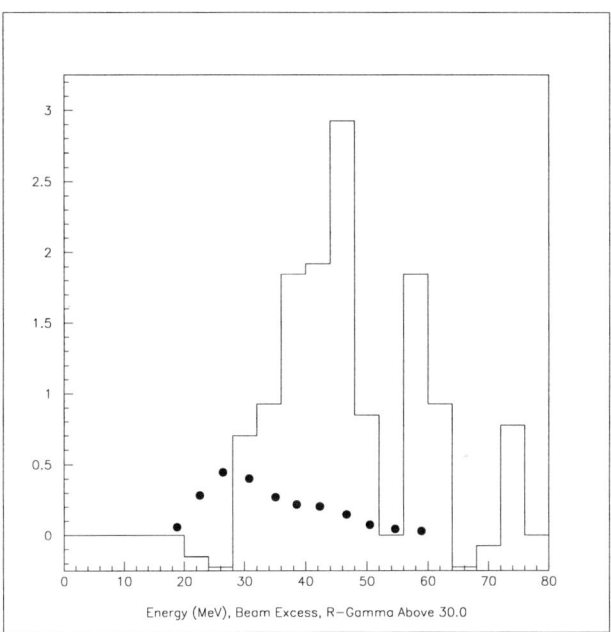

Figure 6. Beam excess spectrum of e^+ candidates in the $\bar{\nu}_e p \to e^+ n$ search, 93 and 94 data, requiring the presence of a coincident gamma ray with $R_\gamma > 30$. The solid dots are the calculated number of events from conventional neutrino sources.

30 cut. In the 36-60 MeV window, there are 10 beam-on candidate $e^+\gamma$ coincidences, 0.7 scaled beam off, and a calculated conventional background of 0.67 events. The appearance of the spectrum in Figure 6, with three events near 60 MeV and one above 70 MeV, may be rather puzzling, as the e^+ endpoint in $\bar{\nu}_e p \to e^+ n$ is ~50 MeV. Perhaps some process not considered here is contributing beam-on events in the higher energy region, but certainly the statistics are too low to draw this conclusion.

If *all* of the $\bar{\nu}_\mu$ produced by μ^+ decay at rest were to have converted to $\bar{\nu}_e$ before entering the detector, we would have detected 4250 beam-on e^+ and 999 (=0.235 × 4250) beam-on $e^+\gamma$ coincidences. If the beam-on excesses above the calculated conventional backgrounds are owing to $\bar{\nu}_\mu \to \bar{\nu}_e$ oscillation, then one computes, from the excess of e^+ candidates, an oscillation probability of $37.4/4250 = (8.8 \pm 2.1_{stat})$ × 10^{-3}; the oscillation probability computed from the excess of $e^+\gamma$ coincidences is $8.6/999 = (8.6 \pm 3.2_{stat}) \times 10^{-3}$. The statistical errors are dominated by the fluctuations in the number of beam-on events. The consistency between the two numbers is somewhat deceiving; a wider range appears in computing the oscillation probability at various intermediate cuts in R_γ. The range from 2σ above the largest probability to 2σ below the smallest is 0.9×10^{-3} to 1.5×10^{-2}. More properly, of course, one should fit to the entire e^+ sample, for all values of the R_γ parameter, to find the size of the excess with a correlated gamma; this analysis will appear elsewhere. In the usual two-state model of neutrino mixing [16], the oscillation probability corresponds, in the limit of large Δm^2 – which for LSND is > 20 eV^2 – to $(1/2)\sin^2 2\theta$.

In summary, a preliminary analysis of the 1993 and 1994 LSND data yields

83 beam-on positron candidates in the 36-60 MeV range, with a scaled beam-off background of 30.8 events and a calculated conventional neutrino-induced background of 14.8 events. A tight cut on the presence of a 2.2 MeV capture gamma in coincidence with the positron yields 10 candidate $e^+\gamma$ coincidences beam-on, a scaled beam-off of 0.7, and a calculated conventional background of 0.67. If the observed beam excesses above conventional background are owing to $\bar{\nu}_\mu \to \bar{\nu}_e$ oscillation, they would indicate an oscillation probability (at approximately the 95% confidence level) in the range 0.9 $\times$ 10^{-3} to 1.5×10^{-2}. We hope to illuminate the origin of these events by additional scrutiny and by the accumulation of more data with this instrument.

References

[1] The LSND collaboration consists of the following people: A. Eisner, Y. Wang, M. Sullivan (*University of California, IIRPA*); K. McIlhany, I. Stancu, W. Strossman, G.J. VanDalen (*University of California, Riverside*); W. Vernon (*University of California, San Diego*); D. Bauer, D. Caldwell, A. Lu, S. Yellin (*University of California, Santa Barbara*); D. Smith (*Embry-Riddle Aeronautical University*); I. Cohen (*Linfield College*); R.D. Bolton, R.L. Burman, J. Donahue, F.J. Federspiel, J. Hill, G.T. Garvey, W.C. Louis (Spokesman), V. Sandberg, M. Schillaci, D.H. White, D. Whitehouse (*Los Alamos National Laboratory*); R.M. Gunasingha, R. Imlay, W. Metcalf (*Louisiana State University*); B. Boyd, K. Johnston (*Lousiana Technical University*); B.D. Dieterle, R.A. Reeder (*University of New Mexico*); A. Fazely (*Southern University*); C. Athanassopoulos, L.B. Auerbach, V. Highland (deceased), J. Margulies, D. Works, Y. Xiao (*Temple University*).

[2] R.L. Burman, M.E. Potter, and E.S. Smith, Nucl. Instrum. Methods A **291**, 621 (1990); R.C. Allen, et. al., Nucl. Instrum. Methods A **284**, 347 (1989).

[3] D. Works, Ph.D. thesis, Temple University, 1995, unpublished.

[4] J. Napolitano, et. al., Nucl. Instrum. Methods A **274**, 152 (1989).

[5] R.A. Reeder et. al., Nucl. Instrum. Methods A **334**, 353 (1993).

[6] M. Albert, et. al., Phys. Rev. C **51**, R1065 (1995).

[7] M. Albert, Ph.D. thesis, University of Pennsylvania, 1994, unpublished; C. Athanassopoulos, Ph.D. thesis, Temple University, 1995, unpublished.

[8] KARMEN Collaboration, B.E. Bodmann et. al., Phys. Lett. B **332**, 251 (1994);

[9] Fukugita et. al., Phys. Lett. B **212**, 139 (1988); S.L. Mintz and M. Pourkaviani, Phys. Rev. C **40**, 2458 (1989); T.W. Donnelly, Phys. Lett. B **93**, 43 (1973).

[10] LSND collaboration, to appear.

[11] R.M. Gunasingha, Ph.D. thesis, Louisiana State University, 1995, unpublished.

[12] T.K. Gaisser and J.S. O'Connell, Phys. Rev. D **34**, 822 (1986); W.M. MacDonald, E.T. Dressler, and J.S. O'Connell, Phys. Rev. C **19**, 455 (1979). The ^{12}C nucleus is modeled with 25 MeV binding energy, 220 MeV/c Fermi momentum, and 6 neutrons.

[13] M. Fukugita *et. al.*, Phys. Rev. C **41**, 1359 (1990).

[14] K. McIlhany *et. al.*, *The LSND Monte Carlo Program*, Proceedings of the International Conference on Computing in High Energy Physics **94**, San Francisco, California, 21-27 April 1994.

[15] CHARM Collaboration, F. Bergsma *et. al.*, Physics Letters B **157**, 469 (1985).

[16] B. Kayser, *The Physics of Massive Neutrinos*, World Scientific, Singapore, 1989.

NEUTRINO REACTIONS IN NUCLEI IN THE LARGE AND IN THE SMALL

S.L. Mintz
Physics Department
Florida International University
Miami, Florida 33199

M. Pourkaviani
MP Consulting Associates
Altamonte Springs, FL 32701

INTRODUCTION

The experimental situation in neutrino nuclear physics has if anything in the last year become more exciting than ever. There is an oscillation experiment at LAMPF[1] which may be producing positive results. There is also a muon neutrino experiment on ^{12}C which is in wild disagreement with several calculations. In addition experiments continue at KARMEN on both exclusive and inclusive electron neutrino reactions in ^{12}C. There has also been a preliminary experiment on ^{127}I and the possibility of others following. It is thus a pleasure to report that the field is generally in ferment. The theoretical situation is in similar disarray. This is particularly true for the muon neutrino reaction on ^{12}C where theoretical calculations and experimental results differ by a factor of three. Thus we have a situation where exciting results are being reported and strenuous attempts are being made to understand them. The current confusion is probably healthy and will lead it is to be hoped to a better understanding of nuclear neutrino reactions and to the resolution of fundamental questions as to whether the neutrino has a mass and as to whether it oscillates.

We therefore will organize this paper into two parts. In the first we shall discuss selected experiments which seem particularly interesting. This discussion can be neither detailed nor inclusive but should help to give some impression of what is running and what is planned. In the second part of the paper we shall give a similar discussion of current theoretical efforts. Finally we shall draw some conclusions and discuss some future efforts which may be interesting.

AN EXPERIMENTAL UPDATE

The LSND oscillation experiment[1] at LAMPF will be covered in detail in another paper in these Proceedings but it is necessary to give at least a short discussion of it. Should the results be due to oscillation it is clearly of the highest importance and could be interesting even if an unusual background is involved. The experiment makes use of anti-neutrinos from μ^+ decays which were produced in turn from π^+ decays. Both decays take place in the LAMPF beam stop essentially at rest so that the muon anti-neutrino spectrum is a standard Michel spectrum result. The decay:

$$\mu^+ \longrightarrow e^+ + \bar{\nu}_\mu + \nu_e \tag{1}$$

yields electron neutrinos and muon anti-neutrinos but no electron anti-neutrinos. There are, of course, π^- mesons as well but their numbers should be smaller for several reasons. First because the beam of particles from the accelerator is a proton beam, positively charged particles predominate after collision with the target. Secondly, both π^- and μ^- decays are suppressed because of relatively large capture rates for π^- in the beam stop. The point is that the production of $\bar{\nu}_e$ should be very small. Thus there is the possibility to observe neutrino oscillation via the reaction:

$$\bar{\nu}_\mu \longrightarrow \bar{\nu}_e \tag{2a}$$

$$\bar{\nu}_e + p \longrightarrow e^+ + n. \tag{2b}$$

This reaction is followed by a capture reaction:

$$n + p \longrightarrow {}^2H + \gamma \tag{3}$$

forming a deuteron and releasing 2.2 MeV with the photon. The final state positron and the subsequent photon are both observed thus providing a much clearer signal than is usual in oscillation experiments.

The detector is the Liquid Scintillator Neutrino Detector (LSND). This detector consists of a large cylindrical tank 8.5 meters long and 5.5 meters in diameter filled with 51,000 gallons of liquid scintillator. On the inside surface of the detector 1220 phototubes are mounted. The system allows detection of Čerenkov and scintillation light. The detector is located approximately 29 meters from the beam dump at a 12 degree angle relative to the original proton direction. We have given this description of the LSND because it will figure in several of the experiments which are described here. We should also note that the detector is well shielded but we shall not go into a further details and shall refer readers to reference 1.

The group has seen as of the summer of 1994, eight electron events followed by the characteristic γ signal whereas an expected background would have given them approximately one event. The question, of course is, whether all background has been considered. This is a difficult question and the group expects to accumulate substantial additional data which may help to answer this question. Meanwhile the possibility of observed neutrino oscillation remains exciting.

Another experiment[2] which is very interesting and is relevant to a series of other neutrino reaction experiments, either completed[3] or in progress[4] is the measurement of the reaction:

$$\nu_\mu + {}^{12}C \longrightarrow \mu^- + X. \tag{4}$$

This is an inclusive reaction, i.e. the final state hadron is not observed. Unlike earlier inclusive neutrino reactions in ^{12}C, this particular experiment is not done over the Michel spectrum. Rather a spectrum which peaks at around 50 MeV and is relatively large at 150 MeV is used. This spectrum is small by 200 MeV but continues until approximately 290 MeV. As threshold for the charged current reaction is approximately 124 MeV, this spectrum allows a glimpse at the behavior of higher q^2 reactions of this type than have been observable in the past. The group has obtained this spectrum from decays in flight of pions, i.e. $\pi^+ \longrightarrow \mu^+ + \nu_\mu$. After various cuts taking into account background reactions, they have obtained about 270 events of which approximately 40 may be attributed to cosmic ray background. This is a reasonable number of events for a neutrino process and the experiment is continuing. At present the group has obtained an average cross section of:

$$<\sigma> = (8.3 \pm 0.7 \; stat. \pm 1.6 \; sys.) \times 10^{-40} \; cm^2. \quad (5)$$

This value agrees with some theoretical calculations but not with others and we will discuss this experiment further in the theoretical section.

An experiment which may be important for nuclear solar neutrino detectors is the reaction:

$$\nu_e + {}^{127}I \longrightarrow e^- + X. \quad (6)$$

The ^{127}I nucleus is a particularly appropriate one for use in a solar neutrino detector. It has its giant dipole resonance at a relatively low energy with a center at approximately 15 MeV. The resonance is broad enough that higher energy neutrinos from the reaction:

$$^8B \longrightarrow {}^8Be + e^+ + \nu_e \quad (7)$$

which is responsible for the largest flux of solar neutrinos in the range of energy from 2 MeV to 10 MeV, might have an enhanced reaction rate with an ^{127}I detector. Furthermore ^{127}I begins to emit nucleons for reactions in which neutrinos are at energies above 7.2 MeV. Finally ^{127}I has a large neutron excess,21, which should also cause higher reaction rates. In addition iodine can be combined with sodium and lithium to make suitable targets.

The desirability of constructing better solar neutrino detectors is very clear because in most of the presently used detectors, the target nuclei can be excited to many states some of which decay to the final measured states and some of which do not. This makes theoretical calculations very difficult. Until recently no neutrino detector had actually been calibrated by the use of a terrestrial source. This situation is now changing but it still can be said that there is much additional work which must be done before present neutrino detector results can be used with complete confidence. It is clear that at present theoretical calculations for neutrino nuclear reactions are not reliable enough to draw conclusions concerning neutrino oscillations.

The features which make ^{127}I a potential candidate for a solar neutrino experiment make it an interesting target for a Michel spectrum experiment. The giant dipole resonance can be excited by a very large fraction of the neutrino flux. This simplifies the theoretical analysis since closure may be invoked with more justification than for the ^{12}C over this same energy range. Also for the same reasons as previously discussed cross sections should be much higher than for the ^{12}C case.

A preliminary experiment has already been undertaken at LAMPF[5] for this process. Electron neutrinos with energies given by a Michel spectrum react with sodium iodide which has been disolved in water. The resulting xenon gas is collected from the water having been swept out by a flow of helium gas. The gas is then purified and put into a gas proportional counter. The xenon then decays via orbital electron capture back into ^{127}I. A preliminary result was obtained by the group of the order of:

$$<\sigma> \approx 5.5 \pm 1.1 \times 10^{-40} \; cm^2. \quad (8)$$

We shall discuss this result later in this paper. The group is currently beginning the construction of a 100 ton iodine detector at Homestake. This target will have approximately 5×10^{29} target atoms of iodine (about one quarter of the number of chlorine atoms in the present detector). Because this detector will be capable of separating daytime events from night events, it will be able to look for neutrino oscillation caused by the earth. The group hopes to use a ^{37}Ar source producing 814 KeV neutrinos to calibrate the ^{127}I detector at an energy close to that from ^7Be neutrinos. This would increase the usefulness of an iodine detector greatly as it is very difficult to undertake reliable calculations in this range at present. The KARMEN collaboration[6] is also considering a Michel spectrum measurement for ^{127}I. At present there are discussions between these two groups for a differential cross section experiment for the reaction, $\nu_e + ^{127}I \longrightarrow e + X$, at low energies. This would completely calibrate a solar neutrino detector but clearly this will not happen for some time.

Another low energy detector is being readied by a group at the university of Maryland[7], this detector is meant for solar neutrinos and makes use of $^7Li^{127}I$. Actually the detector makes use of unseparated or ^6Li depleted lithium. Natural lithium is about 7.5 percent ^6Li and about 92.5 percent ^7Li. Iodine is of course virtually 100 percent ^{127}I. This particular target is Europium activated and both the nuclei and electrons are targets. Charged current reactions such as:

$$\nu_e + ^7Li \longrightarrow ^7Be + e^-$$

or as:

$$\nu_e + ^{127}I \longrightarrow ^{127}Xe + e^-.$$

can occur. In addition a corresponding set of neutral current reactions:

$$\nu + ^7Li \longrightarrow ^7Li + \nu$$

and

$$\nu + ^{127}I \longrightarrow ^{127}I + \nu$$

can occur where ν is any neutrino and the final state nucleus may be a ground state or excited state. Finally there are reactions of the kind:

$$\nu + e^- \longrightarrow \nu' + e'^-$$

from atomic electrons. These reactions contribute to the cross section which is therefore substantially larger than usual. The target is expected to be around 33 tons in mass which is one to two orders of magnitude lighter than current detectors. This target would give rise to approximately 2 charged current events per day, about 1 neutral current reaction per day, and approximately 40 electron scatterings per day. This detector is still in the preliminary stages of planning and development but would clearly be a very useful addition to detectors currently in operation. One of the possibilities which the group is considering is a calibration with anti-neutrinos which are available at low energies from reactor sources. Although solar neutrinos are not anti-neutrinos the practice might be very useful. The relationship between neutrino and anti-neutrino cross sections for the ^{127}I nucleus is unfortunately not direct. We shall discus this question further in the theoretical section.

The KARMEN collaboration continues its work on the reaction, $\nu_e + ^{12}C \longrightarrow e^- + X$. The KARMEN experimental[8] arrangements have been described in detail elsewhere. We mention only that it is the longest running nuclear neutrino experiment ever and that Michel spectrum neutrinos are used. The results of this group for the reaction $\nu_e + ^{12}C \longrightarrow e^- + X$ have been falling over time. Their current value is in the neighborhood of $<\sigma> \approx 6.3 \times 10^{-42}$ cm^2 where this number represents just the transitions to the excited states of ^{12}N. The group has recently has noted an anomaly in their time distribution of neutrinos[9]. This anomaly is apparently persistent and may possibly be a rare decay of the π^+. They are currently investigating this matter.

Finally we conclude with a mention of the Sudbury Neutrino Observatory which is head-quartered at Queens University in Kingston, Ontario. This detector is primarily a solar neutrino detector which makes use of liquid deuterium. It is therefore able to observe both charged and neutral current reactions:

$$\nu_e + ^2H \longrightarrow p + p + e^- \quad (9a)$$

and

$$\nu_e + ^2H \longrightarrow p + n + \nu_e \quad (9b)$$

respectively. This is a relatively clean reaction as there are no excited states of the deuteron. There have also been both experiments and calculations at somewhat higher Michel spectrum energies at LAMPF in reasonable agreement with each other[10,11] as well as some lower energy calculations[12]. This group is also considering the installation of a small accelerator at the laboratory to make isotopes which would provide neutrino sources of a variety of energies. This would enable them to calibrate their detector more accurately than any detector has been calibrated to date. It is to be hoped that they will be able to do this.

A THEORETICAL UPDATE

The theoretical situation particularly for inclusive neutrino reactions in nuclei has remained unclear. There are basically two shell model based methods of calculation in use. One makes use of a Fermi gas model and the other is a random phase approximation calculation. There are also two other calculations which are relevant. One is an impulse approximation based model by Kim and Mintz which uses phenomenological imput from total muon capture results. It makes use of the closure approximation and therefore cannot be used for calculations below the giant dipole resonance. The other is a fully phenomenological model by Mintz and Pourkaviani which also makes use of total muon capture data but does not require a closure approximation. It makes use of a random phase approximation. As the papers on this latter method are relatively recent[13,14,15], we give a brief outline of it below. models in what follows.

We consider as an example the reaction, $\nu_e + ^{127}I \longrightarrow e^- + X$. The method however applies equally well to all inclusive neutrino reactions in nuclei where many states contribute and for which total muon capture results exist. Even if no muon capture results exist, it is still possible to use the method, as we shall see, if related results exist. We begin with the matrix element:

$$M_{ki} = \frac{G}{\sqrt{2}} \cos\theta \bar{u}_e \gamma^\lambda (1 - \gamma_5) u_\nu <k|J_\lambda^\dagger(0)|^{127}I> \quad (10)$$

where k is a particular final state and:

$$J_\mu^\dagger(0) = V_\mu^\dagger(0) - A_\mu^\dagger(0) \quad (11)$$

The cross section may be written as:

$$\sigma_c = \sum_k \frac{m_\nu}{2ME_\nu} \int d^3P_e |M_{ki}|^2 \frac{m_e}{E_e(2\pi)^3} \frac{d^3P_k}{2E_k(2\pi)^3} (2\pi)^4 \delta^4(P_k + P_e - P_\nu - P_i). \quad (12)$$

The quantity $|M_{ki}|^2$ is written as:

$$|M_{ki}|^2 = \frac{G^2 \cos^2 \theta_C}{2m_\nu m_\nu} L^{\sigma\lambda} <k|J_\sigma^\dagger(0)|^{12}C><k|J_\lambda^\dagger(0)|^{127}I>^*. \tag{13}$$

The quantity, $L^{\sigma\lambda}$, is the lepton tensor appropriate to this process and is given by:

$$L^{\sigma\lambda} = p_e^\sigma p_\nu^\lambda - p_e \cdot p_\nu g^{\sigma\lambda} + p_\nu^\sigma p_e^\lambda - \epsilon^{\alpha\sigma\beta\lambda} p_{e\alpha} p_{\nu\beta}. \tag{14}$$

In order to work with average quantities, we assume an average nuclear excitation of δ given by:

$$M_x - M_i = \delta \tag{15}$$

where δ is clearly a function of the incoming neutrino energy and must be determined. We also assume on the basis of our knowledge of individual states that the interaction is largely in the forward direction and that:

$$<E_e> \simeq E_\nu - \delta. \tag{16}$$

and:

$$<\vec{p}_e> \simeq \sqrt{(E_\nu - \delta)^2 - m_e^2}. \tag{17}$$

Although the value of δ, will vary with incoming neutrino energy, above the giant dipole resonance at 15 MeV it should increase slowly in our region of interest. We note that the location of the giant dipole resonance varies from nucleus to nucleus. In addition for the the case of ^{12}C we have recently obtained the dependence of δ on the neutrino energy above the giant dipole resonance. As expected the increase is a slow one which does decrease the cross sections by a few percent. In any case we can assume δ to be known. This enables us to easily obtain $<E_e>$ and $<\vec{p}_e>$ over a large part of the relevant range of neutrino energy. We can, using these averages obtain the quantity:

$$\sigma_c = \frac{G^2 \cos^2 \theta_C}{2ME_\nu} \int d\Omega_e \sum_k <k|J_\sigma^\dagger(0)|^{12}C><k|J_\lambda^\dagger(0)|^{127}I>^* L^{\sigma\lambda}$$

$$\times \frac{<|\vec{p}_e|>}{2M - 2E_e + 2E_e \cos\theta_e \frac{<E_e>}{<|\vec{p}_e|>}}. \tag{18}$$

The hadronic part of eq.(17) may be replaced as follows:

$$<k|J_\sigma^\dagger(0)|^{12}C><k|J_\lambda^\dagger(0)|^{127}I>^* \equiv Q_{\lambda\sigma}(P_i, <q>) \tag{19a}$$

$$Q^{\mu\nu} = \alpha g^{\mu\nu} + \frac{\beta}{M^2} P_i^\mu P_i^\nu + \frac{\gamma}{M^2} P_i^\mu <q^\nu>$$
$$+ \frac{\delta}{M^2} <q^\mu> P_i^\nu + \frac{\rho}{M^2} <q^\mu><q^\nu> + \frac{\eta}{M^2} \epsilon^{\mu\nu\lambda\sigma} P_\lambda <q_\sigma>. \tag{19b}$$

which is a tensor. We have previously shown[13,14] that this tensor may be reduced to the form which contains only two unknown functions. This argument relies on a random phase approximation and on the assumption that the quantity q^2 is small. The resulting tensor is:

$$Q^{\mu\nu(3)} = \alpha g^{\mu\nu} + \frac{\beta}{M^2} P_i^\mu P_i^\nu. \tag{20}$$

The cross section is then found to be:

$$\sigma_c = \frac{G^2 \cos^2(\theta_C)}{4\pi} \frac{<|\vec{p}_e|><E_e>D}{M(M+E_\nu)} \tag{21}$$

where:

$$D = \beta - 2\alpha \tag{22}$$

and an impulse approximation based calculation[16] gives $D(q^2)$ as:

$$D = a_o - b_o q^2. \tag{23}$$

We assume this simple q^2 dependence for D, which we may also write as $D = a_o + b_o|q^2|$.

Thus our result depends upon two parameters, a_o and b_o. Low error total muon capture results are available for the process, $\mu^- +{}^{127}I \longrightarrow \nu_\mu + X$. However a little thought shows that this is not the muon capture process which is appropriate to determine the the hadronic part of the matrix element $\nu_e +{}^{127}I \longrightarrow e^- + X$. In the neutrino reaction there is an excess of 21 neutrons all of which serve as targets for the neutrino helping to boost the expected rates. We need a total muon capture rate for a nucleus with 21 excess protons but the same total number of nucleons, as ^{127}I. Such a nucleus would not be stable but we shall see that we can construct a muon capture rate for such a nucleus. We therefore note that by proceeding by a calculation very similar to the neutrino reaction case[13], we find for the total muon capture rate an expression similar to Eq.(21):

$$\Gamma_{TOT} = \frac{C|\Phi(0)|^2 G^2 \cos^2\theta_C <E_\nu>^2 D}{8\pi M_i(M_i+m_\mu)}. \tag{24}$$

As we remarked earlier, the total muon capture rate is needed for a nucleus which has 74 protons, 53 neutrons and a total $A = 127$. We may call this nucleus ^{127}W and note that it does not exist. However we may make use of the fact that there is an extremely accurate semi-empirical formula for the total muon capture rate in nuclei[17]. This formula is given by:

$$\Gamma = Z_{eff}^4 G_1 [1 + G_2 \frac{A}{2Z} - G_3 \frac{A-2Z}{2Z} - G_4(\frac{A-Z}{2A} + \frac{A-2Z}{8AZ})] \tag{25}$$

where $G_1 = 261, G_2 = -0.040, G_3 = -0.26$, and $G_4 = 3.24$. This formula, Eq.(24) fits all exisitng to within 20 percent and in most cases is much better. We therefore use Eq.(25) to produce a muon capture rate for ^{127}W. We take the ratio of the rates obtained by Eq.(25) for ^{127}W to that for ^{127}I and set it equal to the expression in terms of D given by Eq.(24). When we do this we find that:

$$4.1 \times \frac{{}^W Z}{{}^I Z} = \frac{{}^W D}{{}^I D} \tag{26a}$$

or

$$5.7\, {}^I D = {}^W D. \tag{26b}$$

Thus from the total muon capture rate for ^{127}I we are able to proceed. We should note that it is not usually so difficult to make use of the total muon capture rate. In the ^{12}C case the the levels of ^{12}B and ^{12}N are isotopically related and so the total muon capture rate can be used directly for this and similar cases. It must also be remarked that the total muon capture rate in ^{127}I can be used directly to obtain the anti-neutrino reaction cross section for that nucleus. For ^{127}I the total muon capture rate is known[17] and is given by $\Gamma_{TOT} = 11.2 \pm .11 \times 10^6 \; sec^{-1}$. We also need the correction factor,C, in eq.(24). This as is well known,is given by $C = (\frac{Z_{eff}}{Z})^4$ which for the iodine case case yields:

$$C = .093. \tag{27}$$

We note that even a small error in Z_{eff} will have large consequences but the effective values for Z are believed to be well known. Using $E_\nu \simeq 90.62 \; MeV$, we obtain:

$$^I D = 1.244 \times 10^{12} \tag{28a}$$

or

$$^W D = 7.1 \times 10^{12}. \tag{28b}$$

We have at this point a value for $^W D$ at q^2 appropriate for muon capture but we need a piece of additional information to determine $^W D$ completely. In the case of ^{12}C we relied on inclusive electron scattering data and some impulse approximation results to fully obtain D. Here this is not possible. However an impulse approximation result[16] yields a value for $\tilde{D} = \frac{D(q^2)}{D(0)}$ given by:

$$\tilde{D} = \frac{[1 - (\frac{(A-Z)}{2A})\delta(\vec{q}^2)]}{[1 - \frac{(A-Z)}{2A}\delta(0)]} \tag{29a}$$

where:

$$\delta(\vec{q}^2) = (\frac{d}{r_o})^3 (1 - \frac{\vec{q}^2 d^2}{10}). \tag{29b}$$

and where from Eq.(15) we may write:

$$\tilde{D} = (1 - \frac{b_o}{a_o}q^2) \equiv (1 - b'q^2). \tag{27c}$$

This yields:

$$\tilde{D} = 1 + .1898q^2. \tag{29d}$$

and thus with Eq.(26b) determines $^W D$. We note that $\delta(\vec{q}^2)$ is unrelated to the δ of Eq.(15). Still $\delta(E_\nu)$ is needed as a function of neutrino energy. Above the giant dipole resonance we expect closure to be applicable and so for $E_\nu > 30 MeV$ we set $\delta(E_\nu) = 15 \; MeV$. We choose this value from previous experience[13,14] with total muon capture rates where good results are obtained at 15 to 20 MeV above the giant dipole resonance with closure. From our experience with the ^{12}C case, below 30 MeV we use a decreasing δ given by:

$$\delta(E_\nu) = 1.667 \times 10^{-2} E_\nu^2 - 6.0023 \times 10^{-3}. \tag{30}$$

We have tried several different forms for $\delta(E_\nu)$ and the precise form of the function does not have more than an effect of a few percent on the value of $<\sigma_c>$. We are now able to evaluate the cross section. In figure 1 we plot our results this model for the charged current cross section,σ_c. As already noted, at and above 30 MeV we assume $\delta(E_\nu)$ is given by the constant,15 MeV, but below 30 MeV we write it as given in Eq.(30). We note that $<\sigma_c>$ is not greatly affected by any reasonable choice of

functions. From figure 1 and the Michel spectrum we obtain the spectrum averaged value:
$$<\sigma_c> = 636 \times 10^{-42} \ cm^2. \tag{31}$$

where we expect a 35 percent error to be associated with this number. As was noted earlier an experiment was actually performed on ^{127}I. Because only xenon gas was recovered, other final states not decaying to xenon gas were not observed. Thus we expect the experimental results to be smaller than those produced here and thus agreement between this result and Eq.(8). These results are also in line with a number of estimates that have been circulating among researchers in the field which are in the region of $<\sigma> \approx 600 \times 10^{-42} \ cm^2$.

As we remarked earlier the anti-neutrino reaction in ^{127}I:
$$\bar{\nu}_e + ^{127}I \longrightarrow e^+ + X \tag{32}$$

can be calculated making direct use of the total muon capture result for ^{127}I. There is however a problem in that the q^2 dependence obtained for D is relatively large. As we have no electron scattering data as in the ^{12}C case we make use of the result of Eq.(29a),Eq.(29b),and Eq.(29c) to obtain:
$$\tilde{D} = 1 - 1.2 \frac{q^2}{m_\mu^2}. \tag{33}$$

Because we have a large neutron surplus we get this large q^2 dependence. This leads to a case which is different than the ^{12}C cases and the neutrino reaction in ^{127}I. In all of the previous cases, the first term in D, Eq.(23), i.e., a_o, dominated. In this case the contribution from the b_o term is substantial. This makes the results more model dependent for this anti-neutrino reaction than for the corresponding neutrino reaction or for either of the ^{12}C cases. The reason is that if the q^2 dependence is small then the a_o term in D dominates and because $q^2 \approx (.75m_\mu)^2$ for muon capture is also small, a_o is effectively determined by the known muon capture result. Because q^2 is smaller yet for the neutrino reactions we are considering, a_o essentially determines the numerical results for these reactions. However this is not the case for the anti-neutrino reaction in ^{127}I. We also have noted that for the case of ^{12}C the q^2 dependence which we obtained via electron scattering data was smaller than the impulse approximation obtained dependence for D. In this case because the result was small in either case it did not matter much. If we assume a similar scaling for the ^{127}I case, then a $\tilde{D}$ given by:
$$\tilde{D} = 1 - .207 \frac{q^2}{m_\mu^2} \tag{34}$$

is obtained. This is a very large difference in q^2 dependence which leads to a large difference in Michel spectrum averaged cross sections. For the direct impulse approximation q^2 we obtain:
$$<\sigma> = 118.36 \times 10^{-42} \ cm^2 \tag{35a}$$

and for the scaled version we obtain:
$$<\sigma> = 173.23 \times 10^{-42} \ cm^2. \tag{35b}$$

Although these numbers are much less than the estimates for the neutrino reaction in ^{127}I they are still greater by more than a factor of 10 than the cross sections in ^{12}C

which have been sucessfully studied for more than five years. Thus experimentalists should be in no way discouraged from pursuing the anti-neutrino reaction in ^{127}I.

In figure 1 we show the cross section for the neutrino process in ^{127}I. There is a bump in this curve at about 30 MeV. This is where the region below the giant dipole resonance is joined to the region above where closure has been used and thus the bump is just an artifact. It can also be seen that the cross section increases rapdily with nuetrino energy. In figure 2 we show the cross section for the anti-neutrino reaction in ^{127}I. The solid curve is for the D of Eq.(34) and the dotted curve is for the D of Eq.(33).

As we also discussed, there have been recent measurements for the reaction, $\nu_\mu + {}^{12}C \longrightarrow \mu^- + X$. The spectrum averaged result, given by Eq.(5) is in disagreement with two theoretical calculations. One of these is a random phase approximation which seems to work well at lower energies[18]. This result yields:

$$<\sigma> = 20 \times 10^{-40} \; cm^2. \qquad (36)$$

A Fermi gas model calculation also yields a large result[2], namely:

$$<\sigma> = 24 \times 10^{-40} \; cm^2. \qquad (37)$$

These results are all of the order of three times the measured result. There is another calculation for this process by Kim and Mintz[16] which makes use of an impulse approximation result which is normalized via the total muon capture rate. This result, which makes use of the closure approximation, yields a cross section of:

$$<\sigma> = 11 \times 10^{-40} \; cm^2 \qquad (38)$$

which is within the expected error of the experimental result. Recently we have undertaken a calculation using the model which we described above. Our results yield:

$$<\sigma> = 10.8 \times 10^{-40} \; cm^2 \qquad (39a)$$

and a more careful treatment of the final state electromagnetic interaction of the muon with the nucleus yields:

$$<\sigma> = 13.1 \times 10^{-40} \; cm^2. \qquad (39b)$$

Both Eq.(39a) and Eq.(39b) have a 25 per cent error associated with them and therefore they are in agreement with the measured results. A point which has been raised by the the experimental group concerns the contributions of higher l states to the cross section. In their Fermi gas model cross section, they assume in fact that half of the contributions are higher l states. We believe that as the neutrino energy increases, if fact the higher l state contributions increase. In fact we believe that this can be seen in electron scattering data from ^{12}C. However, even at $E_\nu = 175 \; MeV$, where the spectrum has fallen to less than one fifth of its maximal value, and allowing a generous scattering angle of 20 degrees, q^2 = -3166.13 $(MeV/c)^2$ which is still only about half of the q^2 = -6280 $(MeV/c)^2$ appropriate to muon capture. This leads us to believe that by phenomenologically incorporating muon capture into D that we have included the appropriate l state contributions. It may be that these contributions are too heavily counted in the Fermi gas model calculations.

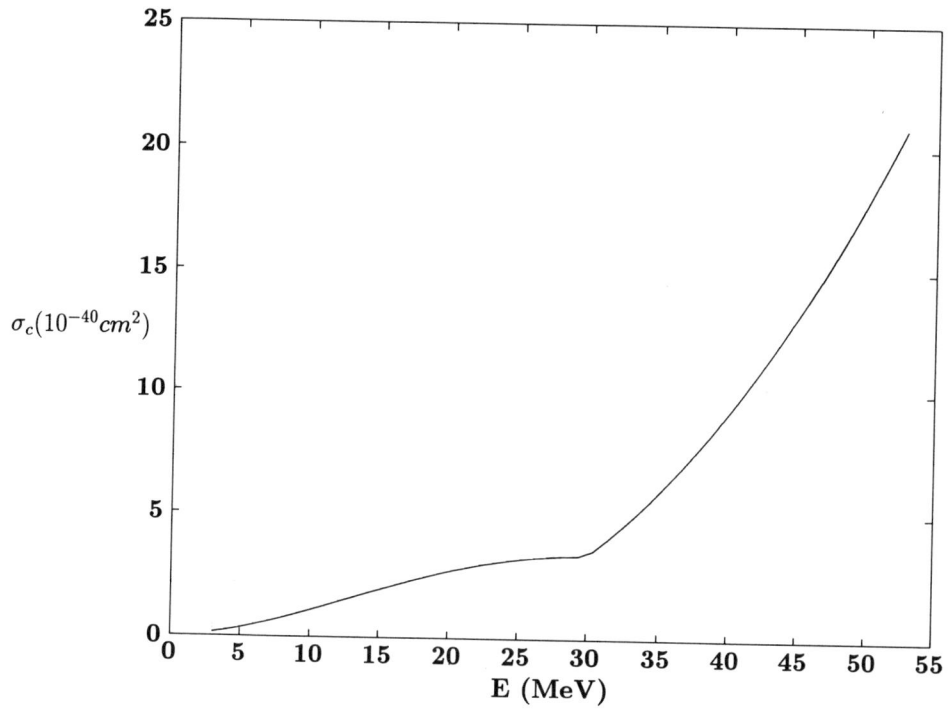

Figure 1 Cross section for the reaction $\nu_e + {}^{127}I \longrightarrow e^- + X$ as a function of neutrino energy. This calculation is via the model described here.

DISCUSSION

Thus we see that much activity is underway both theoretically and experimentally in neutrino nuclear physics. The most potentially interesting work involves the problem of neutrino oscillation. At LAMPF data is still being accumulated. There is also the possible evidence for solar neutrino oscillation and atmospheric oscillation but we have avoided a discussion of them here as they are adequately dealt with in other papers presented at this conference. The best we can say at present is that the long standing problem of neutrino oscillation has not yet been solved but that progress is being made. It is hoped that the Sudbury Neutrino Observatory may provide important data to the solution of this problem.

Very important fundamental experiments for ν_e interactions in ^{12}C and for ν_μ in the same nucleus are providing a body of data which must now be explained by theoreticians. Here this situation is far from satisfactory. Two different models seem to be able to deal with the ν_e data at least within the theoretical and experimental error. However the situation for the higher energy ν_μ experiment is much less clear. Additional calculations must surely be made and further attempts to understand the role of higher l state contributions must also be undertaken.

Intial measurements on larger nuclei,i.e. ^{127}I have been performed. It would be very good if these measurements could be run for long enough so that a substantial body of data could be obtained. This would also be a great aid to thoretical calculations. For what data is available the agreement between experiment and theory is

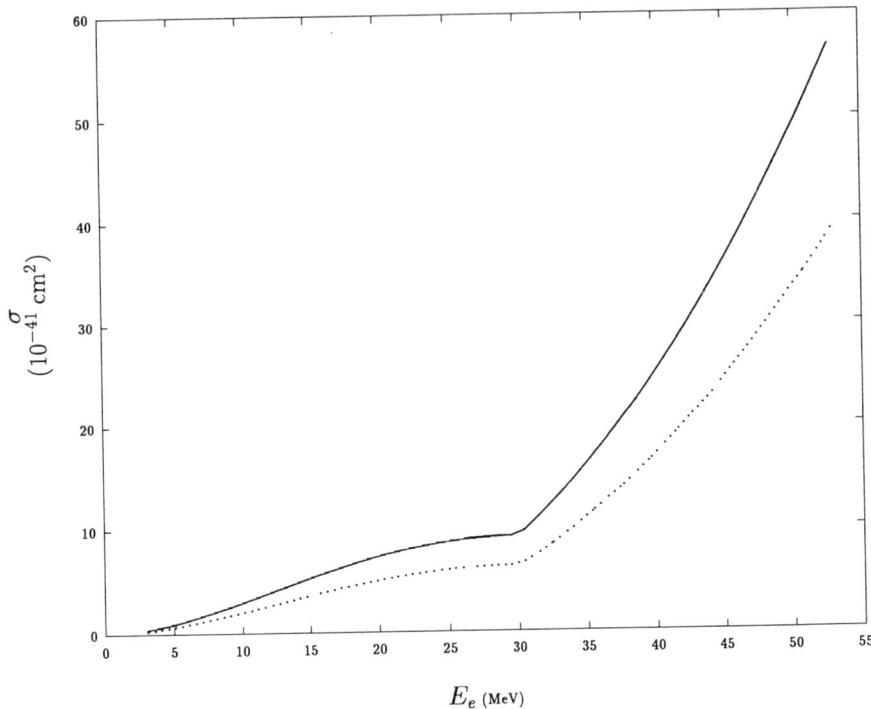

Figure 2 Cross section for the reaction $\bar{\nu}_e + {}^{127}I \longrightarrow e^+ + X$ as a function of neutrino energy. The solid curve is the cross section for the D of Eq.(34) and the dotted curve is for the D of Eq.(33).

better than might have been expected. It would be good to know if this is an artifact or represents a real understanding.

Finally, it would be useful to know how good the random phase approximation is for the theoretical model described in this paper. One possible way of doing this is to compare neutrino and anti-neutrino reactions on a system for which the final states are isotopically related. The difference between these two cross sections would be proportional to a single term in the hadron tensor, the η term in Eq.(19b). It would be very useful to know the size of this term which would be a measure of the correctness of the random phase approximation.

To conclude, the field of neutrino nuclear physics is still in its infancy. This is one of the features that make it a delightful area of study. We can certainly look forward to interesting results for a long time to come.

REFERENCES

1. W.C. Louis et al., Neutrino Oscillation Studies at LAMPF, preprint, LA-UR 94-2761,1994.
2. M. Albert et al., Measurement for the reaction ${}^{12}C(\nu_\mu, \mu^-)X$ Near Threshold, preprint, LAMPF, 1994.
3. R.C. Allen, Phys Rev Lett.**64**,1871(1990)
4. R.Maschuw et al.,**Neutrino Physics Proceedings of the International Workshop, Heidelberg**,eds. H.V. Klapdor and B. Pohv,Springer,Berlin,147(1988).

[5] K. Lande, private communication.
[6] J.A. Edgington, private communication.
[7] G.Chang, private communication.
[8] B.Zeitnitz, Prog. in Part. and Nucl. Phys. **13**,445(1995).
[9] B. Armbruster et al.,Anomaly in the Time Distribution of Neutrinos from a Pulsed Beam Stop,KARMEN preprint,1995.
[10] S.E. Willis et al., Phys. Rev. Lett.**44**,552(1980).
[11] S.L. Mintz,Phys. Rev.**D22**,2918(1980).
[12] K. Kubodera and S. Nozawa,Int. Jour. Mod. Phys.**E3**,101(1994).
[13] S.L. Mintz and M. Pourkaviani, J. Phys. **G20**,925(1994).
[14] M. Pourkaviani and S.L. Mintz, Nuc. Phys.**A573**,501(1994).
[15] S.L. Mintz and M. Pourkaviani, Nuc. Phys.**A584**,665(1995).
[16] C.W. Kim and S.L. Mintz, Phys. Rev. **C31**,274(1985).
[17] T.Suzuki et al.,Phys. Rev. **C35**,2212(1987).
[18] E.Kolbe et al.,Nucl. Phys.**A540**,599(1992).

PHYSICS INTEREST IN $\mu^+\mu^-$ COLLIDERS

V. Barger

Physics Department
University of Wisconsin
Madison, WI 53706

ABSTRACT

Feasibility studies for future $\mu^+\mu^-$ colliders with 500 GeV and 4 TeV center-of-mass energy are currently underway. Here we discuss the physics potential of such colliders, stressing possibilities that are complementary to physics at future e^+e^- linear colliders. Of particular interest are s-channel Higgs resonance production, pair production of heavy supersymmetric particles and a strongly interacting WW sector.

1. INTRODUCTION

Studies are underway to determine the feasibility of constructing high luminosity $\mu^+\mu^-$ colliders.[1] Two scenarios under consideration are a (250 GeV)×(250 GeV) collider with luminosity $\mathcal{L} = 2 \times 10^{33}$ cm^{-2} s^{-1} and a (2 TeV)×(2 TeV) collider with $\mathcal{L} = 10^{35}$ cm^{-2} s^{-1}.[1,2] A schematic design for a muon collider is shown in Fig. 1.[2]

The most pressing fundamental physics issues to be addressed by future colliders are two-fold:[3,4]

- to find the mechanism responsible for electroweak symmetry breaking,

- to see if a low-energy supersymmetry exists.

For both the Higgs mechanism and supersymmetry, new particles are expected and an important goal at future colliders is to determine the mass spectrum, widths, and couplings of these particles. If no Higgs bosons with mass $m_H < 800$ GeV are found, then the study of strong $W_L W_L$ scattering in the TeV regions becomes paramount.[5] As we will see, muon colliders offer significant capabilities in both areas.[3]

Generally speaking, it should be possible to do anything at a $\mu^+\mu^-$ collider that can be done at an e^+e^- collider of the same E_{CM} and luminosity, provided that the detector backgrounds can be managed. Thus we first briefly address the potential of

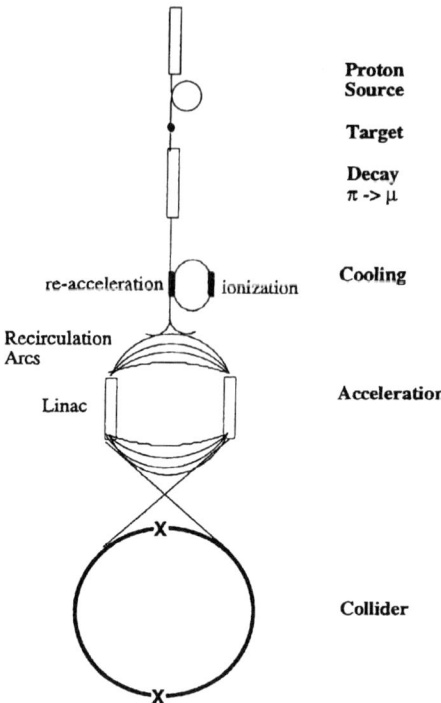

Figure 1. Muon collider schematic (from Ref. 2).

e^+e^- colliders to address the major physics issues above. An e^+e^- collider can search for a Standard-Model-like Higgs boson h via the $Z^* \to Zh$ process for $m_h < 0.7\sqrt{s}$. Since there is a guaranteed upper bound on the lightest Higgs boson in the Minimal Supersymmetric Standard Model (MSSM) of $m_h \lesssim 130$–150 GeV,[6,7] a $\sqrt{s} = 300$ GeV e^+e^- machine could exclude or affirm the light Higgs prediction of supersymmetric Grand Unified Theories. The heavy neutral MSSM Higgs bosons H and A can be discovered only via the process $e^+e^- \to Z^* \to H + A$ since the tree-level ZHH and ZAA couplings are very small or zero, respectively. Representative Higgs boson masses[8] in SUSY GUT models (assuming universal GUT scale scalar and gaugino masses m_0 and $m_{1,2}$) are shown in Fig. 2 versus m_0. Typically $m_A \simeq m_H \gtrsim 200$–$250$ GeV. Hence e^+e^- colliders can discover these Higgs states only for $m_H \lesssim \sqrt{s}/2$.

An e^+e^- linear collider can also pair produce sparticles with mass $< \sqrt{s}/2$. The highest energy expected for an e^+e^- linear collider is $\sqrt{s} = 2$ TeV. This energy reach is likely adequate for producing the lightest chargino ($\chi_1^\pm \chi_1^\mp$), the lightest neutralinos ($\chi_1^0 + \chi_2^0$) and possibly also for pair producing the sleptons $\tilde{e}, \tilde{\mu}, \tilde{\tau}$ and the highest stop $\tilde{t}_1$. However, such an energy could be inadequate for producing the heavier chargino and neutralinos and the other squarks.

If no Higgs bosons exist with $m_h < 800$ GeV, the study of strong $W_L W_L$ scattering could begin at a 1.5–2 TeV e^+e^- machine.[9,10] However the expected signal rates are only a few hundred events per year for a luminosity of 2×10^{34} cm^{-2} sec^{-1}.

A $\mu^+\mu^-$ collider has several advantages for the search and study of new physics signals. First, since $m_\mu \gg m_e$ and the Higgs coupling to fermions is proportional to the mass, appreciable s-channel resonance production of Higgs bosons can occur in $\mu^+\mu^-$ collisions. This may allow discovery of Higgs bosons and precision measurements of their masses and widths. Also the higher energy reach of a 4 TeV muon collider

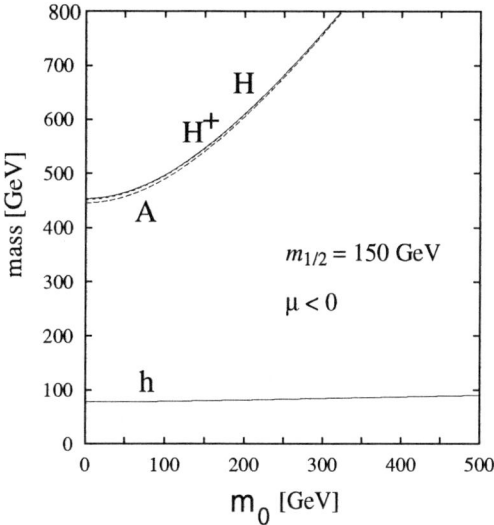

Figure 2. MSSM Higgs mass spectrum versus a universal scalar mass m_0 for fixed gaugino mass $m_{1/2} = 150$ GeV (from Ref. 8.)

over that of e^+e^- colliders could be a crucial advantage. An extended kinematical range would be opened for sparticle pair production. Alternatively, if there are no Higgs bosons with $m_H < 800$ GeV, detailed study of strong $W_L W_L$ scattering would be possible due to higher signals at higher energies with fewer backgrounds due to photon exchange contributions.

2. s-CHANNEL HIGGS RESONANCE PRODUCTION

The novelty of muon colliders is to be able to study Higgs bosons as s-channel resonances in the cross section; see the diagram in Fig. 3. The s-channel Higgs resonance cross section is

$$\sigma_h = \frac{4\pi \Gamma(h \to \mu\mu) \Gamma(h \to X)}{(s - m_h^2)^2 + m_h^2 \Gamma_h^2}, \tag{1}$$

where h denotes a generic Higgs boson, Γ_h is the total width and X denotes a final state. The widths of the MSSM Higgs bosons are typically[3,4] in the range 0.005 GeV to a few GeV, as illustrated in Fig. 4. The machine $\sqrt{s}$ resolution is expected to be about 0.04 GeV with resolutions in the range 0.01 GeV to 1 GeV possible.[11] Thus an energy scan could give a precision determination of m_h and also Γ_h of the MSSM Higgs bosons. For $m_A \gtrsim 200$ GeV, typical in SUSY GUT models, the couplings of the Higgs bosons are proportional to the following factors[12]

$$\begin{array}{cccc} & \mu^+\mu^-, b\bar{b} & t\bar{t} & ZZ, W^+W^- \\ h & 1 & -1 & 1 \\ A & \tan\beta & -1/\tan\beta & 0 \\ A & -i\gamma_5 \tan\beta & -i\gamma_5/\tan\beta & 0 \end{array} \tag{2}$$

where $\tan\beta = v_2/v_2$ is the ratio of vacuum expectation values of the two Higgs bosons. Thus the production rates for H and A will be strongly dependent on the value of the $\tan\beta$. Figure 5 shows the $\tan\beta = 5$ result for the H signal rate ($S_H = \int \sigma_H d\sqrt{s}$) integrated over $\pm 1\% \sqrt{s}$ at $\sqrt{s} = m_H$ along with the corresponding backgrounds in the

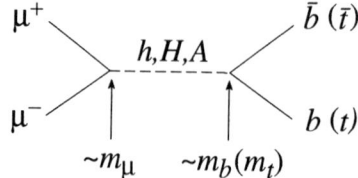

Figure 3. s-channel diagram for production of h, H, A MSSM Higgs bosons with decays to $b\bar{b}$ and $t\bar{t}$ final states.

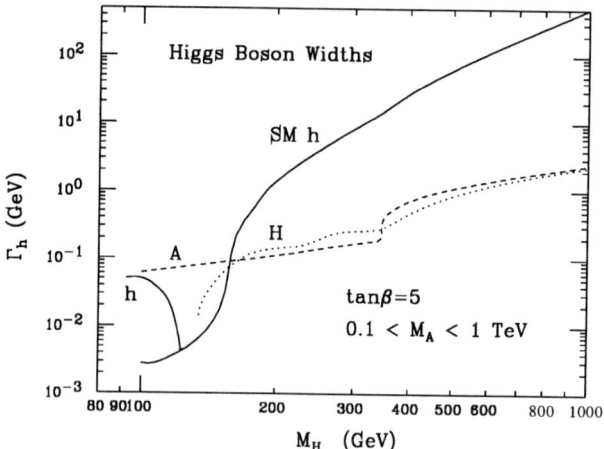

Figure 4. Total widths of the SM and MSSM Higgs bosons, with $\tan\beta = 5$ in the MSSM (from Ref. 3).

$b\bar{b}$ and $t\bar{t}$ channels.[3] For $\tan\beta \gtrsim 5$, the $b\bar{b}$ signal is dominant and $> 5\sigma$ discovery is possible with reasonable luminosities, as illustrated in Fig. 6. Direct s-channel production would allow H, A discovery up to the machine kinematic limit, $\sqrt{s} = m_h$, so long as $\tan\beta$ is not small. This is an important extension to the search range at e^+e^- colliders.

For the lightest supersymmetric Higgs boson, the s-channel resonance search is made more difficult by the presence of large Z-resonance backgrounds. In this case the Higgs boson would presumably first be discovered via the $Z^* \to Zh$ mode at an e^+e^- or $\mu^+\mu^-$ collider. Then a precision m_h and Γ_h determination could be made by a scan over a few energies.[4] Measurements of Γ_h and $\Gamma(h \to \mu\mu)BF(h \to X)$ from the s-channel process could distinguish a Standard Model Higgs boson from the lightest MSSM Higgs.

High longitudinal polarization of both beams would be beneficial for s-channel Higgs physics since

$$\sigma^H(P) = (1 + P^2)\sigma^H(P = 0) \qquad (3)$$
$$\sigma^{bgd}(P) = (1 - P^2)\sigma^{bgd}(P = 0) . \qquad (4)$$

A polarization of $P = 0.84$ can compensate for a factor of ten less luminosity.[13]

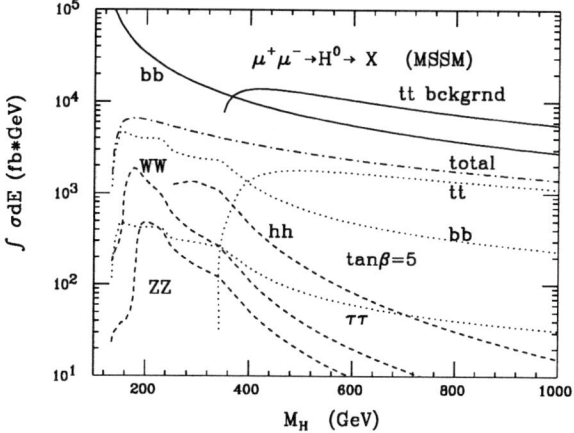

Figure 5. The MSSM H-boson signal rate $S_H = \int \sigma_H d\sqrt{s}$ integrated over a resolution $\pm 1\% m_H$ at $\sqrt{s} = m_H$ and the corresponding integrated backgrounds (from Ref. 3).

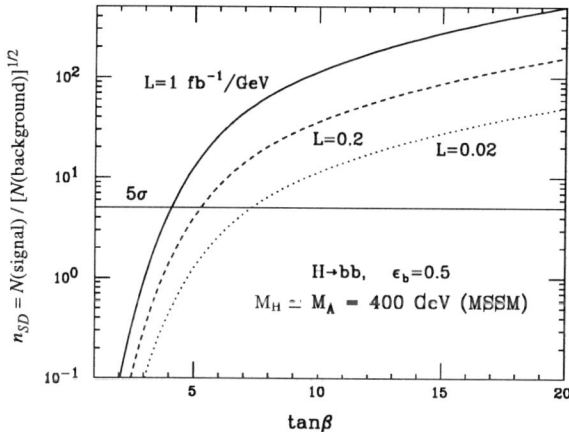

Figure 6. Statistical significance $n_\sigma = \sqrt{L}\, S_H/\sqrt{B_H}$ of the MSSM H-boson signal versus $\tan\beta$ for several luminosities (from Ref. 3).

3. SUSY FACTORY

If $M_{\rm SUSY} \sim 1$ TeV, many sparticles could be beyond the reach of a $\sqrt{s} = 2$ TeV e^+e^- collider. Due to the p-wave suppression of spin-0 production at a e^+e^- or $\mu^+\mu^-$ machine, energies well above threshold are required, as illustrated in Fig. 7. The LHC can produce the squarks and sleptons, but disentangling the spectrum will be a real challenge at a hadron collider, due to the complex cascade decays and the QCD backgrounds.

A 4 TeV $\mu^+\mu^-$ collider with luminosity of 10^{35} cm^{-2} s^{-1} would be a factory for the pair production of SUSY particles. Consider the production of 1 TeV sparticles. The cross sections and annual event rates for squarks (of one flavor in the approximation of L, R degeneracy), charginos and top are respectively

$$\sigma_{\tilde{u}_{L,R}} = 4\beta^3 \text{ fb} \to 2500 \text{ events} \qquad (5)$$
$$\sigma_{\tilde{d}_{L,R}} = 1\beta^3 \text{ fb} \to 600 \text{ events} \qquad (6)$$

$$\sigma_{\chi^\pm} = 6\beta \text{ fb} \to 5000 \text{ events} \qquad (7)$$
$$\sigma_t = 8 \text{ fb} \to 8000 \text{ events}. \qquad (8)$$

The production of heavy SUSY particles will give spherical events near threshold characterized by multijets, missing energy (associated with the lightest SUSY particle) and leptons. There should be no problem separating the signals from the Standard Model backgrounds.

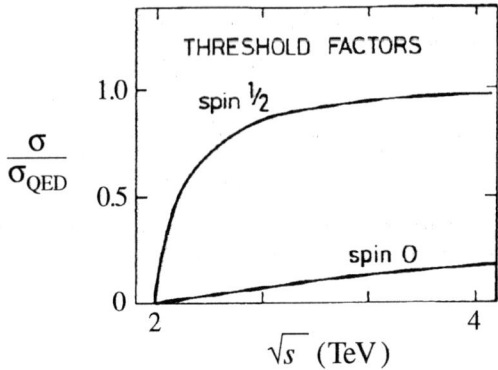

Figure 7. Kinematic suppresion of fermion pair production and squark pair production near threshold in $\mu^+\mu^-$ collisions, for 1 TeV particles.

4. STRONGLY INTERACTING ELECTROWEAK SECTOR (SEWS)

In the event that the electroweak sector is strongly interacting, the energy reach of a 4 TeV $\mu^+\mu^-$ collider would be extremely valuable, since the SEWS effects increase with energy. Figure 8 shows the WW fusion graphs of interest, where the W bosons are longitudinally polarized.[5] The size of the SEWS signals[3,10] can be estimated by calculating the cross section with a 1 TeV Higgs and subtracting the result with a massless Higgs (which removes the scattering due to transversely polarized W bosons)

$$\begin{array}{cccc} & & \Delta\sigma(W_L W_L) & \Delta\sigma(Z_L Z_L) \\ \mu^+\mu^- & 4 \text{ TeV} & 80 \text{ fb} & 50 \text{ fb} \\ e^+e^- & 1.5 \text{ TeV} & 8 \text{ fb} & 6 \text{ fb} \end{array} \qquad (9)$$

At a 4 TeV muon collider the SEWS cross section could be a factor of 10 higher and the luminosity a factor of 5 higher than at a 1.5 TeV e^+e^- collider.

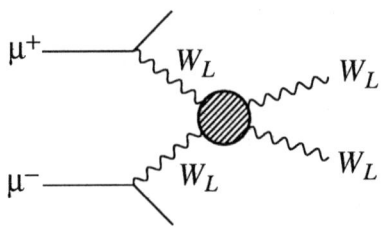

Figure 8. Strong $W_L W_L$ scattering diagram.

5. CONCLUSION

Muon colliders would offer new opportunities at both the low energy ($\sqrt{s} \simeq 500$ GeV) and high energy ($\sqrt{s} \simeq 4$ TeV) frontiers to discover SUSY particles and measure their properties, discover Higgs bosons in the s channel with masses up to $\sqrt{s}$ and measure their masses, widths and couplings, or in their absence study a strongly interacting electroweak sector with higher signal rates at higher energies.

ACKNOWLEDGMENTS

I thank the members of the physics goals working group at the Sausalito $\mu^+\mu^-$ workshop for collaboration on the results reported here. This work was supported in part by the U.S. Department of Energy under Grant No. DE-FG02-95ER40896. Further support was provided by the University of Wisconsin Research Committee, with funds granted by the Wisconsin Alumni Research Foundation.

REFERENCES

1. *Proceedings of the Second Workshop on the Physics Potential and Development of $\mu^+\mu^-$ Colliders*, Sausalito, California, 1994, D. Cline, ed. (to be published); see also First Workshop proceedings in Nucl. Inst. and Meth. **A350**, 24–26 (1994).
2. R. B. Palmer and A. Tollestrup, unpublished report (1995).
3. V. Barger *et al.*, Physics Goals Working Group Report in Ref. 1.
4. V. Barger, M. S. Berger, T. Han and J. F. Gunion, University of Wisconsin-Madison report MADPH-95-884.
5. M.S. Chanowitz and M.K. Gaillard, Nucl. Phys **B261**, 379 (1985); J. Bagger et al., Phys. Rev. **D49**, 1246 (1994).
6. M. Drees, Int. J. Mod. Phys. **A4**, 3635 (1989); J. Ellis *et al.*, Phys. Rev. **D39**, 844 (1989); L. Durand and J.L. Lopez, Phys. Lett. **B217**, 463 (1989); J.R. Espinosa and M. Quirós, Phys. Lett. **B279**, 92 (1992); P. Binétruy and C.A. Savoy, Phys. Lett. **B277**, 453 (1992); T. Morori and Y. Okada, Phys. Lett. **B295**, 73 (1992); G. Kane, *et al.*, Phys. Rev. Lett. **70**, 2686 (1993); J.R. Espinosa and M. Quirós, Phys. Lett. **B302**, 271 (1993); U. Ellwanger, Phys. Lett. **B303**, 271 (1993); J. Kamoshita, Y. Okada, M. Tanaka *et al.*, Phys. Lett. **B328**, 67 (1994).
7. V. Barger, *et al.*, Phys. Lett. **B314**, 351 (1993); P. Langacker and N. Polonshy, University of Pennsylvania preprint UPR-0594-T, hep-ph 9403306.
8. V. Barger, M.S. Berger, and P. Ohmann, Phys. Rev. **D49**, 4908 (1994).
9. K. Hagiwara, J. Kanzaki and H. Murayama, DTP/91/18.
10. V. Barger, K. Cheung, T. Han, R.J.N. Phillips, University of Wisconsin preprint MADPH-95-865.
11. G. P. Jackson and D. Neuffer, private communication.
12. J.F. Gunion and H.E. Haber, Nucl. Phys. **B272**, 1 (1986).
13. Z. Parsa (to be published).

SECTION V

STRINGS AND SUPERSTRINGS

SPIN FIELD VERTICES AND GAUGE SYMMETRY

L. Dolan

Department of Physics, University of North Carolina
Chapel Hill, North Carolina 27599-3255, USA

INTRODUCTION

Non-perturbative analysis of the superstring suggests enhanced gauge symmetry at special values of the moduli of the internal space which determines a particular compactification to four spacetime dimensions. In this talk we present a perturbative analysis involving locality properties of the spin conformal fields and BRST covariant quantization, which enhances the gauge symmetry as described by tree level amplitudes. The operator products of the vertices with the superVirasoro currents are given, expressions for bispinor states describing gauge bosons are computed, and perturbative unitarity of the theory is discussed.

LOCALITY OF SPIN FIELDS AND SUPERCONFORMAL FIELDS

In four spacetime dimensions, the vertex operators associated with massless Ramond states can be described by

$$V^{(1)}_{-\frac{3}{2}}(k,z) = v^{1\dot\alpha}(k) S_{\dot\alpha}(z) e^{ik \cdot X(z)} f^\ell \Sigma_\ell(z) e^{-\frac{3}{2}\phi(z)}$$
$$V^{(2)}_{-\frac{3}{2}}(k,z) = v^{2\alpha}(k) S_\alpha(z) e^{ik \cdot X(z)} f^{\bar\ell} \Sigma_{\bar\ell}(z) e^{-\frac{3}{2}\phi(z)}. \quad (1)$$

In order to maintain locality for the string vertices, we choose the following operator product expansions:

$$S_\alpha(z) S_\beta(\zeta) = (z-\zeta)^{-\frac{1}{2}} C^{-1}_{\alpha\beta} + \ldots$$
$$S_\alpha(z) S_{\dot\beta}(\zeta) = (z-\zeta)^0 \gamma^\mu_{\alpha\dot\beta} \frac{1}{\sqrt{2}} \psi_\mu(\zeta) + \ldots$$
$$S_{\dot\alpha}(z) S_\beta(\zeta) = (z-\zeta)^0 \gamma^\mu_{\dot\alpha\beta} \frac{1}{\sqrt{2}} \psi_\mu(\zeta) + \ldots$$
$$S_{\dot\alpha}(z) S_{\dot\beta}(\zeta) = (z-\zeta)^{-\frac{1}{2}} C^{-1}_{\dot\alpha\dot\beta} + \ldots \quad (2a)$$

$$\Sigma_\ell(z)\Sigma_{\dot n}(\zeta) = (z-\zeta)^{-\frac{3}{4}} C^{-1}_{\ell\dot n} + \ldots$$

$$\Sigma_{\dot\ell}(z)\Sigma_n(\zeta) = (z-\zeta)^{-\frac{3}{4}} C^{-1}_{\dot\ell n} + \ldots$$

$$\Sigma_\ell(z)\Sigma_n(\zeta) = (z-\zeta)^{-\frac{1}{4}} \Gamma^a_{\ell n} \frac{1}{\sqrt{2}} \psi_a(\zeta) + \ldots$$

$$\Sigma_{\dot\ell}(z)\Sigma_{\dot n}(\zeta) = (z-\zeta)^{-\frac{1}{4}} \Gamma^a_{\dot\ell\dot n} \frac{1}{\sqrt{2}} \psi_a(\zeta) + \ldots \tag{2b}$$

where the subleading terms are less singular by integer powers of $(z-\zeta)$, and the field $\psi^a(z)$ is understood to represent $\tilde\gamma^5 \otimes \psi^a(z)$, in order to maintain anticommutativity of $\psi^a(z)$ and $\psi^\mu(z)$. The operator relations above are not single-valued, but when taken in appropriate combinations with each other and the ghost fields, the resulting string vertices are local (in the sense of meromorphic operator product expansions), at least at zero momentum; and the momentum-dependent fields are local when the complete closed string holomorphic and antiholomorphic expressions are considered.

In (2b), we have chosen $1 \leq a \leq 6$ and a Weyl representation for the internal gamma matrices given for $1 \leq a \leq 3$, $1 \leq \ell, \dot\ell \leq 4$ by

$$\Gamma^a = \begin{pmatrix} 0 & (\alpha^a)^\ell_{\dot m} \\ -(\alpha^a)^{\dot\ell}_m & 0 \end{pmatrix}; \quad \Gamma^{a+3} = i\begin{pmatrix} 0 & (\beta^a)^\ell_{\dot m} \\ (\beta^a)^{\dot\ell}_m & 0 \end{pmatrix}; \quad C = \begin{pmatrix} 0 & (I_4)^{\ell\dot m} \\ (I_4)^{\dot\ell m} & 0 \end{pmatrix}. \tag{3}$$

Also $\tilde\gamma^5 \equiv \gamma^5 (-1)^{\sum_{n>0} \psi^\mu_{-n}\psi^\mu_n}$ and $\tilde\Gamma^7 \equiv (i\Gamma^1\Gamma^2\Gamma^3\Gamma^4\Gamma^5\Gamma^6)(-1)^{\sum_{n>0} \psi^a_{-n}\psi^a_n}$ where $i\Gamma^1\Gamma^2\Gamma^3\Gamma^4\Gamma^5\Gamma^6 = \begin{pmatrix} (I_4)^\ell_m & 0 \\ 0 & -(I_4)^{\dot\ell}_{\dot m} \end{pmatrix}$. The matrices α^a, β^a are real antisymmetric, and form representations of $SU(2)$. In the Weyl representation, the charge conjugation and gamma matrices have the following properties: for four dimensions,

$$C_{\alpha\beta} = -C_{\beta\alpha}, \qquad \gamma^\mu_{\alpha\dot\beta} = \gamma^\mu_{\dot\beta\alpha}, \quad \text{etc.} \tag{4a}$$

for six dimensions,

$$C_{\ell\dot n} = C_{\dot n\ell}, \qquad \Gamma^a_{\ell n} = -\Gamma^a_{n\ell}, \quad \text{etc.} \tag{4b}$$

for ten dimensions,

$$C_{A\dot B} = -C_{\dot B A}, \qquad \Gamma^M_{AB} = \Gamma^M_{BA}, \quad \text{etc.} \tag{4c}$$

To check the locality of the vertex operators in (1) at zero momentum, we use (2,4) to show

$$V^{(1)}_{-\frac{3}{2}}(z) V^{(1)}_{-\frac{3}{2}}(\zeta) = (z-\zeta)^{-3} v^{1\dot\alpha}(k_1) v^{1\dot\beta}(k_2) C^{-1}_{\dot\alpha\dot\beta} f^\ell f^n \Gamma^a_{\ell n} \frac{1}{\sqrt{2}} \psi_a(\zeta) e^{-3\phi(\zeta)} + \ldots$$

$$\cong -V^{(1)}_{-\frac{3}{2}}(\zeta) V^{(1)}_{-\frac{3}{2}}(z) \tag{5a}$$

where in (5a) the operator product $V^{(1)}_{-\frac{3}{2}}(z) V^{(1)}_{-\frac{3}{2}}(\zeta)$ is defined for $|z| > |\zeta|$, the product $V^{(1)}_{-\frac{3}{2}}(\zeta) V^{(1)}_{-\frac{3}{2}}(z)$ is defined for $|\zeta| > |z|$ and $\cong$ denotes equal in the sense of analytic continuation[1,2]. Similary, we find local fermionic fields

$$V^{(i)}_{-\frac{3}{2}}(z) V^{(j)}_{-\frac{3}{2}}(\zeta) \cong -V^{(j)}_{-\frac{3}{2}}(\zeta) V^{(i)}_{-\frac{3}{2}}(z) \tag{5b}$$

for all $1 \leq i, j \leq 2$, using (1,2).

We see in (6) that by selecting a particular supercurrent $F(z)$, we can define $q = -\frac{1}{2}$ ghost picture fermion vertex operators. We know that any supercurrent whose operator product with a fermion vertex operator has at most a $(z-\zeta)^{-\frac{3}{2}}$ singularity ensures BRST invariance for the vertex operator. In this section, we choose a supercurrent such that the contribution from the internal degrees of freedom has at most a $(z-\zeta)^{-\frac{1}{2}}$ singularity with (1). Then

$$V^{(1)}_{-\frac{1}{2}}(k,\zeta) = \lim_{z\to\zeta} e^{\phi(z)} F(z) V^{(1)}_{-\frac{3}{2}}(k,\zeta)$$
$$= u^{1\alpha}(k) S_\alpha(\zeta) e^{ik\cdot X(\zeta)} f^\ell \Sigma_\ell(\zeta) e^{-\frac{1}{2}\phi(\zeta)}. \tag{6a}$$

$$V^{(2)}_{-\frac{1}{2}}(k,\zeta) = \lim_{z\to\zeta} e^{\phi(z)} F(z) V^{(2)}_{-\frac{3}{2}}(k,\zeta)$$
$$= -u^{2\dot\alpha}(k) S_{\dot\alpha}(\zeta) e^{ik\cdot X(\zeta)} f^\ell \Sigma_{\dot\ell}(\zeta) e^{-\frac{1}{2}\phi(\zeta)}. \tag{6b}$$

BRST invariance of the vertex operators in the $q = -\frac{1}{2}$ picture (6) follows from BRST invariance of the picture changed $q = -\frac{3}{2}$ operators. Locality holds from (2) for these operators as well:

$$V^{(1)}_{-\frac{1}{2}}(z) V^{(1)}_{-\frac{1}{2}}(\zeta) = (z-\zeta)^{-1} u^{1\alpha}(k_1) u^{1\beta}(k_2) C^{-1}_{\alpha\beta} f^\ell f^n \Gamma^a_{\ell n} \frac{1}{\sqrt{2}} \psi_a(\zeta) e^{-\phi(\zeta)} + \ldots$$
$$\cong -V^{(1)}_{-\frac{1}{2}}(\zeta) V^{(1)}_{-\frac{1}{2}}(z) \tag{7a}$$

where we note that unlike the Neveu-Schwarz case, the statisitics of the fields associated with Ramond states does not change from one picture to another. In general, $V^{(i)}_{-\frac{1}{2}}(z) V^{(j)}_{-\frac{1}{2}}(\zeta) \cong -V^{(j)}_{-\frac{1}{2}}(\zeta) V^{(i)}_{-\frac{1}{2}}(\zeta)$. Also, locality holds between fields in different pictures:

$$V^{(1)}_{-\frac{1}{2}}(z) V^{(1)}_{-\frac{3}{2}}(\zeta) = (z-\zeta)^{-1} u^{1\alpha}(k_1) v^{1\dot\beta}(k_2) \gamma^\mu_{\alpha\dot\beta} \psi_\mu(\zeta) f^\ell f^n \Gamma^a_{\ell n} \psi_a(\zeta) \tfrac{1}{2} e^{-\phi(\zeta)} + \ldots$$
$$\cong -V^{(1)}_{-\frac{3}{2}}(\zeta) V^{(1)}_{-\frac{1}{2}}(z) \tag{7b}$$

To establish (7b), we note as previously mentioned below (2) that the field $\psi^a(z)$ is understood to represent $\tilde\gamma^5 \otimes \psi^a(z)$, and we use $\gamma^5 u^1 \sim u^1$, $\gamma^5 v^1 \sim -v^1$, etc. In general, $V^{(i)}_{-\frac{1}{2}}(z) V^{(j)}_{-\frac{3}{2}}(\zeta) \cong -V^{(j)}_{-\frac{3}{2}}(\zeta) V^{(i)}_{-\frac{1}{2}}(\zeta)$.

The vertex operators for the massless Neveu-Schwarz states in the canonical $q = -1$ superconformal ghost picture are

$$V_{-1}(k,z,\epsilon) = \epsilon \cdot \psi(z) e^{ik\cdot X(z)} e^{-\phi(z)} \tag{8a}$$
$$V^a_{-1}(k,z) = \tilde\gamma^5 \otimes \psi^a(z) e^{ik\cdot X(z)} e^{-\phi(z)}. \tag{8b}$$

States in the Neveu-Schwarz matter system (i.e. without ghosts) form[1] superconformal fields $V(z,\theta) = V_q(z) + \theta V_{q+1}(z)$ with upper and lower components related by

$$G(z) V_q(\zeta) = (z-\zeta)^{-1} V_{q+1}(\zeta)$$
$$G(z) V_{q+1}(\zeta) = (z-\zeta)^{-2} 2h_q V_q(\zeta) + (z-\zeta)^{-1} \partial V_q(\zeta). \tag{9}$$

For the BRST charge given by $Q = Q_0 + Q_1 + Q_2$, we have that BRST invariance holds for both the vertices in (8) since

$$[Q_0, V_{-1}(k,z)] = \tfrac{d}{dz}[c(z)V_{-1}(k,z)]; \qquad [Q_2, V_{-1}(k,z)] = 0 \qquad (10)$$

and

$$\begin{aligned}
Q_1(z)V_{-1}(k,\zeta) &= -\tfrac{1}{2} : e^{-\chi(z)} : e^{\phi(z)} G(z) V_{-1}(k,\zeta) \\
&= -\tfrac{1}{2} : e^{-\chi(z)} : (z-\zeta)^1 G(z) V_{-1}^{\text{matter}}(k,\zeta) \\
&= -\tfrac{1}{2} : e^{-\chi(z)} : V_0^{\text{matter}}(k,\zeta) \\
&= \text{regular terms}.
\end{aligned} \qquad (11a)$$

Therefore

$$[Q_1, V_{-1}k,z)] = 0 \qquad (11b)$$

and $[Q, V_{-1}(k,z)]$ is a total derivative. In the $q = 0$ superconformal ghost picture, the vertices in (8) become

$$\begin{aligned}
V_0(k,z,\epsilon) &= \lim_{z\to\zeta} e^{\phi(z)} G(z) V_{-1}(k,\zeta\epsilon) \\
&= [k\cdot\psi(\zeta)\epsilon\cdot\psi(\zeta) + \epsilon\cdot a(\zeta)] e^{ik\cdot X(\zeta)} \qquad (12a) \\
V_0^a(k,z) &= \lim_{z\to\zeta} e^{\phi(z)} G(z) V_{-1}^a(k,\zeta) \\
&= [k\cdot\psi(\zeta)\tilde\gamma^5 \otimes \psi^a(\zeta) + (\lim_{z\to\zeta}(z-\zeta)\bar F(z)\tilde\gamma^5 \otimes \psi^a(\zeta))] e^{ik\cdot X(\zeta)}.
\end{aligned}$$
$$(12b)$$

BISPINOR STATES AS GAUGE BOSONS

In a Weyl representation, the four-dimensional γ matrix Clifford algebra given by $\{\gamma^\mu,\gamma^\nu\} = 2\eta^{\mu\nu}$ with $\eta^{\mu\nu} = \text{diag}(-1,1,1,1)$ can be represented as

$$(\gamma^\mu)^A{}_B = \begin{pmatrix} 0 & (\bar\sigma^\mu)^\alpha{}_{\dot\beta} \\ (\sigma^\mu)_{\dot\alpha}{}^\beta & 0 \end{pmatrix}; \qquad C^{AB} = \begin{pmatrix} (i\sigma^2)^{\alpha\beta} & 0 \\ 0 & (i\sigma^2)^{\dot\alpha\dot\beta} \end{pmatrix} \qquad (13)$$

where $\sigma^\mu = (\sigma^0, \sigma^i)$ and $\bar\sigma^\mu = (-\sigma^0, \sigma^i)$ are given by $\sigma^0 \equiv \begin{pmatrix} 1 & 0 \\ 0 & 1 \end{pmatrix}$ and the Pauli matrices σ^i. We define $\gamma^5 = (i\gamma^0\gamma^1\gamma^2\gamma^3)^A{}_B = \begin{pmatrix} (\sigma^0)^\alpha{}_\beta & 0 \\ 0 & -(\sigma^0)_{\dot\alpha}{}^{\dot\beta} \end{pmatrix}$. The charge conjugation matrices C^{AB} and $(C^{-1})_{AB}$ are tensors used to raise and lower indices: $C_{AD}^{-1}(\gamma^\mu)^D{}_B \equiv (\gamma^\mu)_{AB}$ and $C^{BD}(\gamma^\mu)^A{}_D \equiv (\gamma^\mu)^{AB}$. The transpose relation which defines C^{AB} is $C_{AB}^{-1}(\gamma^\mu)^B{}_C C^{CD} = -(\gamma^{\mu T})^D{}_A$ and it implies the matrices $(\gamma^\mu)^{AB}$ and $(\gamma^\mu)_{AB}$ are symmetric.

Since we are in the Weyl representation, we can use van der Waerden notation[3,4] for spinor indices $1 \leq \alpha, \dot\alpha \leq 2$. The two linearly independent solutions to the massless Dirac equation $k\cdot\gamma u^\ell(k) = 0$ are now given by solutions of the Weyl equations

$$k_\mu \sigma^{\mu\dot\alpha}{}_\beta u^{1\beta} = 0 \qquad k_\mu \bar\sigma^{\mu\alpha}{}_{\dot\beta} u^{2\dot\beta} = 0 \qquad (14a)$$

as

$$u^{1\beta}(k) = \begin{pmatrix} k^0 + k^3 \\ k^1 + ik^2 \end{pmatrix} (k^0 + k^3)^{-\frac{1}{2}}; \qquad u^{2\dot\beta}(k) = \begin{pmatrix} -k^1 + ik^2 \\ k^0 + k^3 \end{pmatrix} (k^0 + k^3)^{-\frac{1}{2}}. \quad (14b)$$

We define two additional spinors $v^\ell(k)$ by $k \cdot \gamma v^\ell \sim u^\ell$ i.e.

$$\frac{1}{\sqrt{2}} k_\mu \bar\sigma^{\mu\alpha}{}_{\dot\beta} v^{1\dot\beta} = u^{1\alpha} \qquad \frac{1}{\sqrt{2}} k_\mu \sigma^{\mu\dot\alpha}{}_\beta v^{2\beta} = -u^{2\dot\alpha} \qquad (15a)$$

as

$$v^{1\dot\beta}(k) = \begin{pmatrix} k^0 + k^3 \\ k^1 + ik^2 \end{pmatrix} (2(k^0)^2(k^0+k^3))^{-\frac{1}{2}}; v^{2\beta}(k) = \begin{pmatrix} -k^1 + ik^2 \\ k^0 + k^3 \end{pmatrix} (2(k^0)^2(k^0+k^3))^{-\frac{1}{2}}.$$
(15b)

Note that formally $v^{1\dot\beta} = \frac{1}{\sqrt{2k^0}} u^{1\beta}$ and $v^{2\beta} = \frac{1}{\sqrt{2k^0}} u^{2\dot\beta}$. In (14,15) we have $k_\mu k^\mu = 0$. The spin decomposition of the Weyl bispinors into the two helicity states of the massless vector is as follows:

$$u^{1\alpha} u^{1\beta} = -\epsilon_\lambda^+ k_\kappa (\bar\sigma^\lambda \sigma^\kappa \sigma^2)^{\alpha\beta}$$
$$u^{2\dot\delta} u^{2\dot\gamma} = \epsilon_\lambda^- k_\kappa (\sigma^\lambda \bar\sigma^\kappa \sigma^2)^{\dot\delta\dot\gamma}. \quad (16)$$

We also find that

$$u^{1\alpha} v^{1\dot\beta} = \epsilon_\lambda^+ (\bar\sigma^\lambda \sigma^2)^{\alpha\dot\beta} \sqrt{2}; \qquad u^{2\dot\delta} v^{2\gamma} = \epsilon_\lambda^- (\sigma^\lambda \sigma^2)^{\dot\delta\gamma} \sqrt{2}$$
$$v^{1\dot\alpha} u^{1\beta} = \epsilon_\lambda^+ (\sigma^\lambda \sigma^2)^{\dot\alpha\beta} \sqrt{2}; \qquad v^{2\delta} u^{2\dot\gamma} = \epsilon_\lambda^- (\bar\sigma^\lambda \sigma^2)^{\delta\dot\gamma} \sqrt{2} \quad (17)$$

where the expressions in (17) are defined only up to a gauge transformation $\epsilon_\lambda \to \epsilon_\lambda + k_\lambda$. In the Lorentz gauges, which are defined by $k \cdot \epsilon^\pm(k) = 0$, we have that

$$i\epsilon^{\mu\nu\lambda\rho} \epsilon_\lambda^\pm k_\rho = \pm(\epsilon^{\mu\pm} k^\nu - \epsilon^{\nu\pm} k^\mu) \qquad (18)$$

holds generally for the circularly-polarized polarization vectors[5]. The expressions (16,17) satisfy (14),(15) with use of (18). We note that (16) describes only two polarizations since we can show that

$$\epsilon_\lambda^- k_\kappa (\bar\sigma^\lambda \sigma^\kappa \sigma^2)^{\alpha\beta} = 0; \qquad \epsilon_\lambda^+ k_\kappa (\sigma^\lambda \bar\sigma^\kappa \sigma^2)^{\dot\delta\dot\gamma} = 0. \qquad (19)$$

The proof of (19) is as follows:

Consider

$$\sigma_\mu \epsilon_\lambda^- k_\kappa (\bar\sigma^\lambda \sigma^\kappa \sigma^2)^{\alpha\beta} = (\epsilon_\mu^- k_\sigma - \epsilon_\sigma^- k_\mu + i\epsilon_\lambda^- k_\kappa \epsilon^{\lambda\kappa}{}_{\mu\sigma})(\sigma^\sigma \sigma^2)^{\alpha\beta} = 0 \qquad (20)$$

which for $\mu = 0$ is the left equation in (19).

The spin decomposition of the Weyl bispinors into the two spin zero states is

$$u^{1\alpha} u^{2\dot\beta} = k_\kappa (i\bar\sigma^\kappa \sigma^2)^{\alpha\dot\beta}$$
$$u^{2\dot\delta} u^{1\gamma} = k_\kappa (i\sigma^\kappa \sigma^2)^{\dot\delta\gamma}. \qquad (21)$$

TREE LEVEL NON-ABELIAN COUPLING

In this section, we discuss a set of tree amplitudes involving massless gauge boson states coming from the superstring's Ramond-Ramond sector, which exhibit a non-abelian structure. In the spirit of previous investigations [6], the gauge symmetry is enlarged by modifying the choice of the worldsheet supercurrent. Although such states are forbidden by the conventional analysis of [7], nonetheless we find their kinematic structure interesting, and suggest one possible mechanism by which they could be incorporated into an interacting string model satisfying perturbative space-time unitarity. This mechanism involves a non-hermitian piece of the internal supercurrent $\tilde{F}(z)$. The supercurrent is given by

$$F(z) = a_\mu(z)\psi^\mu(z) + \bar{F}(z) \tag{22}$$

where $0 \leq \mu \leq 3$ and we choose

$$\bar{F}(z) = \tilde{\gamma}^5 \otimes \widehat{F}(z) + \tilde{\gamma}^5 \otimes \tilde{\Gamma}^7 \otimes \tilde{F}(z) \tag{23}$$

where $\widehat{F}(z) \equiv (-\frac{i}{6})\frac{1}{\sqrt{\frac{c_\psi}{2}}} f_{abc}\psi^a(z)\psi^b(z)\psi^c(z)$ and f_{abc} given by the totally antisymmetric structure constants of $SU(2) \otimes SU(2)$ with $f_{abc}f_{abe} = c_\psi \delta_{ce}$. For $\mathcal{S}(z) \equiv f^\ell \Sigma_\ell(z)$ and $\tilde{\mathcal{S}}(z) \equiv \frac{1}{2}e^{-\frac{i\pi}{4}} f^\ell \Sigma_{\tilde{\ell}}(z)$, we have

$$\widehat{F}(z)\mathcal{S}(\zeta) = (z-\zeta)^{-\frac{3}{2}} \tilde{\mathcal{S}}(\zeta) + \ldots$$
$$\widehat{F}(z)\tilde{\mathcal{S}}(\zeta) = (z-\zeta)^{-\frac{3}{2}} \tfrac{1}{4}\mathcal{S}(\zeta) + \ldots \tag{24}$$

where we use operator products of the form

$$f_{abc}\psi^a(z)\psi^b(z)\psi^c(z)\Sigma_\ell(\zeta) = (z-\zeta)^{-\frac{3}{2}} f_{abc}\frac{1}{2\sqrt{2}}(\Gamma^a\Gamma^b\Gamma^c)_\ell^{\tilde{\ell}}\Sigma_{\tilde{\ell}}(\zeta) + \ldots.$$

$\tilde{F}(z)$ corresponds to the remaining internal degrees of freedom with $c = 6$. We assume there exists an anti-periodic supercurrent, and a pair of states in the Ramond sector of this piece of the internal system corresponding to weight zero conformal spin fields $V(z)$ and $U(z)$ such that

$$\tilde{F}(z)V(\zeta) = (z-\zeta)^{-\frac{3}{2}} U(\zeta) + \ldots$$
$$\tilde{F}(z)U(\zeta) = (z-\zeta)^{-\frac{3}{2}} (-\tfrac{c}{24})V(\zeta) + \ldots. \tag{25}$$

(25) corresponds to a non-hermitian choice for the operator $\tilde{F}_0$, that is to say non-hermitian with respect to a vector space of states with positive-definite inner product. This is because in such a space, the eigenvalues of the square of a hermitian operator are non-negative, but from (25) we see $\tilde{F}_0^2 V(0)|0\rangle = -\tfrac{1}{4}V(0)|0\rangle$, i.e., a negative eigenvalue. Note that the definition of the hermitian adjoint for any operator A, which is

$$(A\psi_a, \psi_b) = (\psi_a, A^\dagger \psi_b), \tag{26a}$$

depends on the definition of the inner product

$$(\psi_a, \psi_b), \tag{26b}$$

since (26b) is used to evaluate (26a). One example of a conformal field theory where F_0^2 can take on negative eigenvalues is the b, c, β, γ superconformal ghost system[1,8];

here F_0 is not hermitian with respect to a non-vanishing inner product, given for example by the adjoint state defined as $(c_1|0\rangle)^\dagger = \langle 0|c_{-1}c_0$. (Although F_0 is hermitian with respect to a vanishing norm, given for example by the adjoint state defined as $(c_1|0\rangle)^\dagger = \langle 0|c_{-1}$.) Another example is the combined system $(T_m,(b_1,c_1))$ with $N=1$ superconformal symmetry with $c=15$ of Berkovits and Vafa[9] representing the matter system of an $N=1$ fermionic string whose states physical states are in one-to-one correspondence with the states of the bosonic string with $c=26$.

We also remark that since the norm of a state $\tilde{F}_0|\Psi\rangle$ is given by $||\tilde{F}_0|\Psi\rangle|| = \langle\Psi|\tilde{F}_0^\dagger\tilde{F}_0|\Psi\rangle$, that therefore non-hermitian $\tilde{F}_0$ does not imply $\tilde{F}_0|\Psi\rangle$ is a negative norm state, even though $\langle\Psi|\tilde{F}_0^2|\Psi\rangle < 0$.

The states of the Ramond sector are created by conformal fields called spin fields. The spin fields are nonlocal with respect to the worldsheet supercurrent T_F because they make states in the Ramond sector of the theory, whereas superconformal fields corresponding to states in the Neveu-Schwarz sector are local with respect to T_F. In general, the operator product of a spin field with any fermionic part of a superfield is nonlocal (i.e. double-valued) since spin fields change the boundary condition on fermion fields between periodic and anti-periodic.

In the Ramond sector, the operator products of the spin fields with T_F, have a $(z-\zeta)^{-\frac{3}{2}}$ singularity; all except for the "ground states", (i.e. those for which $h=\frac{c}{24}$), which have only a $(z-\zeta)^{-\frac{1}{2}}$ singularity with T_F. Since L_0 and F_0 commute, all excited states are in pairs $S^\pm(z)|0\rangle$ related by F_0, only the "ground states" need not be paired. We have

$$|h^+\rangle = S^+(0)|0\rangle$$
$$|h^-\rangle = S^-(0)|0\rangle = F_0|h^+\rangle$$
$$(h-\tfrac{c}{24})|h^+\rangle = F_0|h^-\rangle$$
$$L_0|h^\pm\rangle = h|h^\pm\rangle$$

or in terms of the fields:

$$F(z)S^+(\zeta) = (z-\zeta)^{-\frac{3}{2}}S^-(\zeta)$$
$$F(z)S^-(\zeta) = [h-\tfrac{c}{24}](z-\zeta)^{-\frac{3}{2}}S^+(\zeta).$$

If $h=\frac{c}{24}$, global worldsheet supersymmetry is unbroken in the Ramond sector; where F_0 is the global supersymmetry charge satisfying the global supersymmetry algebra $F_0^2 = L_0 - \frac{c}{24}$. In this case the states need not come in pairs, i.e. only $|h^-\rangle$ survives since if F_0 is hermitian, then for $h=\frac{c}{24}$ we have $\langle h^-|h^-\rangle = \langle h^+|F_0^2|h^+\rangle = \langle h^+|L_0 - \frac{c}{24}|h^+\rangle = 0$, i.e. $|h^-\rangle$ is null, i.e. has zero norm.

The operator product expansions for the pair of spin fields $U(z), V(z)$ are chosen to be

$$V(z)V(\zeta) = (z-\zeta)^0 C_0(\zeta) + \ldots$$
$$U(z)U(\zeta) = (z-\zeta)^0 C'_0(\zeta) + \ldots$$
$$V(z)U(\zeta) = \ldots + (z-\zeta)^{-\frac{1}{2}} C_{-\frac{1}{2}}(\zeta) + \ldots$$
$$U(z)V(\zeta) = \ldots - (z-\zeta)^{-\frac{1}{2}} C_{-\frac{1}{2}}(\zeta) + \ldots \qquad (27)$$

where the subleading terms are less singular by integer powers of $(z-\zeta)$, and for example $C_{-\frac{1}{2}}(z)$ is some dimension $-\frac{1}{2}$ operator whose coefficient may be zero apriori, and which is understood to represent $\tilde{\gamma}^5 \otimes \tilde{\Gamma}^7 \otimes C_{-\frac{1}{2}}(z)$; and $C_0(z)$ has weight zero, etc. (27) together with (2) are consistent with locality of the string vertices given in (28).

For $\Sigma^{(1)}(z) \equiv \mathcal{S}(z)V(z)$ and $\tilde{\Sigma}^{(1)}(z) \equiv \tilde{\mathcal{S}}(z)V(z) + \mathcal{S}(z)U(z)$, BRST invariant vertex operators for the Ramond states in the canonical $q = -\frac{1}{2}$ superconformal ghost picture are given by

$$V^{(1)}_{-\frac{1}{2}}(k,\zeta) = [u^{1\alpha}(k)S_\alpha(\zeta)\Sigma^{(1)}(\zeta) - v^{1\dot\alpha}(k)S_{\dot\alpha}(\zeta)\tilde{\Sigma}^{(1)}(\zeta)]\, e^{ik\cdot X(\zeta)}\, e^{-\frac{1}{2}\phi(\zeta)} \quad (28a)$$

and for $\Sigma^{(2)}(z) \equiv \tilde{\mathcal{S}}(z)V(z)$ and $\tilde{\Sigma}^{(2)}(z) \equiv \frac{1}{4}\mathcal{S}(z)V(z) - \tilde{\mathcal{S}}(z)U(z)$,

$$V^{(2)}_{-\frac{1}{2}}(k,\zeta) = [u^{2\dot\alpha}(k)S_{\dot\alpha}(\zeta)\Sigma^{(2)}(\zeta) - v^{2\alpha}(k)S_\alpha(\zeta)\tilde{\Sigma}^{(2)}]\, e^{ik\cdot X(\zeta)}\, e^{-\frac{1}{2}\phi(\zeta)}. \quad (28b)$$

We note that within the fermion scattering amplitudes, the modified fermion emission vertex given above effectively normalizes the Ramond-Ramond bosons to 1 not k^0, i.e. the same normalization as the conventional gauge bosons found in the NS-NS sector, a required feature when both kinds of bosons are used together to form the adjoint representation of a bigger group. Consider the following amplitude. On the left-moving side we have

$$\langle 0|V^{(1)}_{-\frac{1}{2}}(k_1,z_1)\, V_{-1}(k_2,z_2,\epsilon_2)\, V^{(1)}_{-\frac{1}{2}}(k_3,z_3) c(z_1)c(z_2)c(z_3)|0\rangle$$
$$= \langle 0|[u^{1\gamma}(k_1)S_\gamma(z_1)\Sigma^{(1)}(z_1) - v^{1\dot\gamma}(k_1)S_{\dot\gamma}(z_1)\tilde{\Sigma}^{(1)}(z_1)]\, e^{ik_1\cdot X(z_1)} e^{-\frac{1}{2}\phi(z_1)}$$
$$\cdot [\epsilon_{2\mu}\psi^\mu(z_2) e^{ik_2\cdot X(z_2)}] e^{-\phi(z_2)}$$
$$\cdot [u^{1\alpha}(k_3)S_\alpha(z_3)\Sigma^{(1)}(z_3) - v^{1\dot\alpha}(k_3)S_{\dot\alpha}(z_3)\tilde{\Sigma}^{(1)}(z_3)]\, e^{ik_3\cdot X(z_3)} e^{-\frac{1}{2}\phi(z_3)}|0\rangle$$
$$\cdot \langle 0|c(z_1)c(z_2)c(z_3)|0\rangle$$
$$= -[u^{1\gamma}(k_1)\epsilon_{2\mu}\gamma^\mu_{\gamma\dot\alpha}v^{1\dot\alpha}(k_3) + v^{1\dot\gamma}(k_1)\epsilon_{2\mu}\gamma^\mu_{\dot\gamma\alpha}u^{1\alpha}(k_3)]\, \frac{1}{\sqrt{2}}$$
$$\cdot f^n f^\ell C^{-1}_{n\ell}\, \frac{1}{2} e^{-\frac{i\pi}{4}} \quad (29)$$

On the right-moving side we consider

$$\langle 0|V^{(1)}_{-\frac{1}{2}}(k_1,z_1)\, V^a_{-1}(k_2,z_2)\, V^{(1)}_{-\frac{1}{2}}(k_3,z_3) c(z_1)c(z_2)c(z_3)|0\rangle$$
$$= \langle 0|[u^{1\delta}(k_1)S_\delta(z_1)\Sigma^{(1)}(z_1)$$
$$- v^{1\dot\delta}(k_1)S_{\dot\delta}(z_1)\tilde{\Sigma}^{(1)}(z_1)]\, e^{ik_1\cdot X(z_1)} e^{-\frac{1}{2}\phi(z_1)}$$
$$\cdot [\tilde{\gamma}^5 \otimes \psi^a(z_2) e^{ik_2\cdot X(z_2)}] e^{-\phi(z_2)}$$
$$\cdot [u^{1\beta}(k_3)S_\beta(z_3)\Sigma^{(1)}(z_3)$$
$$- v^{1\dot\beta}(k_3)S_{\dot\beta}(z_3)\tilde{\Sigma}^{(1)}(z_3)]\, e^{ik_3\cdot X(z_3)} e^{-\frac{1}{2}\phi(z_3)}|0\rangle$$
$$\cdot \langle 0|c(z_1)c(z_2)c(z_3)|0\rangle$$
$$= u^{1\delta}(k_1) C^{-1}_{\delta\beta} u^{1\beta}(k_3) f^k f^m (-\alpha^a_{km}) \frac{1}{\sqrt{2}} \quad (30)$$

where in (30) we have assumed $C'_0 = \frac{i}{4} C_0$. Combining the left and right-moving pieces (29) and (30) we find up to an overall normalization constant

$$A_3 = if^k f^n C_{n\ell}^{-1} f^\ell f^m \alpha_{mk}^a \frac{1}{2}$$
$$\cdot \epsilon_{2\mu}^-[u^{1\delta}(k_1)u^{1\gamma}(k_1)(-i\sigma^2\bar{\sigma}^\mu)_{\gamma\dot{\alpha}} v^{1\dot{\alpha}}(k_3)u^{1\beta}(k_3)(i\sigma^2)_{\beta\delta}\frac{1}{\sqrt{2}}$$
$$+ u^{1\delta}(k_1)v^{1\dot{\gamma}}(k_1)(-i\sigma^2\sigma^\mu)_{\dot{\gamma}\alpha} u^{1\alpha}(k_3)u^{1\beta}(k_3)(i\sigma^2)_{\beta\delta}\frac{1}{\sqrt{2}}]$$
$$= if^k f^n C_{n\ell}^{-1} f^\ell f^m \alpha_{mk}^a \frac{1}{2}$$
$$\cdot \text{tr}[(-\epsilon_{1\rho}^+ k_{1\sigma}\bar{\sigma}^\rho\sigma^\sigma\sigma^2)(-i\sigma^2\bar{\sigma}^\mu)(\epsilon_{3\lambda}^+\sigma^\lambda\sigma^2)(i\sigma^2)$$
$$+ (\epsilon_{1\rho}^+\bar{\sigma}^\rho\sigma^2)(-i\sigma^2\sigma^\mu)(-\epsilon_{3\lambda}^+ k_{3\kappa}\bar{\sigma}^\lambda\sigma^\kappa\sigma^2)(i\sigma^2)]$$
$$= if_{IaJ}[\epsilon_1^+\cdot\epsilon_2^- \epsilon_3^+\cdot k_1 + \epsilon_2^-\cdot\epsilon_3^+ \epsilon_1^+\cdot k_2] \quad (31)$$

which is the three-gluon tree coupling for this set of polarizations, as $\epsilon_1^+\cdot\epsilon_3^+ \epsilon_2^-\cdot k_3 = 0$. The structure constants $f_{IaJ} = 2\,\text{tr}(M_I^\dagger M_J \alpha^a)$ form the (2,2,2,2) representation of $SU(2)^4$ and suggest that the symmetry group is enhanced to $SO(8)$ from its symmetric subgroup $SU(2)^4$.

PERTURBATIVE UNITARITY

To describe consistent interacting strings, one must in general check that all the scattering amplitudes are modular invariant, finite, and unitary. A guide to this program is the calculation of the partition function, i.e. the one-loop cosmological constant Λ, which can be checked for modular invariance, albeit a quantity equal to zero. For closed strings, the one-loop cosmological constant is defined by

$$\Lambda \equiv \tfrac{1}{2}\text{tr}\ln\Delta^{-1} \quad (32)$$

where

$$\Delta^{-1} = \alpha'(p^2 + m^2); \quad \tfrac{1}{2}\alpha'm^2 = \alpha'm_L^2 + \alpha'm_R^2 \quad (33)$$

so in D space-time dimensions, for $\omega = e^{2\pi i\tau}$,

$$\Lambda = -\tfrac{1}{2}(2\pi)^{-1}(\alpha')^{-\frac{D}{2}}\int_F d^2\tau (Im\tau)^{-2-\frac{1}{2}(D-2)}$$
$$\cdot \sum_{\text{all sectors}} \text{tr}[\bar{\omega}^{\alpha'm_L^2}\omega^{\alpha'm_R^2}\text{possible projections}]. \quad (34)$$

F is a fundamental region of the modular group: $\tfrac{1}{2} \leq Re\tau \leq \tfrac{1}{2}$; $|\tau| > 1$. The unitary $D = 4$, $N = 8$ superstring model considered in [10], which incorporates (2b) as *part* of the internal field theory, has its one-loop modular invariant partition function given by

$$\Lambda = -(4\pi\alpha'^2)^{-1}\int_F d^2\tau (Im\tau)^{-3}|f(\omega)|^{-12}|\omega|^{-\frac{1}{2}}$$
$$\cdot \tfrac{1}{4}[\theta_3^4 - \theta_4^4 - \theta_2^4][\bar{\theta}_3^4 - \bar{\theta}_4^4 - \bar{\theta}_2^4]$$
$$\cdot [\tfrac{1}{2}(|\theta_3|^{12} + |\theta_4|^{12} + |\theta_2|^{12})|f(\omega)|^{-12}|\omega|^{-\frac{1}{2}}]. \quad (35)$$

For the $D = 4$, $N = 8$ model whose couplings are considered in (31), the partition function computed for the degrees of freedom denoted by a_μ, ψ^μ and ψ^a with $\alpha'm_L^2 = L_0 - \frac{c}{24}$, etc. is

$$\Lambda = -(4\pi\alpha'^2)^{-1} \int_F d^2\tau (Im\tau)^{-3} |f(\omega)|^{-12} |\omega|^{-\frac{1}{2}}$$
$$\cdot \tfrac{1}{4}[\theta_3^4 - \theta_4^4 - \theta_2^4][\bar\theta_3^4 - \bar\theta_4^4 - \bar\theta_2^4]. \tag{36}$$

We see (36) is already modular invariant using the transformation properties under $SL(2, I)$ given by

$$\tau \to \tau + 1 : \quad \theta_3 \to \theta_4;\ \theta_4 \to \theta_3;\ \theta_2 \to e^{i\frac{\pi}{4}}\theta_2\,;\ Im\tau \to Im\tau$$
$$\tau \to -\frac{1}{\tau} : \quad \theta_2 \to (-i\tau)^{\frac{1}{2}}\theta_4;\ \theta_4 \to (-i\tau)^{\frac{1}{2}}\theta_2;\ \theta_3 \to (-i\tau)^{\frac{1}{2}}\theta_3;\ Im\tau \to |\tau|^{-2}Im\tau;$$
$$\omega^{\frac{1}{24}} f(\omega) \to (-i\tau)^{\frac{1}{2}} \omega^{\frac{1}{24}} f(\omega) \tag{37}$$

where $f(\omega) = \prod_{n=1}(1-\omega)$ is related to the Dedekind eta function $\eta(\omega) = \omega^{\frac{1}{24}} f(\omega)$, $\omega = e^{2\pi i\tau}$ and $\theta_i(0|\tau)$ are the Jacobi theta functions. We suggest that the inclusion of the remaining internal conformal field theory will leave the modular invariance of (36) unchanged.

CONCLUSIONS

We have developed a formalism for studying four-dimensional massless spin fields, and computed their BRST invariant tree amplitudes. The physical states in the $q = -\frac{1}{2}$ ghost picture, given by say (28a),

$$V_{-\frac{1}{2}}^{(1)}(k,0)|0\rangle \tag{38}$$

satisfy, for F_0 From (23):

$$F_0 V_{-\frac{1}{2}}^{(1)}(k,0)|0\rangle = 0, \tag{39}$$

which is the physical state condition in the old covariant formalism. We can either check (39) explicitly, or recall that it follows from the BRST invariance of (38). These states have the additional feature that

$$F_0^{s.t.} V_{-\frac{1}{2}}^{(1)}(k,0)|0\rangle = -u^{1\alpha}(k) S_\alpha(0) \tilde\Sigma(0) e^{ik\cdot X(0)} |0\rangle \neq 0 \tag{40a}$$
$$\bar F_0\, V_{-\frac{1}{2}}^{(1)}(k,0)|0\rangle = u^{1\alpha}(k) S_\alpha(0) \tilde\Sigma(0) e^{ik\cdot X(0)} |0\rangle \neq 0, \tag{40b}$$

where $F^{s.t.}(z) \equiv a_\mu(z)\psi^\mu(z)$, even though the sum $F_0 = F_0^{s.t.} + \bar F_0$ does annihiliate the states as is shown in (39). Of course (40a) still describes massless states since although $k\cdot\gamma v = u \neq 0$, we have

$$-m^2 v = k^2 v = k\cdot\gamma u = 0, \tag{41}$$

so that

$$(F_0^{s.t.})^2 V_{-\frac{1}{2}}^{(1)}(k,0)|0\rangle = 0. \tag{42}$$

Now, $F_0 V^{(1)}_{-\frac{1}{2}}(k,0)|0\rangle = 0$ implies $F_0^2 V^{(1)}_{-\frac{1}{2}}(k,0)|0\rangle = 0$. Using $\{F_0^{s.t.}, \bar{F}_0\} = 0$ and (42), we find

$$F_0^2 V^{(1)}_{-\frac{1}{2}}(k,0)|0\rangle = ((F_0^{s.t.})^2 + \bar{F}_0^2) V^{(1)}_{-\frac{1}{2}}(k,0)|0\rangle = 0, \quad (43)$$

and thus

$$\bar{F}_0^2 V^{(1)}_{-\frac{1}{2}}(k,0)|0\rangle = 0. \quad (44)$$

So we have $\bar{F}_0^2 V^{(1)}_{-\frac{1}{2}}(k,0)|0\rangle = 0$, but from (40b) that $\bar{F}_0 V^{(1)}_{-\frac{1}{2}}(k,0)|0\rangle \neq 0$. This is consistent with the observation that $\tilde{F}_0$ which forms part of $\bar{F}_0$ is not hermitian. (40b) indicates that global worldsheet supersymmetry generated by the charge $\bar{F}_0$ is broken in the Ramond sector. Tree level amplitudes with other polarizations, and consistency of the one-loop amplitudes are being investigated[11].

The mechanism we have suggested yields a string model with the gauge group derived by enhancing the symmetric subgroup $SU(2)^4$ to $SO(8)$, which although providing non-abelian Ramond-Ramond vector mesons, does not contain a realistic theory. We propose a similar mechanism can be used to generate the gauge symmetry $SU(3) \otimes SU(2) \otimes U(1)$ from a symmetric subgroup such as $SU(2)^2 \otimes U(1)^2$.

REFERENCES

1. D. Friedan, E. Martinec, and S. Shenker, Nucl. Phys. **B271** (1986) 93.

2. L. Dolan, P. Goddard and P. Montague, hep-th/9410029 preprint.

3. B. L. van der Waerden, *Group Theory and Quantum Mechanics*. New York: Springer Verlag, (1974).

4. J. Wess and J. Bagger, *Supersymmetry and Supergravity*. Princeton: Princeton University Press (1992).

5. L. Dolan and S. Horvath, preprint hep-th/9503210.

6. R. Bluhm, L. Dolan and P. Goddard, Nucl. Phys. **B289** 364 (1987).

7. L. Dixon, V. Kaplunovsky and C. Vafa, Nucl. Phys. **B294** (1987) 43.

8. D. Lust and S. Theisen, *Lectures on String Theory*. New York: Springer-Verlag, 1989.

9. N. Berkovits and C. Vafa, Mod. Phys. Lett. **A9** (1994) 653: hep-th/9310170.

10. R. Bluhm, L. Dolan and P. Goddard, Nucl. Phys. **B338** (1990) 529.

11. L. Dolan and S. Horvath, in preparation.

DYNAMICAL SUPERSYMMETRY BREAKING: SOME RECENT DEVELOPMENTS

Michael Dine

Santa Cruz Institute for Particle Physics
University of California
Santa Cruz, CA 95064

ABSTRACT

Recent work has made clear that we have far more control over the dynamics of supersymmetric than non-supersymmetric theories. Here, I discuss some recent developments in dynamical supersymmetry breaking both in ordinary field theory and in string theory. I briefly mention some new examples of theories (developed with Ann Nelson) with dynamical supersymmetry breaking, and describe some features which may be phenomenologically relevant. In string theory, I explain how one can show, in some circumstances, that stringy non-perturbative effects are *smaller* than effects visible in the low energy field theory. Observations of this sort suggest a general approach to string phenomenology. I also describe some string models (developed with Tom Banks) in which one can show that perturbatively and non-perturbatively, at least some of the vacuum degeneracy is not lifted.

INTRODUCTION

Over the last year or so, the beautiful work of Seiberg and collaborators[1,2] and Seiberg and Witten,[3,4] has brought home the point that supersymmetric theories are far more tractable than more conventional field theories. This fact has even been noted in the *New York Times Week In Review*, which devoted a recent article to the subject.[5] The main tool in these analyses consists in understanding the constraints imposed by supersymmetry on the low energy effective theory. In the past, these ideas have been applied to problems of dynamical supersymmetry breaking, both in field theory and in string theory. These are the issues I will focus on in this note. The dramatic new development is the application of this tool to explore a whole set of previously inaccessible problems in field theory, including aspects of quark confinement, weak and strong coupling duality, and certain aspects of the moduli spaces of strongly coupled supersymmetric theories, and, most startling in my view, the discovery of non-trivial duality relations between quantum field theories,[3,4] as well as non-trivial conformal field theories in four dimensions.[2] While I won't have time to even touch on most of these developments, the message which I hope to convey in this talk is that we do understand a great deal about these theories. Indeed, even in string theory, where we barely know what the theory is, we can say a great deal!

In my talk at the Coral Gables Meeting, I presented material quite similar to that in a talk I gave at the Fourth International Conference on Physics Beyond the Standard Model.[6] Rather than repeat all of that material here, I would like to focus on the recent developments I alluded to above. In section 2, after briefly reviewing (as in my earlier lecture) some essential features of supersymmetry breaking, I will list some new models with dynamical supersymmetry breaking, and explain there possible relevance to model building. This represents work in progress with Ann Nelson and Yuri Shirman. I believe it is essentially correct, but the reader is warned that the discussion may not be entirely error free. In section 3, I will discuss some issues in string theory. Again, after recapitulating some aspects of strong coupled field theory, I will turn to some new developments. In particular, I will show that in many instances in string theory, it is possible to show that there are no interesting stringy non-perturbative contributions to the superpotential; supersymmetry breaking in these models must be understood entirely within the low energy effective field theory of the light particles. As a result, in some instances, one can argue that there are exact string moduli, even at the non-perturbative level. It should also be possible to make definite statements about supersymmetry breaking in many cases.

SUPERSYMMETRY AND SUPERSYMMETRY BREAKING

Two themes run through any discussion of supersymmetry and supersymmetry breaking:

a. Flat directions: many supersymmetric models possess directions in field space in which the energy vanishes classically. The analysis of the dynamics in these directions is often very simple.

b. Holomorphy in the fields: the low energy effective theory (in both global and local supersymmetry) is described by three functions of the chiral fields: the superpotential, W, the gauge coupling function, f, and the Kahler potential, K. The crucial point is that W and f are "holomorphic" (this means, in essence, analytic) functions of the chiral fields: $W = W(\phi), f = f(\phi)$. Their form is thus highly restricted. On the other hand, K is essentially unrestricted, $K = K(\phi, \phi^\dagger)$.

c. Holomorphy in the couplings: Much of the recent progress has followed from the realization that the couplings of a supersymmetric theory may themselves be viewed as expectation values of chiral fields. Thus the superpotential and gauge coupling functions are holomorphic functions of these parameters.[7,8,9] Thus W and f (but not K) are holomorphic functions of the couplings as well.

These ideas are illustrated by a simple model: an SU(2) gauge theory with one "flavor," i.e., a doublet and antidoublet, Q and $\bar{Q}$. We will first suppose that the classical superpotential vanishes This might occur, for example, if the underlying theory possessed a suitable discrete symmetry. In this case, the scalar potential is given by (here and in what follows we denote the scalar components of the chiral field by the same letter as the chiral field itself)

$$V = \tfrac{1}{2}g^2(Q^*T^aQ - \bar{Q}T^a\bar{Q}^*)^2. \tag{1}$$

If we take the expectation values of Q and $\bar{Q}$

$$Q = \bar{Q} = \begin{pmatrix} v \\ 0 \end{pmatrix}, \qquad (2)$$

then the scalar potential vanishes (for any choice of v). This means that classically the theory possesses a continuous set of physically inequivalent ground states. In each of these states, the gauge symmetry is completely broken and the spectrum consists of a massive gauge multiplet (the massive gauge bosons and their superpartners). In addition there is one massless state. The imaginary (CP-odd) part of this field is a conventional Goldstone boson; the real part arises because it costs no energy to change the expectation value v. This massless state can be written in terms of a gauge invariant field, $\Phi = \bar{Q}Q$. Expanding about the expectation values of the fields:

$$\Phi = \bar{Q}Q = v^2 + v(\delta\bar{Q} + \delta Q) + \ldots . \qquad (3)$$

The effective coupling describing the low energy theory depends on v; it is basically $g^2(M_V)$, since the mass of the gauge bosons provides a lower cutoff on all Feynman integrals. By taking v large we can make the coupling arbitrarily small.

This theory has one non-anomalous global symmetry. This symmetry is an "R" symmetry, which means that it does not commute with supersymmetry. The supercharges rotate by a phase under this transformation,

$$Q_\alpha \to e^{i\omega} Q_\alpha. \qquad (4)$$

The scalar components of the fields Q and $\bar{Q}$ transform as

$$Q \to e^{-i\omega} Q \qquad \bar{Q} \to e^{-i\omega} \bar{Q}. \qquad (5)$$

The fermionic components have charge -2 under the symmetry.

It is crucial to what follows that, because supersymmetry is unbroken classically, the low energy effective action describing the light field will be supersymmetric, even if supersymmetry is spontaneously broken. Heuristically, this follows from the fact that supersymmetry breaking requires the existence of a massless fermion, in the case of global supersymmetry, and this must arise (at weak coupling) from the classically massless fields.[10] In the case of supergravity, it follows from the fact that, if supersymmetry is unbroken at tree level there is a massless spin-$\frac{3}{2}$ field. By general theorems, consistent coupling of such a field requires supersymmetry. As for any supersymmetric theory, then, the goal is to determine the superpotential, $W(\Phi)$, as well as the Kahler potential and the gauge coupling function, after integrating out the massive fields.

The effective lagrangian is constrained by the symmetries of the underlying theory, as well as by the holomorphy of the superpotential. In particular, it will respect the $U(1)_R$ of the underlying theory. Under an R symmetry, the superpotential transforms as

$$W \to e^{2i\omega} W. \qquad (6)$$

This fixes the forms of the superpotential *uniquely*:[11]

$$W_{np} = c \frac{\Lambda^5}{\bar{Q}Q} = c \frac{\Lambda^5}{\Phi}. \qquad (7)$$

Here

$$\Lambda = e^{-8\pi^2/b_o g^2} \qquad (8)$$

and c is a constant of order one. Λ is the non-perturbative scale of the $SU(2)$ theory.

Even before attempting to evaluate the constant c, this is an extraordinary result. First, it implies that there is no renormalization of the superpotential in perturbation theory. Even more than that, it guarantees that non-perturbative corrections are not random, but are highly constrained. In fact, the constant c can be obtained from a completely straightforward instanton computation, and shown to be non-zero.[12] This result means that supersymmetry is broken, but not in precisely the way one might like. The potential corresponding to Eq. (7) falls to zero at infinity – the region of weak coupling – while blowing up in the region of strong coupling, where one does not have control of the computation. This sort of runaway behavior is not uncommon, and will typify the situation in string theory.

One might hope to obtain a theory with a "nice vacuum" by adding a mass term, $m\bar{Q}Q$ so that the potential would rise at infinity. However, in this case, there are N *supersymmetric* ground states, in accord with the Witten index theorem.[13] This can be seen by taking

$$W = mQ\bar{Q} + W_{np} \qquad (9)$$

and solving

$$\frac{\partial W}{\partial Q} = \frac{\partial W}{\partial \bar{Q}} = 0.$$

What sorts of models exhibit supersymmetry breaking with a stable ground state? The following two conditions appear to be sufficient:
  a. The classical theory should possess no flat directions.
  b. The theory should possess a global symmetry, which is spontaneously broken.

It is easy to see why one expects supersymmetry breaking under these circumstances: Suppose supersymmetry were unbroken. Then, the Goldstone boson would be accompanied by a (scalar) superpartner. However, because there is no potential for the Goldstone boson, there is no potential for the superpartner. This implies the existence of a flat direction, which violates the original assumption (unless the partner is itself a Goldstone boson, or the flat direction is compact). This argument is rather heuristic, but in many cases one can calculate the non-perturbative potential, and show that this "theorem" is satisfied. The simplest known such model has gauge group $SU(3) \times SU(2)$, and a set of chiral fields similar to that of one generation of the standard model, without the positron:[14]

$$Q\,(3,2) \qquad \bar{u}\,(\bar{3},1) \qquad \bar{d}\,(\bar{3},1) \qquad L\,(1,2). \qquad (10)$$

At the classical level, one also adds the most general superpotential allowed by the gauge symmetries:

$$W = \lambda Q L \bar{d}. \qquad (11)$$

It is not hard to check that this model has no flat directions. It does possess two global $U(1)$ symmetries. If λ and the $SU(2)$ gauge coupling are small, then the theory reduces to $SU(3)$ gauge theory with two flavors, a theory in which instantons generate a superpotential. Adding this to the classical superpotential, one finds that the true minimum has broken supersymmetry. The vacuum energy, spectrum, and other features of the theory can then be computed systematically.[14,15]

Until recently, only a few such models were known. In the last few months, however, many additional models exhibiting dynamical supersymmetry breaking have been discovered.
  a. Intriligator, Seiberg and Shenker, exploiting recent developments in understanding the moduli spaces of supersymmetric theories, have discovered a very simple

model, $SU(2)$ gauge theory with a single chiral field in the spin 3/2 representation, which breaks supersymmetry. This model has a number of features which are different than previously known models. For example, in the limit that the classical superpotential is set to zero, there are exact flat directions even non-perturbatively.[16]

b. Ann Nelson, Yuri Shirman and I have constructed (inspired by a suggested of Poppitz and Randall) many new models with dynamical supersymmetry breaking.

These latter models will be described in some detail in a forthcoming paper,[17] but let me mention a few features here. In ref. 11, an infinite set of models which break supersymmetry was described. These were models with gauge group $SU(N+4)$, where N was odd, and with N chiral fields, $\bar{F}^a$, in the antifundamental representation and one in the antisymmetric tensor representation. Adding the most general superpotential,

$$W = \lambda_{ab} A \bar{F}^a \bar{F}^b \qquad (12)$$

led to a model without flat directions and with a non-anomalous R symmetry. One strategy for constructing generalizations of these models is to take a particular one, and simply discard some of the gauge multiplets while keeping the chiral multiplets. One might then add the most general superpotential allowed in the reduced theory. This procedure is guaranteed to yield chiral models which are free of anomalies. As we will see, the resulting theories often possess non-anomalous R symmetries, and also have no flat directions.

The simplest such model is given by the case $N = 1$, i.e. an $SU(5)$ theory with a $\bar{5}$ and 10. In this case, the superpotential vanishes. One can now modify this theory by taking the gauge group to be the $SU(3) \times SU(2)$ subgroup. Under this group, the $\bar{5}$ and 10 decompose as a $(3,2)$, two $(\bar{3},1)$'s, and a $(1,2)$. If we add the most general superpotential, we obtain the 3 − 2 model of dynamical supersymmetry breaking. We obtain something new if we retain instead an $SU(4) \times U(1)$ subgroup, where the $U(1)$ generator is

$$Y = diag(1,1,1,1,-4). \qquad (13)$$

Now the ten decomposes as an antisymmetric tensor, A_2, (the subscript indicates the $U(1)$ charge), a fundamental, 4_{-3}, a $\bar{4}_{-1}$ and a singlet, S_4. The most general allowed superpotential is

$$W = \lambda S_4 \bar{4}_{-1} 4_{-3}. \qquad (14)$$

With the superpotential, it is easy to show that there is no flat direction. To see this note first that the most general flat direction of the $SU(4)$ D term has the form

$$A = \begin{pmatrix} a\sigma_2 & 0 \\ 0 & a\sigma_2 \end{pmatrix} \quad 4 = \bar{4} = \begin{pmatrix} b \\ 0 \\ 0 \\ 0 \end{pmatrix} \quad S = c. \qquad (15)$$

The $U(1)$ D term requires

$$2|a^2| + 4|c^2| - 4|b|^2 = 0. \qquad (16)$$

But combined with the vanishing of the F terms, one sees that $a = b = c = 0$. In addition to the absence of flat directions, this model also possesses a non-anomalous R symmetry. So one expects that supersymmetry is broken.

To see this in detail, we can ask about the form of the non-perturbative superpotential, in the limit that the classical superpotential vanishes. There is, in fact, a unique superpotential consistent with the symmetries. Call

$$\mathcal{O} = \bar{4}_i 4^j A^{ik} A^{lm} \epsilon_{jklm}. \tag{17}$$

Then

$$W_{np} = \frac{\Lambda_4^5}{\sqrt{\mathcal{O}}}. \tag{18}$$

where Λ_4 is the scale of the $SU(4)$ theory.

To see how this term is generated, suppose that the coupling, λ is small compared to all of the gauge couplings. In this case, we might expect the system to sit far out in the flat direction of the D terms,

$$b = c \gg a. \tag{19}$$

In this direction, $SU(4) \times U(1)$ is broken to $SU(3) \times U(1)$. In the low energy theory, there is one light 3 and $\bar{3}$, i.e. one has supersymmetric QCD with one flavor. This theory is well-known to develop a superpotential,

$$W_{np} = \frac{\Lambda_3^4}{\sqrt{q\bar{q}}} \tag{20}$$

where Λ_3 is the scale of the $SU(3)$ theory. It is not hard to see that this corresponds precisely to the superpotential above. For example, $\Lambda_3^3 = \Lambda_4^5/b$, so that numerically the superpotentials coincide. In addition, if the $U(1)$ coupling is small, $\lambda \ll g_1 \ll 1$, the low energy theory has approximate flat directions in which $SU(3)$ is broken to $SU(2)$; gluino condensation than generates the required superpotential. Alternatively, one can consider the hierarchy, $g_a \ll \lambda \ll 1$. In this case, one expects $a \gg b$. Then at the first stage, the gauge symmetry is broken to $Sp(4) \approx SO(5)$, with two 5's. Again, the appropriate superpotential is generated.

This model has many generalizations. For example, there are a set of models with gauge group $SU(n) \times U(1)$ (n even). Start with the theory with gauge group $SU(n+1)$, an antisymmetric tensor and $n-3$ antifundamentals. Throw out those generators of $SU(n+1)$ which do not lie in an $SU(n) \times U(1)$ subgroup, where the $U(1)$ generator is $\tilde{T} = diag(1, 1, \ldots, 1, -n)$. The original chiral fields decompose as

$$A_2 \quad n_{1-n} \quad (n-3)\bar{n}_{-1} \quad (n-3)S_n. \tag{21}$$

(Here A is an antisymmetric tensor and S_n a singlet of the $SU(n)$.) At the classical level, one can add to this model a superpotential,

$$W = \Gamma_{ab} A \bar{n}^a \bar{n}^b + \lambda_{ab} n \bar{n}^a S^b. \tag{22}$$

It is not hard to check that for general matrices Γ and λ, there are no flat directions; on the other hand, there is a non-anomalous R symmetry, and supersymmetry is broken.

There are infinite variations which one can make on this construction. For example, start with the $SU(N)$ theory and break to $SU(N-2) \times SU(2) \times U(1)$. In 17, we will present other sets of models in which supersymmetry is only broken when one includes non-renormalizable terms in the superpotential.

These theories open up many new possibilities for model building. For example, one of the difficulties in the phenomenological models of ref. 18was the appearance of Fayet-Iliopoulos D terms for the $U(1)$ responsible for feeding supersymmetry breaking to "ordinary fields" ("messenger $U(1)$"). With these new theories, it is possible to forbid such D terms by discrete symmetries. The models with supersymmetry broken in the presence of non-renormalizable operators offer the possibility that the supersymmetry breaking is at a higher energy scale than in earlier models. These possibilities are currently being explored.

STRONGLY COUPLED STRING THEORY

As we have explained, two themes run through any discussion of dynamical supersymmetry breaking: flat directions and holomorphy. These permit one to make powerful statements about the non-perturbative dynamics of the theory, even in some cases at strong coupling. String theories are an obvious arena in which to explore these ideas. String theories are notorious for flat directions. Moreover, in string theory not only is non-perturbative physics difficult – we do not even possess a non-perturbative definition of the theory! As we will see, the methods we have described earlier permit one to make powerful statements about these theories. They lead, for example, to a two line argument for the finiteness of string theory.[19] More strikingly, they permit us in many instances to place strong bounds on the size of any inherently stringy non-perturbative effects. Perhaps most important of all, they suggest how, even if string theory is strongly coupled, it may yet make contact with reality. The most simple and dramatic prediction of this framework is that – even at strong coupling – the theory, at low energies, should look like a supersymmetric theory with small, explicit soft breakings.

One of the disturbing features of string theory is that if it does describe nature, one can argue (though not quite prove) that it must be strongly coupled.[20] This is troubling, for many reasons, not least of all that we do not know what the non-perturbative theory is.

We have hinted above that the constraints imposed by supersymmetry itself may provide a way out of these dilemmas. To understand why holomorphy is such a powerful tool in dealing with string theory, it is necessary remember that the dimensionless coupling constant of string theory may itself be thought of as the expectation value of a dynamical scalar field. In four dimensional string models which preserve $N = 1$ supersymmetry, this field is a component of a chiral superfield, usually denoted by the letter S:

$$S = \frac{8\pi^2}{g^2} + D + ia + \ldots . \qquad (23)$$

Here D represents the fluctuating part of the dilaton field. a is an axion, which couples universally to the $F\tilde{F}$ of each gauge group. (The decay constant of this axion is of order the M_P.) g here represents the gauge coupling at the string scale. In the real world, then, one might imagine that S is a number of order 300 or so.

In string perturbation theory, it is not hard to show that the axion is truly an axion, i.e., the theory is symmetric under shifts[21]

$$a \to a + \delta \qquad (24)$$

or

$$S \to S + i\delta. \qquad (25)$$

On the other hand, the superpotential must be a holomorphic function of S. This means that the superpotential is independent of S, i.e., the superpotential is not renormalized![19]

Since the superpotential is independent of S, we have also learned that the theory possesses, classically and in perturbation theory, an exact flat direction; there are a continuum of physically inequivalent ground states, characterized by different values of the coupling. As we have mentioned, in string theory the degeneracy is typically much larger, extending to other "moduli." In what follows, we will focus, for simplicity, on the dilaton; all of our remarks are readily extended to these other moduli.

Just as in field theory, we want to investigate what happens beyond perturbation theory. In particular, as in field theory, we expect that the Peccei-Quinn symmetry is broken, and that the non-renormalization theorem may break down. All existing work on this problem involves examining the low energy effective theory for effects which might break supersymmetry. One sort of breakdown which has been widely discussed in the literature is "gluino condensation in a hidden sector."[22] The point here is that if one has a pure gauge sector of the theory (no chiral matter) with gauge group, say, SU(N), then condensation of the gluinos of this sector gives rise to a superpotential which behaves as

$$W(S) \sim e^{-S/N}. \tag{26}$$

Before dealing with any phenomenological issues, a natural question to ask is whether even for weak coupling (large S), gluino condensation (we use this term as a code word for any supersymmetry-breaking effects visible in the low energy effective theory) is the largest non-perturbative effect. After all, who says that there can't be much larger effects (behaving, say, as $e^{-1/g}$), arising from integrating out massive string modes?

In [6], I discussed how discrete symmetries of string theory[23] can be used to show that at weak coupling, gluino condensation gives a larger contribution to the superpotential than any stringy non-perturbative effect.[24] Here, I would like to briefly describe a different approach to this problem, currently being explored with Tom Banks. The idea is to consider theories in which there is a $U(1)$ for which a Fayet-Iliopoulos term is generated.[25] As an example, consider the O(32) heterotic string compactified on the familiar Calabi-Yau theory based on the quintic in CP^4. The low energy gauge group is then $O(26) \times U(1)$. Before any modding out, there are 101 26's with charge $+1$, ϕ^{+1}, paired with 101 singlets, N^{-2}, of charge -2. There is also one 26 and of charge -1 and a singlet of charge 2. In order to cancel the Fayet-Iliopoulos term, some singlet of charge -2 must obtain a vev. In addition, the dilaton multiplet transforms under the symmetry. In the present case, e^{-S} behaves as a field of charge -400. If one considers now the effective lagrangian at scales a little below the string scale and above $<N^{-2}>$, one sees that there can be no term in the superpotential involving S alone (times vev's of $<N>$). So non-perturbative stringy effects cannot break supersymmetry at all! In order to decide what happens in this theory, one needs, then, to understand the low energy dynamics. This can be done along the lines of section 2. In the present example, one can show that supersymmetry is broken at generic points in the moduli space, but that it is restored at points of high symmetry: some of the moduli are truly moduli even non-perturbatively! Other models can be analyzed similarly.

One would like to apply this eventually to some sort of string phenomenology. In our present state, we still will need to understand what stabilizes the dilaton. The ideas discussed in ref. 24 and in 6 still probably represent the best we can presently do in thinking about how supersymmetry might be broken in string theory in the real world.

CONCLUSIONS

Theories with low energy supersymmetry are astoundingly tractable. The examples discussed here represent only a small sample of the questions which have been attacked by these methods. It is almost certain that they further shed light on many issues, both in string theory and field theory. Perhaps they will yet provide some insight into some of the hardest questions we face, such as the problem of the cosmological constant and the existence of a strongly coupled string ground state.

REFERENCES

1. N. Seiberg, *Phys. Rev.* **D49** (1994); K. Intriligator, R.G. Leigh and N. Seiberg, *Phys. Rev.* **D50** (1994) 1092; K. Intriligator and N. Seiberg *Nucl. Phys.* **B431** (1994) 551.
2. N. Seiberg, *Nucl. Phys.* **B345** (1995) 129.
3. N. Seiberg and E. Witten, *Nucl. Phys.* **B426** (1994) 19.
4. N. Seiberg and E. Witten, *Nucl. Phys.* **B431** (1994) 484.
5. G. Johnson, *New York Times, Week In Review*, Dec. 10, 1994, p. 3.
6. M. Dine, SCIPP 95/11 (1995).
7. M.A. Shifman and A.I. Vainshtein, *Nucl. Phys.* **B277** (1986) 456 and *Nucl. Phys.* **B359** (1991) 571.
8. D. Amati, K. Konishi, Y. Meurice, G.C. Rossi and G. Veneziano, *Phys. Rep.* **162** (1988) 169.
9. N. Seiberg, *Phys. Lett.* **B318** (1993) 469.
10. E. Witten, *Nucl. Phys.* **B188** (1981) 513.
11. I. Affleck, M. Dine and N. Seiberg, *Nucl. Phys.* **B241** (1984) 493.
12. S. Cordes, *Nucl. Phys.* **B273** (1986) 629; D. Finnell and P. Pouliot, SLAC-PUB-95-6768.
13. E. Witten, *Nucl. Phys.* **B202** (1982) 253.
14. I. Affleck, M. Dine and N. Seiberg, *Nucl. Phys.* **B256** (1985) 557.
15. J. Bagger, E. Poppitz and L. Randall, *Nucl. Phys.* **B426** (1994) 3.
16. K. Intriligator, N. Seiberg, and S.H. Shenker, RU-94-75 (1994), hep-ph-9410203.
17. M. Dine, A. Nelson and Y. Shirman, to appear.
18. M. Dine, A. Nelson and Y. Shirman, SCIPP-94-21 (1994), hep-ph-9408384.
19. M. Dine and N. Seiberg, *Phys. Rev. Lett.* **57** (1986) 2625.
20. M. Dine and N. Seiberg, *Phys. Lett.* **162B** (299) 1985 and in *Unified String Theories*, M. Green and D. Gross, Eds. (World Scientific, 1986).
21. M. Green, J. Schwarz and E. Witten, *Superstring Theory*, Cambridge University Press, Cambridge (1987).
22. J.P. Derendinger, L.E. Ibanez and H.P. Nilles, *Phys. Lett.* **155B** (1985) 65; M. Dine, R. Rohm, N. Seiberg and E. Witten, *Phys. Lett.* **156B** (1985) 55.
23. T. Banks and M. Dine, *Phys. Rev.* **D45** (1992) 424; L. Ibanez and G.G. Ross, *Phys. Lett.* **260B** (1991) 291; *Nucl. Phys.* **368** (1992) 3.
24. T. Banks and M. Dine, RU-50-94 (1994), hep-th-9406132.
25. M. Dine, N. Seiberg and E. Witten, *Nucl. Phys.* **B289** (1987) 889.

IDENTIFICATION AS BLACK HOLES OF ALL MASSIVE SUPERSTRING STATES

Paul H. Frampton

Institute of Field Physics
Department of Physics and Astronomy
University of North Carolina
Chapel Hill, NC 27599-3255

INTRODUCTION

In this talk I shall present some observations and results [1] which suggest that the massive states in the superstring spectrum should be identified as black holes. These results are based on properties of dual amplitudes, and comparison of the thermal distributions of strings and black holes. Then I shall consider the fate of mini black holes if present at the Big Bang.

REGGE TRAJECTORIES

The spectrum of the superstring obtained by factoring tree amplitudes, and evident from the partition function, lies on linear Regge trajectories. Taking the Regge slope as α' these may be written in the form

$$\alpha(s) = 2 + \alpha's - n/2$$

where s is the squared mass m^2 and n is a non-negative integer which is even for bosons and odd for fermions. At m=0 are the graviton, gravitino, and lower-spin states while massive states are at or above the Planck mass $\sim 10^{19}$ GeV.

The slope α' of the leading trajectory can be evaluated in a classical model where the string is a straight rigid rotator of length 2ℓ and mass per unit length ρ with mass M = $2\rho\ell$. The angular momentum J is computed from the boundary condition that the ends move at the speed of light (c) as $J = (cM^2/6\rho)$. The density per unit length $\rho = M_{Planck}/L_{Planck} = c^2/G$ so we estimate that $\alpha' = G/(6c)$; one can easily check that this has the same dimension as h/M^2 where h = Planck's constant.

In scattering, all states with $J \leq \alpha' M^2$ are coupled but it can be shown (see below) that those states with $J \leq (\alpha' M^2)^{1/2}$ are most strongly coupled.

For comparison, consider the radius of the event horizon for an uncharged Kerr black hole given by [2]

$$R(Kerr) = \frac{GM}{c^2} + \sqrt{\frac{G^2M^2}{c^4} - \frac{J^2}{M^2c^2}}$$

This can be a black hole only for $J \leq GM^2/c$. If, on a J versus M^2 plot, we draw this Kerr limit together with the trajectories described by the first equation above we immediately see that all the massive superstring states are below the Kerr line. Backreaction can shift the black hole masses giving some non-linearity as is now under study. But at least for masses large compared to the Planck mass all the massive superstring states may collapse to black holes. These states are physical black holes only on-shell. Off-mass-shell they may contribute virtually to low-energy scattering of massless states, making perturbative superstrings viable as an effective theory.

DUAL AMPLITUDES [3]

Take the scattering of two massless spin-zero states described by the dual amplitude

$$A(s,t) = \frac{\Gamma(1-\alpha_s)\,\Gamma(1-\alpha_t)}{\Gamma(1-\alpha_s-\alpha_t)}$$

$$= \sum_{n=1}^{\infty} \frac{R_n(t)}{n-\alpha_s}$$

and decompose $R_n(t)$ to partial waves

$$R_n(t) = \sum_{\ell=0}^{n} C_n^\ell\, P_\ell(\cos\theta_s)$$

then

$$C_n^\ell = (\ell+\tfrac{1}{2}) \int_{-1}^{+1} R_n(t(z))\, P_\ell(z)\, dz$$
$$= n(\ell+\tfrac{1}{2}) \sqrt{2\pi}\, [(n-a-4\mu^2)/2]^{-1/2} J$$

where

$$J = \frac{1}{2}\pi i \int \frac{d\zeta}{\sqrt{\zeta}} \exp[(3a^2 + 4\mu^2 - 1)\zeta/2]$$

$$I_{\ell+\frac{1}{2}}[(n-a-4\mu^2)\zeta/z]\,[2\sinh\zeta/2]^{-n-1}$$

Using the asymptotic form of the modified Bessel function $I_{\ell+1/2}$ in a saddle-point estimation gives

$$C_n^\ell \sim \sqrt{\tfrac{1}{2}(\ell+\tfrac{1}{2})}\,\Big/[\zeta_0/2\,\cosh(\zeta_0/2) - \sinh((\zeta_0/2)]^{1/2}E$$

with

$$E = [\exp(\cosh\zeta_0/2)\tanh\zeta_0/4]^{\ell+1/2}\Big/[2\sinh(\zeta_0/2)]^n$$
$$2\cosh[(3a+4\mu^2-1)\zeta_0/2]$$

when $n-\ell$ = odd; if $n-\ell$ = even, the final cosh is replaced by sinh.

The conclusion from studying this expression is that $\ell \leq \sqrt{n}$ dominate, as expected from a lever-arm argument for fixed impact parameter. This means that the more strongly-coupled states are even deeper into the allowed region for black holes and, with lower angular momentum, more closely approximate Schwarzschild-like black holes.

THERMAL DISTRIBUTIONS [4,5]

The Helmholtz free energy per unit 4-volume (F/VT) has been computed for a thermal distribution of strings. For temperature T greater than the critical temperature T_c (approximately the Planck temperature) the tree level dominates, and in a 1/n approach the result is

$$(F/VT) \underset{\sim}{\overset{T\to\infty}{}} T^p$$

where the power p = (d–1) for quantum field theory in d space-time dimensions; p = 0 for open strings and p = 1 for closed strings. For the closed string case there is a pathological singularity in the classical limit [4]. The surprising result was that strings have a much lower (F/VT) than fields.

In this spinless (Schwarzschild) limit a thermal distribution of black holes has

$$(F/VT) = \text{constant}$$

corresponding to the open string case. This suggests that the two may be identifiable.

QUANTUM BLACK HOLES

The spectrum of black holes according to the superstring will be $\sqrt{n}$ M_{Planck} where n is an integer. This spectrum has consequences for emitted massless states, entropy and evaporation. Consider the radiative decay $(BH)_n \to (BH)_{n'} + \gamma$ where the initial and final black holes have masses $\sqrt{n}$ M_{Planck} and $\sqrt{n'}$ M_{Planck}. In the rest frame the photon energy is

$$E_\gamma = (M_{Planck}/2)\,(n-n')/\sqrt{n}$$

implying a characteristic spectrum e.g.

$$(2E_\gamma/M_{Planck}) = (k_1/\sqrt{n}),\ (k_2/\sqrt{n-k_1})\ \ldots\ldots\ldots,\ \frac{k_n}{(n - \sum_{i=1}^{n-1} k_i)}.$$

As n goes to infinity, this becomes a continuous spectrum. But for small n the quantum aspect dominates. Defining the entropy (S) of a black hole as the logarithm of the number of different quantum configurations, the latter is given by the number of ordered partitions of n which is 2^{n-1} (this can be seen as the (n–1) spaces between n dots). Then the entropy S is [6]

$$S = (n-1)\,\ln 2$$

FATE OF PRIMORDIAL MINI BLACK HOLES

In the very early universe mini black holes can either evaporate into massless states or, if the collision time is sufficiently short compared to the Hubble time and to the evaporation lifetime, coalesce into successively larger BHs.

If the universe undergoes inflation after the Big Bang the primordial BH abundance would be diluted to at most a few per horizon-size unless the early universe became matter-dominated to the extent that overclosure and recollapse occurred. We show now that that did not happen, at least in a representative toy model.

Let $t_1 = 3 \times 10^{-44}$s represent the cosmic Planck time, μ = Planck mass and $V_1 = 10^{-93}$ cm^3, the horizon volume at time t_1. The mini BH lifetime is $\tau \sim M^3/n_\phi$ where n_ϕ = number of distinct massless modes; the cross-section $\sigma \sim M^2/\beta^2$ where β is the BH velocity.

The critical density at t_1 is $\rho_1 = 2.5 \times 10^{-93}$ g/cm³ (using $\rho \sim t^{-2}$ valid for radiation or matter dominated situations). Coalescence of BHs gives a doubling of mass $\mu \to 2\mu \to 4\mu \to$ etc. so we take $\mu_n = \mu_1 2^{n-1}$ and, in view of the lifetime, take time stages $t_n = 8^{n-1} t_1$. Suppose now that the critical density is valid, formed entirely of mass μ black holes (2×10^{89} BHs). Since the density $\rho \sim t^{-2}$ we have $\rho_n = 64^{1-n} \rho_1$.

To calculate the probability p_n of collision in the nth time slice we consider a cylinder of length $\beta_n t_n$ and cross-sectional area σ_n. Then

$$p_n = \sigma_n \beta_n t_n (\rho_n / \mu_n)$$

and $p_n = p_{n-1} p_{n-1} / (64 \times 2)$ where the factors 64, 2 come from ρ and from coalescence respectively. This yields

$$\frac{p_n}{(p_{n-1})^2} = 2^{-7/4}$$

implying very rapid coalescence. After seven 8-foldings all of the BHs have evaporated. This happens by the GUT scale 10^{16} GeV and so no primordial black holes survive to the inflationary stage.

This work was in collaboration with T.W. Kephart and supported in part by the U.S. Department of Energy under Grant No. DF-FG05-85ER-40219, Task B.

REFERENCES

1. P.H. Frampton and T.W. Kephart, preprint IFP–708–UNC (1995).
2. R. Kerr, *Phys. Rev. Lett.* 11:327 (1963), S. Chandrasekhar, "The Mathematical Theory of Black Holes," Oxford University Press (1983).
3. P.H. Frampton and Y. Nambu, in *Quanta*, University of Chicago Press (1970), beginning on page 403.
4. J.J. Atick and E. Witten, *Nucl. Phys.* B310:291 (1988).
5. J.W. York, *Phys. Rev.*, D33:2092 (1986).
6. V.F. Mukhanov, in *Complexity, Entropy and the Physics of Information*, Addison-Wesley (1990).

EFFECTS OF THE TOP LANDAU POLE ON ELECTROWEAK PHYSICS IN SUSY UNIFICATION

Pran Nath[1] and R. Arnowitt[2]

[1]Department of Physics, Northeastern University
Boston, MA 02115

[2]Department of Physics, Texas A&M University
College Station, TX 77843

Abstract

An analysis is given of the effects of the Landau pole in the top Yukawa coupling on electroweak physics. It is found that proximity to the Landau pole leads to a scaling of the SUSY mass spectrum. It is shown that the Landau pole makes a negative contribution to the stop(mass)^2 which turns tachyonic as one approaches the Landau pole. It is also shown that a magnification of errors can occur in the vicinity of the Landau pole which can affect precision analyses of physics at the electroweak scale.

I. INTRODUCTION

The CDF Collaboration gives a determination of the top mass at $m_t = 176 \pm 8(stat.) \pm 10(sys.)$ GeV. It is conceivable that eventually the mass of the top may go even higher and end up being close to the CDF upper limit or above it[2]. The question of interest to us is to explore the implications of such a heavy top for SUSY unification. The importance of this question arises due to the fact that the top Yukawa coupling has a pseudo fixed point, or a Landau pole which for supersymmetric unification has an upper limit of ≈ 200 GeV[3-6] Thus when the top mass is near the CDF upper limit or near the mean of the D0 value one cannot escape being close to the Landau pole, and even for $m_t = 170$ GeV Landau pole effects are signficant. We shall discuss in this paper a number of interesting phenomena that arise specifically because of the proximity of the top to the Landau pole.

The outline of the paper is as follows: In Sec. II we give a brief discussion of the Landau pole in the top Yukawa coupling. In Sec. III we discuss how the singularity structure of the top Yukawa coupling propagates in the SUSY parameters via renormalization group equations. In Sec. IV we discuss the relationship of the Landau pole and radiative breaking, and show that radiative breaking always occurs as one approaches the Landau pole. In sec. V we show that the Landau pole drives the light stop towards the tachyonic limit and the condition that the SUSY spectrum be free of tachyons limits severely the parameter space of the theory. In Sec. VI we discuss the effect of the Landau pole on precision physics. It is shown that physical quantities which are directly coupled to the top, are sensitive to small variations in the parameters m_t, α_G and $\tan\beta$ (where $\tan\beta = v_2/v_1$, and where v_2 gives mass to the top and v_1 gives mass to the bottom quark) when one is in proximity to the Landau pole. Finally, conclusions are given in Sec. VII.

II. THE TOP LANDAU POLE IN SUPERGRAVITY UNIFICATION

The framework we use for our analysis is supergravity grand unification[7]. This model has a sound theoretical foundation. Let us mention briefly a few basic features of the theory. First the theory is formulated at scales below the Planck scale as an effective theory which arises from coupling of N=1 supergravity with N=1 gauge and N=1 matter chiral multiplets. The theory manufactures its own spontaneous breaking of supersymmetry via a hidden sector and generates soft SUSY breaking terms of $O(M_G/M_{Planck})^n M_G$ and thus the electro-weak physics is protected from the GUT size corrections. In the low energy domain the model depends on 7 parameters aside from the top mass. These consist of two GUT parameters, α_G, and M_G; four SUSY breaking parameters, i.e., the universal scalar mass m_o, the universal gaugino mass $m_{1/2}$, the soft SUSY breaking trilinear parameter A_o and the soft SUSY breaking bilinear parameter B_o, and the Higgs mixing parameter μ_o.

Now one can think of the LEP data on $\alpha_1, \alpha_2, \alpha_3$ as measurement of α_G and M_G when one uses the renormalization group equations. The CDF data gives m_t, and radiative breaking of the electro-weak symmetry fixes μ. Thus one is left with four parameters which one can choose to be $m_o, m_{1/2}, A_o$ and $\tan\beta$. There are 32 SUSY particles in the theory whose masses can be computed in terms of these four parameters. Thus this theory makes 28 predictions which can be tested at supercolliders and in non-accelerator experiment to verify this model.

What we wish to focus on is this talk in the specific issue of the top mass. As pointed out already if the upper limit of the CDF data is realized, then the top will be within striking distance of the Landau pole. We explore here the implications of this result for SUSY unification. The Landau pole phenomenon is quite simple. One can illustrate it by one-loop analysis of the renormalization group equations of the gauge and Yukawa couplings. At the one loop level the solution to the Yukawa coupling of the top is given by[8]

$$Y_t = Y_o E(t)/(1+6Y_o F(t)) \tag{2.1}$$

where Y_o is the top Yukawa coupling at the GUT scale, Y_t is the top Yukawa coupling at the electro-weak scale, $t = \ell n(M_G^2/Q^2)$, and $E(t)$ and $F(t)$ are given by

$$E(t) = (1+\beta_3)^{\frac{16}{3b_3}}(1+\beta_2)^{\frac{3}{b_2}}(1+\beta_1)^{\frac{13}{9b_1}}, \quad F(t) = \int_0^t E(t)dt \tag{2.2}$$

Here $\beta_i = \alpha_i(0)b_i/4\pi$, where $(b_1,b_2,b_3) = (33/5, 1, -3)$, $\alpha_i(0) = \alpha_G$ and $\alpha_1 = (5/3)\alpha_Y$ with Y being the hypercharge in the Standard Model. Eq. (2.1) shows that one achieves a pseudo-fixed point corresponding to $Y_t^f = E/6F$. At this point the solutions at low energy become independent of Y_o which becomes large. Thus the position of the Landau pole is determined completely by the gauge coupling constant, and one gets

$$m_t^f = (8\pi/\alpha_t(t))^{1/2} M_Z \cos\theta_w \sin\beta \tag{2.3}$$

For some typical values, $M_G \approx 10^{16} GeV$, $\alpha_G \approx 1/24$, one has $m_t^f \approx 200\sin\beta\ GeV$. Experimentally the value of $\tan\beta$ is not known. Theoretically, from radiative breaking we know that $\tan\beta \rangle 1$, which implies $m_t^f \rangle 140\ GeV$. Thus the top mass may already lie close to

m_t^f, although this situation is not forced. However, if the final values of the CDF/D0 data give a top mass close to 190 GeV, it would be difficult to avoid the situation of the top mass lying in the vicinity of the Landau pole.

III. SINGULARITY STRUCTURE OF SUSY PARAMETERS

The analysis of Sec. II showed that there is a Landau pole in the top quark Yukawa coupling. Now the renormalization group analysis couples the top Yukawa to SUSY parameters, some of which also become singular. Thus one finds[9] a proliferation of the Landau pole singularity in parameters A_o, B_o, μ_o. We discuss these briefly. Assuming A_t fixed, one can compute A_o at the GUT scale using renormalization group, and one finds $A_o = A_R/D_o$ where $D_o = 1 - y_t/y_t^f$ and

$$A_R = A_t - (m_{1/2}/m_0)(H_2 - 6H_3 Y_t/E) \tag{3.1}$$

where

$$H_2 = (\alpha_G/4\pi)\left(\frac{16}{3}h_3 + 3h_2 + \frac{13}{15}h_1\right) \tag{3.2}$$

and

$$\frac{H_3}{F} = t\frac{E}{F} - 1, \quad h_i = t(1+\beta_i t)^{-1} \tag{3.3}$$

We note that A_o has the same singular structure that Y_o has. Similarly the renormalization group analysis of μ^2 gives

$$\mu^2 = \mu_o^2 \, D_o^{1/2} (1+\beta_2 t)^{3/b_2} (1+\beta_1 t)^{1/b_1} \tag{3.4}$$

Thus μ_o^2 will show a singularity of $D_o^{-1/2}$ if μ^2 were held fixed as we approach the Landau pole. Actually the situation is more complex as discussed in Sec. IV because μ^2 is determined via radiative breaking of the electro-weak symmetry which implies a specific singularity structure for μ^2. Effectively one finds that $\mu^2 \sim D_o^{-3/2}$ as one approaches the Landau pole. The pole structure of B_o can also be calculated and one finds

$$B_o = \frac{s}{q}\frac{A_R}{D_o} + \frac{1}{4}\frac{\tan 2\beta}{\mu}\frac{A_R^2}{D_o} + B_o(NP) \tag{3.5}$$

where B_o (NP) is the non-singular part and s,q are as defined in Ref. 8. Eq. (3.5) implies that $B_o \sim D_o^{-1}$ near the Landau pole.

IV. TOP LANDAU POLE AND RADIATIVE BREAKING

We begin by determining the behavior of the parameter μ as we approach the Landau pole using the radiative breaking equation which is given by

$$\mu^2 = \frac{m_{H_1}^2 - m_{H_2}^2 \tan^2\beta}{\tan^2\beta - 1} - \frac{1}{2}M_Z^2 \quad (4.1)$$

where m_{H_1} (m_{H_2}) are the masses of the H_1 (H_2) Higgs bosons which give masses to the down (up) quarks. Now one can show that $m_{H_1}^2$ is smooth as one approaches the Landau pole. However, $m_{H_2}^2$ develops a pole so that

$$m_{H_2}^2(pole) = -3Y_t(F/E)A_R^2/D_0 + m_{H_2}^2(NP) \quad (4.2)$$

where $m_{H_2}^2(NP)$ is the non-pole term. Since the residue of the pole term in eq. (4.2) is negative definite, we see that $m_{H_2}^2$ would always turn tachyonic if we approach the Landau sufficiently closely. Since a tachyonic $m_{H_2}^2$ leads to radiative breaking of the electro-weak symmetry one finds that radiative breaking of the electro-weak symmetry necessarily occurs as one approaches the Landau pole. The value of μ^2 also simplifies in the vicinity of the Landau pole. From eqs. (4.1) and (4.2) one finds[10,11]

$$\mu^2 = \mu_R^2/D_0 + \mu^2(NP) \quad (4.3a)$$

where

$$\mu_R^2 = 3\frac{\tan^2\beta}{\tan^2\beta - 1}Y_t\frac{F}{E}A_R^2 \quad (4.3b)$$

and $\mu^2(NP)$ is the non-pole piece. From eq. (4.3a) one finds that μ gets large as we approach the Landau pole and one moves into the scaling region where the masses of the charginos and the neutralinos are given by

$$m_{\tilde{W}_1} \simeq \tilde{m}_2, \quad m_{\tilde{W}_2} \simeq \mu \quad (4.4a)$$

$$m_{\tilde{Z}_1} \simeq m_{\tilde{W}_1}/2, \quad m_{\tilde{Z}_2} \simeq m_{\tilde{W}_1} \quad (4.4b)$$

$$m_{\tilde{Z}_3} \simeq \mu, \quad m_{\tilde{Z}_4} \simeq \mu \quad (4.4c)$$

where $\tilde{m}_2 = (\alpha_2/\alpha_G)m_{1/2}$. From the above we see that $m_{\tilde{W}_1}, m_{\tilde{Z}_1}$ and $m_{\tilde{Z}_2}$ will show no variation while $m_{\tilde{W}_2}, m_{\tilde{Z}_3}$ and $m_{\tilde{Z}_1}$ will show a rapid variation as we approach the Landau pole. A similar analysis holds for the Higgs. For the lightest Higgs one does not expect a rapid variation in the vicinity of the Landau pole since the lightest Higgs mass involves the μ-parameter only at the loop level. The other three Higgs states $(A, H^o, H^\pm)$ become degenerate with a common mass of $m_A^2 \approx 2m_3^2/\sin^2\beta$. Renormalization group analysis shows that in the vicinity of the Landau pole one has

$$m_A^2 = -\frac{3}{\cos(2\beta)}Y_t\frac{F}{E}A_R^2\frac{1}{D_0} + m_A^2(NP) \quad (4.5)$$

Thus the three degenerate Higgs states will show a rapid variation in the vicinity of the Landau pole.

V. LANDAU POLE AND THE TACHYONIC LIMIT

As discussed in the preceding sections, approach to the Landau pole is associated with important effects on the SUSY spectrum. A relevant question is how closely one may approach the Landau pole. The strongest constraint here comes from the condition that the SUSY spectrum be free of tachyons. Now the part of the SUSY spectrum that is most sensitive to the Landau pole is the stop mass matrix which has the form.

$$\begin{pmatrix} m_{\tilde{t}_L}^2 & m_t(A_t + \mu \cot\beta) \\ m_t(A_t + \mu \cot\beta) & m_{\tilde{t}_R}^2 \end{pmatrix} \quad (5.1)$$

where $m_{\tilde{t}_L}^2$ and $m_{\tilde{t}_L}^2$ are as the left and the right stop masses. Diagonalization gives the lighter stop mass to be

$$m_{\tilde{t}_1}^2 = \frac{1}{2}\left[\left(m_{\tilde{t}_L}^2 + m_{\tilde{t}_R}^2\right) - \sqrt{\left(m_{\tilde{t}_L}^2 - m_{\tilde{t}_R}^2\right)^2 + 4(A_t + \mu \cot\beta)^2 m_t^2}\right] \quad (5.2a)$$

As we approach the Landau pole we find

$$m_{\tilde{t}_1}^2 = -2x/D_0 + m_{\tilde{t}_1}^2(NP) \quad (5.2b)$$

where $x = Y_t A_R^2 F/E$. Similarly for the heavier stop mass one has

$$m_{\tilde{t}_2}^2 = \frac{1}{2}\left[\left(m_{\tilde{t}_L}^2 + m_{\tilde{t}_R}^2\right) + \sqrt{\left(m_{\tilde{t}_L}^2 - m_{\tilde{t}_R}^2\right)^2 + 4(A_t + \mu \cot\beta)^2 m_t^2}\right] \quad (5.3a)$$

and as we approach the Landau pole we have

$$m_{\tilde{t}_2}^2 = -x/D_0 + m_{\tilde{t}_2}^2(NP) \quad (5.3b)$$

We notice now a couple of features. First although stop1 by definition is the lighter of the two stops, its pole part is larger by a factor of 2. Further we note that the Landau pole contribution to the stop masses is always tachyonic. Thus as we approach the Landau pole, stop1 moves towards its tachyonic limit. Second the sign and the size of the trilinear parameter A_t determines how fast $\tilde{t}_1$ becomes tachyonic. To get an idea of sizes, if we take $M_G = 10^{16} GeV$ and $\alpha_G = 1/24$, which is consistent with the gauge coupling constant unification using the LEP data, we find that

$$A_R \simeq A_t - 0.61 m_{\tilde{g}} \quad (5.4)$$

Thus if A_R is positive, there is a cancellation between the first and the second terms and the approach to the tachyonic limit is slow. This means that for positive A_t you can get relatively close to the Landau pole without turning $\tilde{t}_1$-tachyonic. On the other hand if A_t is negative, A_R is large and there is a rapid approach to the tachyonic limit. This is illustrated in Fig. 1. Here the top curve exhibits the position of the Landau pole as a functin of $\tan\beta$. The middle curve exhibits the tachyonic limit for the light stop for the case $\mu \langle 0$, $m_o = 600 GeV$, $m_{\tilde{g}} = 300 GeV$ and $A_t = 0.5$. The region below the curve is the allowed region while the region above it is disallowed since in this region the light stop is tachyonic. The bottom curve corresponds to $A_t = -0.5$ with all other parameters being the same as the middle curve. Again the region below the curve is allowed while the region above it is disallowed. We thus note that the tachyonic limit is approached much faster for negative A_t than for positive A_t exactly as expected from eq. (5.4). In general the condition that there be no tachyons is found to be very strong and severely limits the allowed parameter space of the model[9,12,13].

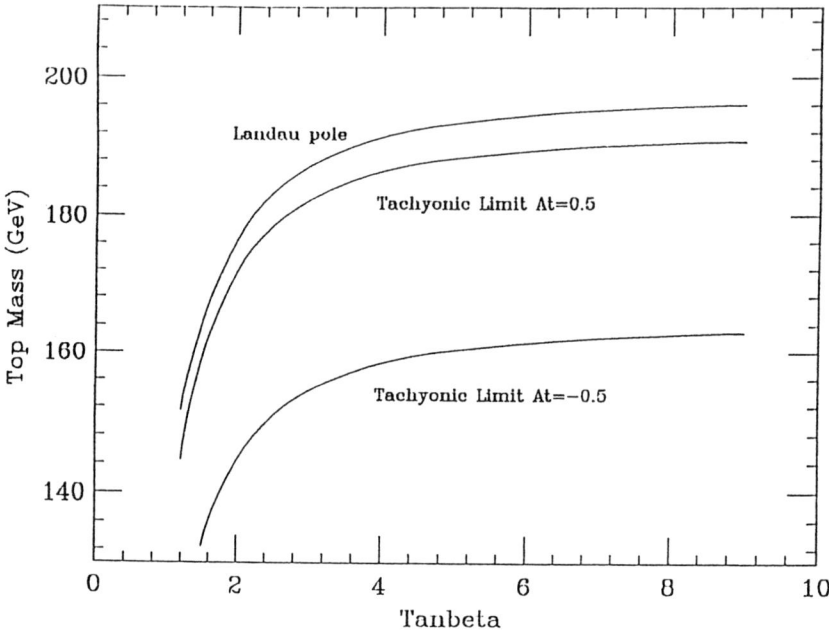

Fig. 1. Approach to the tacchyonic limit is exhibited. The top curve corresponds to the position of the Landau pole as a function of $\tan\beta$. The first curve below the top curve is the tachyonic limit for $A_t = 0.5$ and the second curve below is for $A_t = -0.5$ for the case when $\mu \langle o$, $m_o = 600$ GeV and $m_{\tilde{g}} = 300$ GeV (from Ref. 9).

VI. EFFECT OF THE LANDAU POLE ON PRECISION ANALYSES

Electro-weak physics which depends on the Landau pole may be sensitive to small changes in the input parameters when one is in the vicinity of the Landau pole. Let Q_a be a quantity of interest which exhibits a pole structure so that

$$Q_a = C_a/D_0 + Q_a(NP) \qquad (6.1)$$

Let P_i denote collectively the input parameters. Variations of Q_a with respect to P_i are then given by

$$Q_{a,i} = C_{a,i}/D_0 - C_a D_{0,i}/D_0^2 + Q_a(NP)_{,i} \qquad (6.2)$$

From the above we see that the variations with respect to the initial data involve a term with a single pole, a term with a double pole and a non-pole term. The largest sensitivities arise from variation of those parameters for which $D_{0,i} \neq 0$. There are three such parameters in the theory which are[9]

$$m_t, \alpha_G, \tan\beta \qquad (6.3)$$

We shall label these as order two parameters since their variation leads to a double pole. Currently, there is about 10% ambiguity in the experimental determination of α_3 (and hence of α_G) and m_t. In the vicinity of the Landau pole this uncertainty is enhanced by several fold. Thus the issue of magnification of errors is a very significant one if the top mass eventually ends up being close the to the top Landau pole mass.

VII. CONCLUSION

We have given here a brief summary of results on the analysis of the top Landau pole. In addition to the top Yukawa coupling, other SUSY parameters were found to also exhibit a singular structure. It is seen that radiative electro-weak breaking always occurs as we approach the Landau pole, and in this limit one also finds a scaling of the SUSY mass spectrum. An important effect observed in the analysis is that the Landau pole drives the stop towards its tachyonic limit and the condition that there be no tachyons then severely limits the parameter space of the model. Specifically ones finds that for $\tan\beta = 5.0$, $m_t = 170\ GeV$, $-0.8 m_o \langle A_t \langle 5.0 m_o$. Another effect of the Landau pole arises due to magnification of errors in m_t, α_G, and $\tan\beta_i$ when m_t is in the vicinity of the Landau pole, and this magnification has important consequences for precision analyses such as analyses of $b \to s\gamma$ decay[12,13,15] relic density[16] and event rate analyses[17,18,19,20] and analyses for signals of supersymmetry[21].

ACKNOWLEDGMENTS

This research was supported in part by NSF grant numbers PHY-19306906 and PHY-9411543.

REFERENCES

1. CDF collaboration, F. Abe et al, FERMILAB-PUB-95/022-E; For previous value see Phys. Rev. **D50**, 2996 (1994).
2. The D0 Collaboration S. Abachi et. al, (FERMILAB-PUB-95/028-E) gives a top mass of $199^{+19}_{-21}(stat.) \pm 22(syst.) GeV$.
3. C. Hill, Phys. Rev. **D24**, 691 (1981).
4. V. Barger, M.S. Berger and P. Ohman, Phys. Lett. **B314**, 351 (1993).
5. W. A. Bardeen, M. Carena, S. Pokorski, and C.E.M. Wagner, Phys. Lett. **B20**, 110 (1994); M. Carena, and C.E.M. Wagner, CERN-TH.7393/94 (1994).
6. M. Carena, M. Olechowski, S. Pokorski, and C.E.M. Wagner, Nucl. Phys. **B419**, 213 (1994).
7. A. H. Chamseddine, R. Arnowitt, and P. Nath, Phys. Rev.Lett. **29**, 970 (1982). For reviews, see; P. Nath, R. Arnowitt, and A.H. Chamseddine, "Applied N=1 Supergravity", (World Scientific, Singapore, 1984); H.P. Nilles, Phys. Rep. 110, 1 (1984); R. Arnowitt, and P. Nath, Lectures at VII J.A. Swieca Summer School, Campos de Jordao, Brazil, 1993 (World Scientific, Singapore, 1994).
8. L Ibañez, C. Lopez, and C. Muños, Nucl. Phys. **B256**, 218 (1985).
9. P. Nath, J. Wu and R. Arnowitt, NUB-TH-3116/95; CTP-TAMU-04/94.
10. R. Arnowitt and P. Nath, Phys. Rev. Lett. **69**, 725 (1992).
11. P. Nath and R. Arnowitt, and P. Nath, Phys. Lett. **B289, 368** (1992).
12. J. Wu, R. Arnowitt, and P. Nath, Phys. Rev. **D51**, 1371 (1995).
13. P. Nath, and R. Arnowitt, Phys. Lett. B336, 395 (1994).

14. P. Langacker, Proceedings, PASCOS90 Symposium, Editors, P. Nath and S. Reucroft (World Scientific,Singapore 1990); J. Ellis,S. Kelley, and D. V. Nanopoulos, Phys. Lett. **249B**, 441 (1990), *ibid*, **260B**, 131 (1991); U. Amaldi, W. de Boer, and H. Furstenau, *ibid* **260,** 447 (1991); P. Langacker and N. Polonsky, Phys. Rev. **D47,** 4028 (1993).
15. S. Bertolini, F. Borzumati, A. Masiero, and G. Ridolfi, Nucl. Phys **B353**, 541 (1983); J. Hewett, Phys. Rev. Lett. **70**, 1045 (1993); V. Barger, M. Berger, and R.J.N. Phillips, Phys. Rev. Lett. **70**, 1368 (1993); M. A. Diaz, Phys. Lett. **304B**, 278 (1993). R. Barbieri, and G Guidice, Phys. Lett. **309B**, 86 (1993); J. L. Lopez, D. V. Nanopoulos, and G. Park Phys. Rev. **D48**, R974 (1993); Y. Okada, Phys. Lett. **315B**, 119 (1993); R.Garisto, and J. N. Ng, Phys. Lett. **315B**, 372 (1993); G. Kane, C. Kolda, L. Roszkowski, and D. J. Wells, Phys. Rev. **D49**, 6173 (1994); S. Bertolini, and F. Vissani, SISSA 40/94/EP (1994).
16. R. Arnowitt, and P. Nath, Phys. Lett. **299B**, 58 (1993) and Erratum *ibid* **303B**, 403 (1993); P. Nath and R. Arnowitt, Phys. Rev. Lett. **70** 3696 (1993); J. Lopez, D. Nanopoulos, and K. Yuan, Phys. Rev. **D48**, 2766 (1993); M. Drees and M. M. Nojiri, Phys. Rev. **D48**, 3483 (1993); A. Bottino, V. de Alfero, N. Forengo, G. Mignola, and M. Pignone, Astro. Phys. **2** 67 (1992);T. Falk et. al., Phys. Lett. **B318**, 354 (1993) and the references quoted therein.
17. J. Ellis and R. Flores Phys. Lett. **B263**, 259 (1991); **B300**, 175 (1993); Nucl. Phys. **B400**, 25 (1993); A. Bottino et al., Phys. Lett. **B295**, 330 (1992).
18. M. Drees and M. Nojiri, Phys. Rev. **D48**, 3483 (1993); F. Borzumati, M. Drees and M. Nojiri, Preprint MAD/PH/385.
19. R. Arnowitt and P. Nath, CERN TH.7362/94 (1994)(to appear in Mod. Phys. Lett. A); M. Kamionkowski, K. Griest, G. Jungman, and B. Sadoulet, IASSNS-HEP-94/76 (1994); P. Nath and R. Arnowitt, Proc. of HEP94, Glasgow (1994)..
20. P. Nath and R. Arnowitt, CERN.TH 7363 (1994); E. Diehl, G. Kane, C. Kolda and J.Wells, UM-TH-94-38 (1995).
21. For a recent analysis see H. Baer, C.C. Chen, F. Paige and X. Tata, FSU-HEP-950204/UH-511-817-95

SECTION VI

PROGRESS IN SOME NEW AND OLD IDEAS

INTEGRALS OF MOTION FOR THE SINE-GORDON MODEL WITH BOUNDARY AT THE FREE FERMION POINT

Luca Mezincescu[1,2] and Rafael I. Nepomechie[3]

[1] Physics Department, Bonn University
Nussallee 12, D-53115 Bonn, Germany

[2] Physics Department, University of Miami
Coral Gables, FL 33124 USA (permanent address)

[3] Physics Department, University of Miami
Coral Gables, FL 33124 USA

Abstract: We construct integrals of motion for the sine-Gordon model with boundary at the free Fermion point ($\beta^2 = 4\pi$). We use these integrals of motion to determine the boundary S matrix.

INTRODUCTION

As is well known [1] - [3], if an integrable quantum field theory in 1 + 1 dimensions has integrals of motion of nontrivial dimensions (i.e. different from zero) which do not commute among themselves, these integrals of motion can be used to determine the (nontrivial) momentum dependence of the (bulk) factorizable S matrix, up to a scalar unitarization factor. Such integrals of motion generally are nonlocal and have fractional spin. The use of such integrals of motion constitutes an alternative to the quantum inverse scattering method for solving integrable quantum field theories in 1+1 dimensions [2,3].

Together with A. B. Zamolodchikov, we are engaged in a project [4] to use nonlocal integrals of motion of the sine-Gordon model with boundary [5] to determine the exact relation between the parameters of the action and the parameters of the boundary S matrix given in [5] (for general values of β^2).

We report here some preliminary results in this direction. Specifically, we use such integrals of motion to calculate the boundary S matrix of the sine-Gordon model with boundary at the free Fermion point ($\beta^2 = 4\pi$). Of course, for this particular case, the boundary S matrix can be obtained much more easily by direct integration of the free

Dirac equations. Our goal here is to show that such integrals of motion can exist in an integrable field theory with boundary.

THE BULK THEORY

The bulk Lagrangian density for a free massive Dirac spinor field $\Psi = \begin{pmatrix} \bar{\psi}_+ \\ \psi_+ \end{pmatrix}$, $\Psi^\dagger = \begin{pmatrix} \bar{\psi}_- \\ \psi_- \end{pmatrix}$ in $1+1$ dimensional Minkowski space is[1]

$$\begin{aligned}
\mathcal{L}_0 &= \frac{i}{2}\bar{\Psi}\overleftrightarrow{\partial}\Psi - im\bar{\Psi}\Psi \\
&= -i\left[\bar{\psi}_-\partial_+\bar{\psi}_+ + \bar{\psi}_+\partial_+\bar{\psi}_- - \psi_-\partial_-\psi_+ - \psi_+\partial_-\psi_- - m(\bar{\psi}_-\psi_+ - \psi_-\bar{\psi}_+)\right].
\end{aligned} \quad (1)$$

Evidently, the Lagrangian has a $U(1)$ symmetry; $\psi_+, \bar{\psi}_+$, have charge $+1$ and $\psi_-, \bar{\psi}_-$ have charge -1.

By solving the equations of motion, we are led (following the conventions of [5]) to the mode expansions

$$\begin{aligned}
\psi_+(x,t) &= \sqrt{\frac{m}{4\pi}} \int_{-\infty}^\infty d\theta\, e^{\frac{\theta}{2}}\left[\omega A_-(\theta)e^{ip\cdot x} + \omega^* A_+(\theta)^\dagger e^{-ip\cdot x}\right], \\
\bar{\psi}_+(x,t) &= \sqrt{\frac{m}{4\pi}} \int_{-\infty}^\infty d\theta\, e^{-\frac{\theta}{2}}\left[\omega^* A_-(\theta)e^{ip\cdot x} + \omega A_+(\theta)^\dagger e^{-ip\cdot x}\right],
\end{aligned} \quad (2)$$

where $p\cdot x = p_0 t + p_1 x$, with $p_0 = m\cosh\theta$, $p_1 = m\sinh\theta$, and $\omega = e^{i\pi/4}, \omega^* = e^{-i\pi/4}$. Canonical quantization implies that the only nonvanishing anticommutation relations for the modes are

$$\left\{A_\pm(\theta), A_\pm(\theta')^\dagger\right\} = \delta(\theta - \theta'). \quad (3)$$

We regard $A_\pm(\theta)^\dagger$ as creation operators.

The (bulk) field theory has integrals of motion of nontrivial dimension, e.g., [3,6]

$$Q_\pm = \int_{-\infty}^\infty dx\, \psi_\pm \dot{\psi}_\pm, \qquad \bar{Q}_\pm = \int_{-\infty}^\infty dx\, \bar{\psi}_\pm \dot{\bar{\psi}}_\pm, \quad (4)$$

where the dot (˙) denotes differentiation with respect to time. They form [7] the level 0 $\widehat{sl(2)}$ twisted affine Lie algebra.

We shall consider the field theory (1) on the negative half line $x \leq 0$. Given boundary conditions on the fields at $x = 0$, the boundary S matrix $R_a^b(\theta)$ is defined by [5]

$$A_a(\theta)^\dagger B = R_a^b(\theta) A_b(-\theta)^\dagger B, \quad (5)$$

where B is the boundary operator. In terms of the matrix notation $A = \begin{pmatrix} A_+ \\ A_- \end{pmatrix}$, this relation reads

$$A(\theta)^\dagger B = R(\theta) A(-\theta)^\dagger B. \quad (6)$$

---
[1] Our conventions are $\eta^{00} = -1 = -\eta^{11}$; $\partial_\pm = \frac{1}{2}(\pm\partial_0 + \partial_1)$; $\bar{\Psi} = \Psi^{\dagger T}\gamma^0$; $\gamma^0 = -i\sigma_2$; $\gamma^1 = \sigma_1$, where σ_i are the Pauli matrices.

"FIXED" BOUNDARY CONDITIONS

For simplicity, we first consider the case of "fixed" boundary conditions, for which the value of the sine-Gordon field at the boundary is kept fixed. In the Fermionic description, this corresponds to

$$\left[\psi_+(x,t) + e^{-i(\phi-\varphi)}\bar{\psi}_+(x,t)\right]\Big|_{x=0} = 0,$$
$$\left[\psi_-(x,t) + e^{i(\phi-\varphi)}\bar{\psi}_-(x,t)\right]\Big|_{x=0} = 0, \quad (7)$$

where ϕ and φ are arbitrary real constants. These boundary conditions preserve the $U(1)$ symmetry of the bulk Lagrangian.

By direct computation along the lines of [5], one easily finds that the boundary S matrix is given by

$$R(\theta) = \frac{-1}{\cos(\phi-\varphi) - i\sinh\theta}\begin{pmatrix} \cosh[\theta - i(\phi-\varphi)] & 0 \\ 0 & \cosh[\theta + i(\phi-\varphi)] \end{pmatrix}. \quad (8)$$

We now describe an alternative derivation of this result using certain integrals of motion. Indeed, consider $Q, Q^\dagger$ defined by

$$Q = \int_{-\infty}^{0} dx \left[\bar{\psi}_+\dot{\psi}_+ + e^{2i(\phi-\varphi)}\psi_+\dot{\psi}_+\right],$$
$$Q^\dagger = -\int_{-\infty}^{0} dx \left[\bar{\psi}_-\dot{\psi}_- + e^{-2i(\phi-\varphi)}\psi_-\dot{\psi}_-\right]. \quad (9)$$

One can show using the equations of motion that $Q, Q^\dagger$ are integrals of motion

$$\frac{d}{dt}Q = \frac{d}{dt}Q^\dagger = 0. \quad (10)$$

Substituting the mode expansions (2), we find that[2]

$$Q = \frac{im}{2}\int_0^\infty d\theta \left[e^{-\theta} + e^{2i(\phi-\varphi)}e^\theta\right] A_+(\theta)^\dagger A_-(\theta),$$
$$Q^\dagger = -\frac{im}{2}\int_0^\infty d\theta \left[e^{-\theta} + e^{-2i(\phi-\varphi)}e^\theta\right] A_-(\theta)^\dagger A_+(\theta). \quad (11)$$

The commutation relations of the integrals of motion $Q, Q^\dagger$ with the creation operators $A(\theta)^\dagger$ are therefore given by

$$Q\, A(\theta)^\dagger = A(\theta)^\dagger Q + \frac{im}{2}\left[e^{-\theta} + e^{2i(\phi-\varphi)}e^\theta\right]\sigma_- A(\theta)^\dagger,$$
$$Q^\dagger A(\theta)^\dagger = A(\theta)^\dagger Q^\dagger - \frac{im}{2}\left[e^{-\theta} + e^{-2i(\phi-\varphi)}e^\theta\right]\sigma_+ A(\theta)^\dagger, \quad (12)$$

where $\sigma_\pm = (\sigma_1 \pm i\sigma_2)/2$. Evidently, the asymptotic states are not eigenstates of Q, $Q^\dagger$. Note also that for the particular case of free Fermions, the integrals of motion Q, $Q^\dagger$ are local.

We observe from Eq. (11) that $Q, Q^\dagger$ annihilate the vacuum state $|0>_B$,

$$Q|0>_B = Q^\dagger|0>_B = 0. \quad (13)$$

---

[2]This result can be obtained using certain identities obeyed by the solution (8). A much simpler alternative is to evaluate the integrals in Eq. (9) over the full line $(-\infty, +\infty)$.

Using first the commutation relations (12) and then the definition (6) of the boundary S matrix, we obtain

$$\begin{aligned} QA(\theta)^\dagger|0>_B &= \frac{im}{2}\left[e^{-\theta} + e^{2i(\phi-\varphi)}e^{\theta}\right]\sigma_- A(\theta)^\dagger|0>_B \\ &= \frac{im}{2}\left[e^{-\theta} + e^{2i(\phi-\varphi)}e^{\theta}\right]\sigma_- R(\theta) A(-\theta)^\dagger|0>_B. \end{aligned} \quad (14)$$

On the other hand, reversing the order of operations, we obtain

$$\begin{aligned} QA(\theta)^\dagger|0>_B &= R(\theta) Q A(-\theta)^\dagger|0>_B \\ &= \frac{im}{2} R(\theta)\left[e^{\theta} + e^{2i(\phi-\varphi)}e^{-\theta}\right]\sigma_- A(-\theta)^\dagger|0>_B. \end{aligned} \quad (15)$$

We conclude that the boundary S matrix $R(\theta)$ obeys the constraint

$$\left[e^{-\theta} + e^{2i(\phi-\varphi)}e^{\theta}\right]\sigma_- R(\theta) = R(\theta)\sigma_-\left[e^{\theta} + e^{2i(\phi-\varphi)}e^{-\theta}\right]. \quad (16)$$

A similar analysis using $Q^\dagger$ implies

$$\left[e^{-\theta} + e^{-2i(\phi-\varphi)}e^{\theta}\right]\sigma_+ R(\theta) = R(\theta)\sigma_+\left[e^{\theta} + e^{-2i(\phi-\varphi)}e^{-\theta}\right]. \quad (17)$$

These relations reproduce our previous result Eq. (8), up to a scalar factor.

"FREE" BOUNDARY CONDITIONS

The case of "free" boundary conditions corresponds to

$$\begin{aligned} \left[\bar{\psi}_+(x,t) - e^{-i(\phi+\varphi)}\psi_-(x,t)\right]\Big|_{x=0} &= 0, \\ \left[\bar{\psi}_-(x,t) - e^{i(\phi+\varphi)}\psi_+(x,t)\right]\Big|_{x=0} &= 0. \end{aligned} \quad (18)$$

These boundary conditions break the $U(1)$ symmetry of the bulk Lagrangian.

By direct computation, we obtain the boundary S matrix

$$R(\theta) = \frac{-1}{\cosh^2\theta}\begin{pmatrix} \cosh\theta & -\frac{i}{2}e^{-i(\phi+\varphi)}\sinh 2\theta \\ -\frac{i}{2}e^{i(\phi+\varphi)}\sinh 2\theta & \cosh\theta \end{pmatrix}. \quad (19)$$

This result can also be derived (in the manner explained in the previous section) using the integral of motion

$$Q = \int_{-\infty}^{0} dx\left[\bar{\psi}_+\dot{\psi}_+ + e^{-2i(\phi+\varphi)}\psi_-\dot{\psi}_- - me^{-i(\phi+\varphi)}\left(\bar{\psi}_-\bar{\psi}_+ + \psi_-\psi_+\right)\right], \quad (20)$$

and its Hermitian conjugate $Q^\dagger$.

INTERPOLATING BOUNDARY CONDITIONS

We now consider a one-parameter (h) family of boundary conditions, which interpolates between the "free" $(h=0)$ and "fixed" $(h\to\infty)$ boundary conditions described above. The action is given by[3]

$$S = \int_{-\infty}^{\infty} dt\left\{\int_{-\infty}^{0} dx \mathcal{L}_0 + L_{boundary}\right\}, \quad (21)$$

---

[3] We learned at this conference that a similar action, involving two Fermionic boundary degrees of freedom instead of one, has been found independently by Ameduri, Konik, and LeClair. Their work is reported in the recent preprint [8].

216

where $\mathcal{L}_0$ is given in Eq. (1), and $L_{boundary}$ is given by

$$L_{boundary} = \frac{i}{2}\left[e^{i(\phi+\varphi)}\bar{\psi}_+\psi_+ + e^{-i(\phi+\varphi)}\bar{\psi}_-\psi_- + a\dot{a}\right. \\ \left. - h\left(e^{i\varphi}\bar{\psi}_+ + e^{-i\varphi}\bar{\psi}_- + e^{i\phi}\psi_+ + e^{-i\phi}\psi_-\right)a\right]\Big|_{x=0}, \qquad (22)$$

where $a(t)$ is a Fermionic boundary degree of freedom. This action is a generalization of the one for the off-critical Ising field theory (free massive Majorana field) with a boundary magnetic field [5].

This action implies (upon eliminating a through its equations of motion) the following boundary conditions

$$\left(e^{i\varphi}\bar{\psi}_+ - e^{-i\varphi}\bar{\psi}_- + e^{i\phi}\psi_+ - e^{-i\phi}\psi_-\right)\Big|_{x=0} = 0, \qquad (23)$$

$$\frac{d}{dt}\left(e^{i\varphi}\bar{\psi}_+ + e^{-i\varphi}\bar{\psi}_- - e^{i\phi}\psi_+ - e^{-i\phi}\psi_-\right)\Big|_{x=0} \\ - h^2\left(e^{i\varphi}\bar{\psi}_+ + e^{-i\varphi}\bar{\psi}_- + e^{i\phi}\psi_+ + e^{-i\phi}\psi_-\right)\Big|_{x=0} = 0. \qquad (24)$$

By direct computation, we obtain the boundary S matrix

$$R(\theta) = \frac{-1}{\cos(\phi-\varphi) - i\sinh\theta - \frac{m}{h^2}\cosh^2\theta} \\ \times \begin{pmatrix} \cosh[\theta - i(\phi-\varphi)] - \frac{m}{h^2}\cosh\theta & \frac{im}{2h^2}e^{-i(\phi+\varphi)}\sinh 2\theta \\ \frac{im}{2h^2}e^{i(\phi+\varphi)}\sinh 2\theta & \cosh[\theta + i(\phi-\varphi)] - \frac{m}{h^2}\cosh\theta \end{pmatrix}. \qquad (25)$$

Evidently, this expression interpolates between the "free" and "fixed" boundary S matrices given in Eqs. (19) and (8), respectively.

Eq. (25) is equivalent to the result obtained in [5] for the sine-Gordon model with boundary at the free Fermion point ($\beta^2 = 4\pi$), provided we make the identification

$$e^{2i\xi} = \frac{e^{-i(\phi-\varphi)} - \frac{m}{h^2}}{e^{i(\phi-\varphi)} - \frac{m}{h^2}}. \qquad (26)$$

(The dependence on $\phi+\varphi$ in $R(\theta)$ can be removed by a suitable gauge transformation.) Indeed, the action (21) was formulated specifically to reproduce this result of [5].

We now show that the boundary S matrix (25) can also be derived using integrals of motion. We proceed by solving the constraints (23), (24) to obtain the identity

$$\left[\psi_+ - e^{-i(\phi+\varphi)}\bar{\psi}_- + e^{-i\phi}C(\bar{\psi}_+)\right]\Big|_{x=0} = 0, \qquad (27)$$

where the quantity $C(\psi)$ is defined by

$$C(\psi) = \frac{1}{1 + \frac{\partial_0}{h^2}}\left(e^{i\varphi}\psi + e^{-i\varphi}\psi^\dagger\right). \qquad (28)$$

With the help of this identity, one can show that

$$Q = \int_{-\infty}^{0} dx\left\{\left[\bar{\psi}_+ - e^{-i\varphi}C(\bar{\psi}_+)\right]\left[\dot{\bar{\psi}}_+ - e^{-i\varphi}\dot{C}(\bar{\psi}_+)\right] + e^{-2i(\phi+\varphi)}\psi_-\dot{\psi}_- \right. \\ \left. + me^{-i(\phi+2\varphi)}\left[\psi_- C(\bar{\psi}_+) + \bar{\psi}_- C(\bar{\psi}_+)\right] - me^{-i(\phi+\varphi)}\left(\bar{\psi}_-\bar{\psi}_+ + \psi_-\psi_+\right)\right\}, \qquad (29)$$

and its Hermitian conjugate $Q^\dagger$ are integrals of motion.

Note that these integrals of motion interpolate between the "free" and "fixed" integrals discussed above. Indeed,

$$\text{for } h \to 0, \quad Q \to Q_{free},$$
$$\text{for } h \to \infty, \quad Q \to -e^{-4i\varphi} Q^\dagger_{fixed}, \qquad (30)$$

where Q_{free} and Q_{fixed} are given by Eqs. (20) and (9), respectively.

Substituting the mode expansions (2), we obtain

$$Q = \frac{im}{2} \int_0^\infty d\theta \big[a(\theta) A_+(\theta)^\dagger A_-(\theta) + b(\theta) A_-(\theta)^\dagger A_+(\theta) + c(\theta) A_-(\theta)^\dagger A_-(\theta) + d(\theta) A_+(\theta)^\dagger A_+(\theta) \big], \qquad (31)$$

where

$$a(\theta) = \frac{e^{-\theta} m^2 \cosh^2 \theta}{h^4 + m^2 \cosh^2 \theta},$$

$$b(\theta) = \frac{h^4 e^{-\theta - 4i\varphi}}{h^4 + m^2 \cosh^2 \theta} - \frac{2h^2 m \cosh \theta\, e^{-i(3\varphi + \phi)}}{h^4 + m^2 \cosh^2 \theta} + e^{\theta - 2i(\phi + \varphi)},$$

$$c(\theta) = -\frac{i h^2 m \cosh \theta\, e^{-\theta - 2i\varphi}}{h^4 + m^2 \cosh^2 \theta} - \frac{m \cosh \theta\, e^{-i(\phi + \varphi)}}{h^2 + im \cosh \theta},$$

$$d(\theta) = \frac{i h^2 m \cosh \theta\, e^{-\theta - 2i\varphi}}{h^4 + m^2 \cosh^2 \theta} - \frac{m \cosh \theta\, e^{-i(\phi + \varphi)}}{h^2 - im \cosh \theta}, \qquad (32)$$

(The operator Q has been normal ordered, and an additive constant has been discarded.) The commutation relations of the integrals of motion Q, $Q^\dagger$ with the creation operators $A(\theta)^\dagger$ are therefore given by

$$Q\, A(\theta)^\dagger = A(\theta)^\dagger Q + \frac{im}{2} [a(\theta)\sigma_- + b(\theta)\sigma_+ + e(\theta)\sigma_3 + f(\theta)]\, A(\theta)^\dagger,$$

$$Q^\dagger A(\theta)^\dagger = A(\theta)^\dagger Q^\dagger - \frac{im}{2} [b(\theta)^* \sigma_- + a(\theta)\sigma_+ + e(\theta)^* \sigma_3 + f(\theta)^*]\, A(\theta)^\dagger, \qquad (33)$$

where

$$e(\theta) = \frac{1}{2}(d(\theta) - c(\theta)), \qquad f(\theta) = \frac{1}{2}(d(\theta) + c(\theta)), \qquad (34)$$

and * denotes complex conjugation. By an argument similar to the one used in the case of "fixed" boundary conditions (see Eqs. (13) - (17)), we reproduce the boundary S matrix (25), up to a scalar factor.

DISCUSSION

We observe that the diagonal elements of the boundary S matrix (25) vanish for $h^2 = m$ and $\phi = \varphi$. That is, for these values of the parameters, there is no elastic reflection. There is a similar "critical" value of the boundary magnetic field in the off-critical Ising field theory [5].[4]

We comment briefly on the factor $C(\psi)$ appearing in the integral of motion Q given in Eq. (29). Expanding the definition (28) in a power series in $\frac{1}{h^2}$,

$$C(\psi) = e^{i\varphi} \psi + e^{-i\varphi} \psi^\dagger - \frac{1}{h^2}\left(e^{i\varphi} \dot\psi + e^{-i\varphi} \dot\psi^\dagger\right) + \cdots, \qquad (35)$$

---
[4] A Hermitian integral of motion similar to (29) presumably exists also for the off-critical Ising field theory with boundary magnetic field.

one sees that Q involves an infinite series of higher-spin charges. Such charges have been discussed for the bulk theory in [7]. Also, one can see from the equations of motion for the Fermionic boundary field a that $C(\bar\psi_+)\big|_{x=0}$ is proportional a. Thus, $C(\bar\psi_+)$ seems to be a "continuation" of a away from the boundary.

We have seen that for the free Dirac field theory, a direct calculation of the boundary S matrix $R(\theta)$ is possible, and therefore the alternative calculation using the integrals of motion Q, $Q^\dagger$ is superfluous. However, for an interacting theory, while the former approach is no longer an option, the latter approach may still be viable. An investigation along these lines of the sine-Gordon model with boundary away from the free Fermion point is now in progress [4]. The integrals of motion which we have constructed are valid also at the semiclassical level. Off the free Fermion point, the semiclassical integrals of motion have not been constructed even for the bulk theory. Such integrals of motion may have a geometric origin along the lines of [1].

ACKNOWLEDGMENTS

The work presented here was initiated by A. B. Zamolodchikov, and represents a preliminary account of a joint ongoing investigation with him. We are grateful to him for his invaluable advice, kind hospitality, and for giving us the permission to present some of our joint results here. This work was supported in part by DFG Ri 317/13-1, and by the National Science Foundation under Grant PHY-92 09978.

REFERENCES

[1] M. Lüscher, Nucl. Phys. *B135* (1978) 1.

[2] A.B. Zamolodchikov, "Fractional-spin integrals of motion in perturbed conformal field theory," in Fields, Strings and Quantum Gravity, eds. H. Guo, Z. Qiu and H. Tye, (Gordon and Breach, 1989).

[3] D. Bernard and A. LeClair, Commun. Math. Phys. *142* (1991) 99.

[4] L. Mezincescu, R.I. Nepomechie and A.B. Zamolodchikov, in preparation.

[5] S. Ghoshal and A.B. Zamolodchikov, Int. J. Mod. Phys. *A9* (1994) 3841.

[6] R. K. Kaul and R. Rajaraman, Int. J. Mod. Phys. *A8* (1993) 1815.

[7] A. LeClair, Nucl. Phys. *B415* (1994) 734.

[8] M. Ameduri, R. Konik and A. LeClair, "Boundary sine-Gordon interactions at the free Fermion point," hep-th/9503088.

REFLECTION MATRICES AND POLYMERS AT A SURFACE

Murray T. Batchelor and C. Ming Yung

Department of Mathematics
School of Mathematical Sciences
Australian National University
Canberra ACT 0200, Australia

Abstract: We discuss self-avoiding polymer chains at a boundary from the point of view of boundary reflection matrices and exactly solved lattice models. This approach leads to new exact results for the polymer adsorption transition.

INTRODUCTION

It is well known that the O(n) or n-vector model describes self-avoiding polymer chains in the limit $n = 0$.[1] This relationship with phase transitions and critical phenomena in magnetic spin systems has led to a wealth of exact information on the critical properties of two-dimensional polymers, both in the bulk and at a surface.[2,3] These results have chiefly been obtained via mappings between various models and the Coulomb gas, and identifications based on conformal invariance arguments.[2,3]

Another approach is to seek explicit Bethe Ansatz solutions of the O(n) loop model on the honeycomb lattice,[4] and then to derive the critical properties, such as the central charge and scaling dimensions, from the finite-size behaviour of the eigenspectrum. Starting from Baxter's solution,[5] this programme has been carried out for the bulk critical behaviour.[6-9]

More recently, a solution has been found via the co-ordinate Bethe Ansatz for the O(n) loop model on the honeycomb lattice with *open* boundary conditions, yielding surface critical behaviour.[10] This work raised a number of interesting questions. As pointed out originally by Korepin,[5] the integrable bulk vertex weights follow in a special limit from the Izergin-Korepin R-matrix.[11] On the other hand, it is known that the integrability of models with open boundaries is governed by the reflection matrix (or K-matrix) satisfying the boundary version of the Yang-Baxter equation.[12,13] Thus there should be some connection between the integrable boundary weights of the honeycomb O(n) loop model and the known reflection matrices of the Izergin-Korepin model.[14]

Building on the earlier work of Sklyanin,[12] Mezincescu and Nepomechie,[13] Destri and de Vega,[15] and others, we recently gave such a prescription for obtaining integrable vertex models via reflection matrices on the square lattice with open boundaries and explicitly carried out the procedure for a number of models.[16] In particular, the approach leads to a further integrable set of boundary weights for the O(n) loop model,[16] which in turn provides new exact results for the polymer adsorption transition.[17]

Here we give a brief account of these developments with an emphasis on the related polymer problems.

INTEGRABLE O(n) LOOP MODELS WITH OPEN BOUNDARIES

K-matrices $K^-(u)$ are solutions of the reflection equation

$$R_{12}(u-v) \overset{1}{K^-}(u) R_{21}(u+v) \overset{2}{K^-}(v) = \overset{2}{K^-}(v) R_{12}(u+v) \overset{1}{K^-}(u) R_{21}(u-v). \tag{1}$$

Sklyanin[12] and Mezincescu and Nepomechie[13] showed, in the context of integrable open spin chains, how to build a family of commuting transfer matrices $t(u, \vec{\omega})$ out of $K^-(u)$ and a partner $K^+(u) = K^-(-u-\eta)^t M$; namely

$$t(u, \vec{\omega}) = \text{tr}_a \left[\overset{a}{K^+}(u) R_{a1}(u+\omega_1) \cdots R_{aN}(u+\omega_N) \overset{a}{K^-}(u) R_{Na}(u-\omega_N) \cdots R_{1a}(u-\omega_1) \right] \tag{2}$$

where η and M are the crossing parameter and crossing matrix specific to the R-matrix $R(u)$ in question. The commutativity of $t(u, \vec{\omega})$ allows its eigenvalues to be determined by, e.g., the algebraic or the analytic Bethe Ansatz. Destri and de Vega[15] looked at $t(u, \vec{\omega})$ with $\omega_j = (-)^{j+1}u$ and showed that it pertains to vertex models on a diagonal lattice with open boundary conditions. We further clarified this relation and looked at the mapping of these vertex models onto loop models on the same lattice.[16] Conditions on the K-matrices were found for the feasibility of this mapping and, when possible, expressions were found relating them to the boundary weights.

For the Izergin-Korepin model Mezincescu and Nepomechie showed that there are three inequivalent K-matrices $K^-(u)$.[14] We showed that of these three solutions, two of them – interestingly, not the solution $K^-(u) = 1$ leading to an $U_q(su(2))$-invariant spin chain[18] – are relevant to loop models. Building on their analytic Bethe ansatz calculation for the $K^-(u) = 1$ case,[19] we found the eigenvalue expression and associated Bethe Ansatz equations for the two cases of relevance.[16]

The integrable O(n) loop models we find have partition function

$$Z_{\text{loop}} = \sum_{\mathcal{G}} \rho_1^{m_1} \cdots \rho_{13}^{m_{13}} n^P \tag{3}$$

where the sum is over all configurations $\mathcal{G}$ of non-intersecting closed loops which cover some (or none) of the edges of the square lattice depicted in figure 1. The possible configurations at each vertex are shown in figure 2, with a vertex of type i carrying a Boltzmann weight ρ_i. In the configuration $\mathcal{G}$, m_i is the number of occurrences of the vertex of type i, while P is the total number of closed loops of fugacity n.

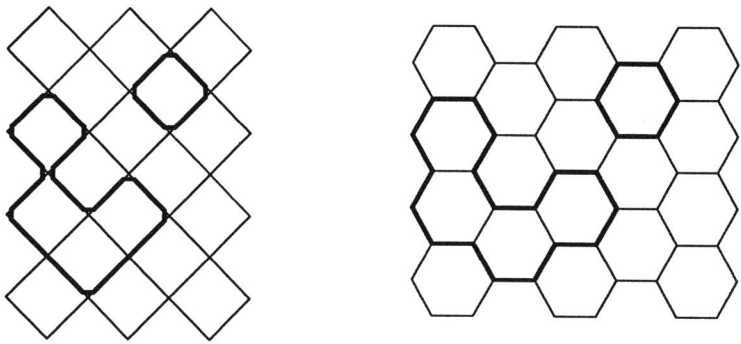

Figure 1: The open square and honeycomb lattices.

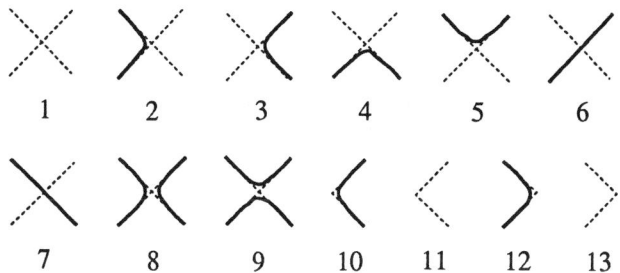

Figure 2: The allowed vertices of the square lattice loop model.

The integrable bulk loop weights are[20,21]

$$\begin{aligned}
\rho_1 &= \sin(3\lambda - u)\sin(u) + \sin(2\lambda)\sin(3\lambda) \\
\rho_2 = \rho_3 &= \epsilon_1 \sin(3\lambda - u)\sin(2\lambda) \\
\rho_4 = \rho_5 &= \epsilon_2 \sin(u)\sin(2\lambda) \\
\rho_6 = \rho_7 &= \sin(3\lambda - u)\sin(u) \\
\rho_8 &= \sin(3\lambda - u)\sin(2\lambda - u) \\
\rho_9 &= -\sin(\lambda - u)\sin(u).
\end{aligned} \quad (4)$$

where $\epsilon_1^2 = \epsilon_2^2 = 1$ and $n = -2\cos(4\lambda)$. The two sets of integrable boundary loop weights are

(A)
$$\begin{aligned}
\rho_{10} = \rho_{13} &= \sin[\tfrac{1}{2}(3\lambda - u)] \\
\rho_{11} = \rho_{12} &= \epsilon_1 \sin[\tfrac{1}{2}(3\lambda + u)]
\end{aligned} \quad (5)$$

(B)
$$\begin{aligned}
\rho_{10} = \rho_{13} &= \cos[\tfrac{1}{2}(3\lambda - u)] \\
\rho_{11} = \rho_{12} &= \epsilon_1 \cos[\tfrac{1}{2}(3\lambda + u)].
\end{aligned} \quad (6)$$

For both cases the eigenvalues of the transfer matrix, acting in the vertical direction, are given by

$$\Lambda(u) = \prod_{j=1}^{m} \frac{\sinh[u_j - i\lambda - \tfrac{1}{2}iu]\sinh[u_j + i\lambda + \tfrac{1}{2}iu]}{\sinh[u_j - i\lambda + \tfrac{1}{2}iu]\sinh[u_j + i\lambda - \tfrac{1}{2}iu]}. \tag{7}$$

For case (A), the Bethe Ansatz roots u_j satisfy the equations ($1 \leq k \leq m$)

$$\left[\frac{\sinh(u_k - i\lambda - \tfrac{1}{2}iu)\sinh(u_k - i\lambda + \tfrac{1}{2}iu)}{\sinh(u_k + i\lambda + \tfrac{1}{2}iu)\sinh(u_k + i\lambda - \tfrac{1}{2}iu)}\right]^N =$$

$$\left[\frac{\sinh(u_k + \tfrac{1}{2}i\lambda)}{\sinh(u_k - \tfrac{1}{2}i\lambda)}\right]^2 \prod_{j\neq k}^{m} \frac{\sinh(u_k + u_j - 2i\lambda)\sinh(u_k - u_j - 2i\lambda)}{\sinh(u_k + u_j + 2i\lambda)\sinh(u_k - u_j + 2i\lambda)}$$

$$\times \frac{\sinh[u_k + u_j + i\lambda]\sinh(u_k - u_j + i\lambda)}{\sinh(u_k + u_j - i\lambda)\sinh(u_k - u_j - i\lambda)}. \tag{8}$$

The effect of the different boundary weights in case (B) is to replace the sinh functions in the squared prefactor on the rhs by cosh functions. This change is sufficient to alter the finite-size behaviour of the eigenspectrum, and thus the critical behaviour.

Integrable O(n) loop models on the honeycomb lattice are obtained by taking $u = \lambda$, as for the bulk case.[20] At this value of the spectral parameter the weight ρ_9 vanishes, allowing the square lattice vertices to be "pulled apart" horizontally. Here this procedure results in a loop model on the open honeycomb lattice depicted in figure 1. The partition function reduces to[17]

$$Z_{\text{loop}} = \sum_{\mathcal{G}} x^L y^{L_s} n^P \tag{9}$$

where the sum is now over loop configurations on the honeycomb lattice. In the limit $n = 0$ the contribution from all closed loops vanishes and the partition sum reduces to a generating function for self-avoiding walks where x is the fugacity of a step in the bulk and y is the fugacity of a step along the surface, in this case either edge of the strip. Here L is the length of the walk in the bulk and L_s is the length of a walk along the surface.

The values of x and y in (9) are not arbitrary, but are fixed by the integrability of the underlying vertex model. The well known Nienhuis result, $x = x^*$, where[22]

$$1/x^* = \sqrt{2 \pm \sqrt{2-n}} \tag{10}$$

follows from the bulk weights (4). The value of the surface coupling y differs for each of the cases (A) and (B). For case (A) we have $y^* = x^*$, i.e., the critical surface coupling is equal to the bulk coupling. In the language of surface critical phemonena,[23] this corresponds to an integrable point on the *ordinary* transition line, where the surface orders at the same temperature as the bulk. However, the boundary weights for case (B) are seen to give[17]

$$1/y^* = \sqrt{\pm\sqrt{2-n}}. \tag{11}$$

In this case the surface orders at a different temperature to the bulk. This integrable point corresponds to the O(n) model at the *special* transition. The positive sign in (11) holds in the dilute phase, where $0 \leq \lambda \leq \pi/4$ ($-2 \leq n \leq 2$). Here $y^* = 1$ at $n = 1$, which corresponds to infinitely strong surface couplings in the Ising model, as expected at the special transition, since the surface (i.e. the one-dimensionsal Ising model) only orders at $T = 0$. We discuss the interpretation of y^* at $n = 0$ further below.

SURFACE CRITICAL BEHAVIOUR

The surface critical behaviour for case (A), at the *ordinary* transition, had been derived previously in the honeycomb limit via the finite-size corrections in the Bethe Ansatz eigenspectra.[10] The central charge is as obtained by a number of techniques, namely[2,3,6-9]

$$c = 1 - 6(g-1)^2/g \qquad (12)$$

where $\pi g = 2\pi - 4\lambda$. The geometric scaling dimensions were found to be given by

$$X_\ell = h_{\ell+1,1} \qquad (13)$$

where $\ell = 1, 2, 3, \ldots$ This is in agreement with an earlier conformal invariance result[24] for X_1 and an identification in terms of the Kac formula,[25] where

$$h_{p,q} = \tfrac{1}{4}gp^2 - \tfrac{1}{2}pq + \frac{q^2 - (g-1)^2}{4g}. \qquad (14)$$

From a further analysis of the exact solution, we see there is a 2-string solution to the Bethe Ansatz equations leading to a further scaling dimension $X_\epsilon = 1$, also in agreement with conformal invariance based arguments.[26]

The surface critical behaviour for case (B), at the *special* transition, can be derived in a similar manner. Again in the honeycomb limit, we find[17,27] that the central charge is as given in (12), with scaling dimensions

$$X_\ell = h_{\ell+1,3} \qquad (15)$$

where again $\ell = 1, 2, 3, \ldots$ This result has an interesting history, being first conjectured by Guim and Burkhardt from their finite-size scaling estimates of X_1 and X_2 at $n = 0$,[28] and more recently by Fendley and Saleur.[29] A further scaling dimension, $X_\epsilon = 2/g - 1$, again follows from an elementary 2-string excitation in the Bethe Ansatz root distribution. This result is also in agreement with a conformal invariance based result.[30]

POLYMERS AT A BOUNDARY

In the lattice model of the polymer adsorption transition at a boundary, the self-avoiding walk has energy[31,28]

$$E = -\epsilon L_s \qquad (16)$$

where ϵ is a positive constant and L_s is the number of steps along the adsorbing boundary. The polymer adsorption transition corresponds to the $n = 0$ limit of the O(n) model at the *special* surface transition. At the self-avoiding walk point at $n = 0$ we have

$$1/x^* = \sqrt{2 + \sqrt{2}} \quad \text{and} \quad 1/y^* = \sqrt{\sqrt{2}}. \qquad (17)$$

The exact critical adsorption temperature T_a is thus given by

$$\exp\left(\frac{\epsilon}{kT_a}\right) = y^*/x^* = \sqrt{1 + \sqrt{2}} = 1.553\ldots \qquad (18)$$

Despite intensive effort on other lattices,[3] this quantity does not appear to have been estimated before for the honeycomb lattice. By way of comparison, corresponding estimates for other two-dimensional lattices are

$$2.041 \pm 0.002 \quad \text{(square lattice)}^{28}$$
$$2.85 \pm 0.07 \quad \text{(triangular lattice)}.^{32}$$

The above model exhibits a desorbed phase for $T > T_a$ and an adsorbed phase for $T < T_a$. The number of configurations of walks with one end attached to the boundary has the asymptotic form[31,28]

$$Z_1 \sim \begin{cases} \mu^L L^{\gamma_1^o - 1} & T > T_a \\ \mu^L L^{\gamma_1^{sp} - 1} & T = T_a \end{cases} \tag{19}$$

where for the honeycomb lattice, $\mu = 1/x^*$. The surface critical exponents γ_1^o and γ_1^{sp} follow from the usual scaling relations, with[23,28]

$$\gamma_1 = (2 - X_1 - X_1^{\text{bulk}})\nu \tag{20}$$

where $X_1^{\text{bulk}} = \frac{5}{48}$ and $\nu = \frac{3}{4}$.[22] From the above results, the exact scaling dimensions at $n = 0$ are $X_1^o = \frac{5}{8}$ and $X_1^{sp} = -\frac{1}{24}$, from which the universal surface critical exponents follow as

$$\gamma_1^o = \frac{61}{64} \quad \text{and} \quad \gamma_1^{sp} = \frac{93}{64}. \tag{21}$$

The value for γ_1^o is the conformal invariance result of Cardy.[24]

The remaining quantity of interest that we discuss here is the crossover exponent ϕ which governs the average number of steps in contact with the boundary at $T = T_a$[31,28]

$$\langle L_s \rangle \sim L^\phi. \tag{22}$$

Again from scaling arguments, $X_\epsilon = 1 - \phi/\nu$.[23,28] Here the relevant scaling dimension is $X_\epsilon = X_2 = \frac{1}{3}$, from which it follows that $\phi = \frac{1}{2}$. More generally this result holds for $-2 \leq n \leq 2$.

ACKNOWLEDGMENTS

M.T.B. thanks J.B. McGuire, L. Mezincescu and R.I. Nepomechie for their kind hospitality and support during his visit to Florida. The work presented here was supported by the Australian Research Council under Large Grant A69231368 and the award of a Senior Research Fellowship to M.T.B.

REFERENCES

1. P.G. de Gennes, "Scaling Concepts in Polymer Physics," Cornell University Press, Ithaca (1979).

2. B. Duplantier, *Phys. Rep.* 184:229 (1989).

3. K. De'Bell and T. Lookman, *Rev. Mod. Phys.* 65:87 (1993).

4. E. Domany, D. Mukamel, B. Nienhuis and A. Schwimmer, *Nucl. Phys.* B 190:279 (1981).

5. R.J. Baxter, *J. Phys. A* 19:2821 (1986).

6. M.T. Batchelor and H.W.J. Blöte, *Phys. Rev. Lett.* 61:138 (1988).

7. J. Suzuki, *J. Phys. Soc. Japan* 57:2966 (1988).

8. M.T. Batchelor and H.W.J. Blöte, *Phys. Rev. B* 39:2391 (1989).

9. J. Suzuki, T. Nagao and M. Wadati, *Int. J. Mod. Phys. B* 6:1119 (1992).

10. M.T. Batchelor and J. Suzuki, *J. Phys. A* 26:L729 (1993).

11. A.G. Izergin and V.E. Korepin, *Commun. Math. Phys.* 79:303 (1981).

12. E.K. Sklyanin, *J. Phys. A* 21:2375 (1988).

13. L. Mezincescu and R.I. Nepomechie, *J. Phys. A* 24:L17 (1991).

14. L. Mezincescu and R.I. Nepomechie, *Int. J. Mod. Phys. A* 7:5231 (1991).

15. C. Destri and H.J. de Vega, *Nucl. Phys. B* 374:692 (1992).

16. C.M. Yung and M.T. Batchelor, *Nucl. Phys. B* 435:430 (1995).

17. M.T. Batchelor and C.M. Yung, *Phys. Rev. Lett.* 74:2026 (1995).

18. L. Mezincescu and R.I. Nepomechie, *Mod. Phys. Lett.* 6:2497 (1991).

19. L. Mezincescu and R.I. Nepomechie, *Nucl. Phys. B* 72:597 (1992).

20. B. Nienhuis, *Int. J. Mod. Phys. B* 4:929 (1990).

21. S.O. Warnaar and B. Nienhuis, *J. Phys. A* 26:2301 (1993).

22. B. Nienhuis, *Phys. Rev. Lett.* 49:1062 (1982).

23. K. Binder, *in:* "Phase Transitions and Critical Phenomena," vol 8, C Domb and J L Lebowitz, eds, Academic, London (1983).

24. J.L. Cardy, *Nucl. Phys. B* 240:514 (1984).

25. B. Duplantier and H. Saleur, *Phys. Rev. Lett.* 57:3179 (1986).

26. T.W. Burkhardt and J.L. Cardy, *J. Phys. A* 20:L233 (1987).

27. C.M. Yung and M.T. Batchelor, in preparation.

28. I. Guim and T.W. Burkhardt, *J. Phys. A* 22:1131 (1989).

29. P. Fendley and H. Saleur, *J. Phys. A* 27:L789 (1994).

30. T.W. Burkhardt, E. Eisenriegler and I. Guim, *Nucl. Phys. B* 316:559 (1989).

31. E. Eisenriegler, K. Kremer and K. Binder, *J. Chem. Phys.* 77:6296 (1982).

32. D. Zhao, T. Lookman and K. De'Bell, *Phys. Rev. A* 42:4591 (1990).

The Classical Space-Time from The Chern Simons Gauge Theory of Gravity

Freydoon Mansouri

Physics Department, University of Cincinnati, Cincinnati, OH 45221

Abstract

The exact dynamics of the 2+1 dimensional Chern Simons gauge theory of the Poincaré group coupled to sources carrying any spin is analyzed. The two body problem is solved exactly. It is shown how the classical space-time with correct asymptotic observables emerges from this gauge theory. Application is made to two particle scattering, obtaining its exact S-matrix in terms of Wilson loops without any explicit reference to time or time ordering. The supersymmetric extension of these results are briefly disscussed.

I. Introduction

In recent years, 2+1 dimensional gravity has been used as a theoretical laboratory to gain insight in the workings of quantum gravity in 3+1 dimensions. One reason for this is that in 2+1 dimensions, the classical gravity action can be seen to arise from the *bona fide* Chern Simons gauge theory of the Poincaré group under certain conditions [1]. The central aim of the present work is to study, non-perturbatively, 2+1 dimensional gravity as a Chern Simons gauge theory of the Poincaré group coupled to sources.

The dynamics of sources coupled to 2+1 dimensional gravity has been studied from a variety of points of view. In the context of metrical Einstein theory, this problem has been analyzed by a number of authors[2-7]. The dynamics of 2+1 dimensional gravity has also been studied in the framework of the Chern Simons gauge theory of the Poincare' group[1,8-14]. This approach, in which the topological aspects of the theory is more manifest, turns out to be very convenient for the study of the metric (gauge) independent features of the 2+1 dimensional gravity. This is the point of view which will be explored in the present manuscript. In section (II), I begin with a brief review of the formalism to establish our notaion. In section (III), I discuss the coupling of sources with any spin to the Chern Simons theory. In section (IV), I present the exact solution of the two body problem for sources of any spin by reducing it to an equivalent one body problem. In section (V), I show how the space-time emerges from the Poincaré gauge theory. In section (VI), I give the exact S- matrix for two particle scattering amplitude. Finally, In section (VII), I briefly discuss the extension of our results to supergravity theories coupled to supersources.

II. Chern Simons Theory of the Poincaré Group

A typical Chern Simons action in 2+1 dimensions is given by

$$I_{cs} = \int_M \gamma_{bc} A^b \wedge (dA^c + \frac{1}{3} f^c_{de} A^d \wedge A^e) \tag{1}$$

where A^a are components of the Lie algebra valued connection

$$A = A^a G_a; \quad A^a = A^a_\mu dx^\mu \tag{2}$$

The quantities G^a are elements of the Lie algebra with structure constants f_{abc}. The quantities γ_{ab} are the components of a suitable non-degenerate metric on the Lie algebra [1]. For Poincaré algebra with elements P_a, J_a, a=0,1,2, the connection can be written as

$$A_\mu = e^a_\mu P_a + \omega^a_\mu J_a; \quad \mu = 0, 1, 2 \tag{3}$$

where e^a_μ and ω^a_μ are gauge fields of the Poincaré group. From this we can compute the components of the field strength in the usual way.

$$F^a_{\mu\nu}[A] = P_a[\partial_\mu e^a_\nu - \partial_\nu e^a_\mu + \epsilon^a_{bc}(\omega^b_\mu e^c_\nu + \omega^c_\nu e^b_\mu)] \tag{4}$$

$$+ J_a[\partial_\mu \omega^a_\nu - \partial_\nu \omega^a_\mu + \epsilon^a_{bc} \omega^b_\mu \omega^c_\nu] \equiv P_a F^a_{\mu\nu}[e] + J_a F^a_{\mu\nu}[\omega] \tag{5}$$

With these preliminaries the Chern Simons action with the Poincaré group as gauge group can be written as [1,15]

$$I_{cs} = \frac{1}{2} \int_M d^3x \, \eta_{ab} \, \epsilon^{\mu\nu\lambda} \, e^a_\mu \, F^b_{\nu\lambda}[\omega] \tag{6}$$

Gauge theoretic metric independent actions of this type can be written down in any number of dimensions[16,17]. What distinguishes the 2+1 dimensional theory from all others is that in this case it becomes a Chern-Simons theory, indicating that the theory is locally trivial and is invariant under the full Poincaré gauge transformations. It is worth emphsizing that at this stage the manifold M in Eq. 6 is not to be identified with metrical space-time. It is specified in terms of its topology and is not endowed with clocks and measuring rods. It is not even clear whether the Poincaré symmetry of Eq 6 is a space- time symmetry. It will be seen in section V that the physical space-time is an output of this gauge theory and that the corresponding Poincaré symmetry is the symmetry of space-time. One can of course by-pass such conceptual subtleties by simply assuming that e^a_μ is invertible. For then one can define a non-singular metric with components

$$g_{\mu\nu} = \eta_{ab} e^a_\mu e^b_\nu \tag{7}$$

and reduce Eq. 6 to the standard Einstein action. Although this can be done [1,16,17], it would destroy the Poincaré gauge symmetry and would defeat the very purpose of approaching this problem from the gauge theory point of view.

To set up a canonical formalism, let the manifold M have the topology R $\times \Sigma$, where R is the real line representing x^0, and Σ is a two-dimensional manifold. Then, Eq. 6 reduces, up to a total derivative term [14], to

$$I_{cs} = \int dx^0 \int_\Sigma -\epsilon^{ij} e^a_i \partial_{x^0} \omega_{aj} - \eta_{ab} (e^a_0 F^b[\omega] + \omega^a_0 F^b[e]) \tag{8}$$

where

$$F^a[A] = \epsilon^{ij} F^a_{ij}[A] \tag{9}$$

It is easy to see from Eq. 8 that e_i^a and ω_i^a are canonically conjugate dynamical variables. On the other hand the x^0 derivatives of e_0^a and ω_0^a do not appear in the action, so that they are Lagrange multipliers. Their coefficients can then be viewed as constraints:

$$F^a[e] = F^a[\omega] = 0 \tag{10}$$

Their closure under Poisson brackets yields a representation of the Poincaré algebra. The constraint Eqs. 10 allow us to formally write down the Hamiltonian of the theory according to the prescription of Dirac:

$$G = -\int_\Sigma (\, \rho^a F_a[e] + \tau^a F_a[\omega] \,) \tag{11}$$

It coincides with the Hamiltonian obtained from the Eq. 8. This is not the whole story, however, and the consistency of the Hamiltonian formalism as well as the proper definition of energy later on require, in general, the addition of a boundary term to Eq. 11. We will obtain such a term for the Chern Simons action coupled to sources in section IV.

In contrast to 3+1 dimensions, the constraints of the 2+1 dimensional Poincaré gauge theory are solvable [1,8]. More specifically, they determine ω_μ^a to be a flat SO(2,1) connection on Σ, modulo gauge transformations.

III. Coupling to Sources (Particles)

The Poincaré invariance of the Chern Simons gauge theory suggests that we can introduce the notion of a particle into this theory in the same way as we do in particle physics in 3+1 dimensions. To this end, we take a particle to be an irreducible representation of the Poincaré group. Its mass and spin are the eigenvalues of the two *independent* Casimir operators of this group. As discussed in references [12-14], a spinless source of mass m carries a Poincaré charge $p^a = (p^0, \vec{p})$ with the condition that in the rest frame of the source $p^a = (m, 0)$, where m^2 is an eigenvalue of the first Casimir operator, p^2, of the Poincaré group. A source with spin carries another Poincaré charge j^a, i.e., the total (Lorentz) angular momentum given by

$$j^a = \epsilon^a_{\ bc} q^b p^c + s^a \tag{12}$$

The quantity q^a is the Poincaré coordinate canonically conjugate to the charge (momentum) p^a. From this, we can obtain the eigenvalues of the second Casimir operator W^2, of the Poincaré group. Denoting the eigenvalue by the same symbol W, we have

$$W = p \cdot j = p \cdot s \tag{13}$$

Since W is a Poincaré invariant, it can be evaluated in any convenient frame. In particular, in the rest frame of the source, $s^a = (s, 0)$, so that $W = ms$. Following Wigner's classic work, we identify s with the intrinsic spin of the source.

The simplest coupling of a spinless source to the Chern Simon action was first given by Witten [1]. For each source with canonical phase space variable p^a and q^a, the coupling is accomplished by means of the first order action

$$I = \int_C d\tau \, \eta_{ab} \, p^a [\partial_\tau q^b + t^\mu (e_\mu^b + \epsilon^b_{\ cd} q^c \omega_\mu^d)] + \lambda(p^2 - m^2) \tag{14}$$

where $t^\mu = dx^\mu/d\tau$. It is clear from the action that the quantities p^a, and q^a are canonically conjugate to each other and satisfy Poisson brackets. For more than one source, one can add an action of this type for each source. For sources with spin, we generalize [12-14] this action to

$$I = \int_C d\tau \; \eta_{ab} \left[p^a \partial_\tau q^b + t^\mu (p^a e^b_\mu + j^a \omega^b_\mu) \right] + \lambda_1 (p^2 - m^2) + \lambda_2 (W^2 - m^2 s^2) \quad (15)$$

where j^a and W are given, respectively, by Eqs. 12 and 13.

In the presence of sources, the constraint equations take the form

$$\tilde{F}^a[\omega] \equiv F^a[\omega] - \sum_{k=1}^{N} P^a_{(k)} \delta^2(x, x_{(k)}) = 0 \quad (16)$$

$$\tilde{F}^a[e] \equiv F^a[e] - \sum_{k=1}^{N} J^a_{(k)} \delta^2(x, x_{(k)}) = 0 \quad (17)$$

where $J_{(k)}$ is the orbital (Lorentz) angular momentum of the kth source. As mentioned above, these constraint equations are solvable and determine ω^a_μ to be a flat SO(2,1) connection on Σ modulo gauge transformations. The analog of the constraint Hamiltonian given by Eq. 11 is now

$$H_0 = - \int_\Sigma \rho^a \tilde{F}_a[e] + \tau^a \tilde{F}_a[\omega] \quad (18)$$

It is still neccessary to add a boundary integral to this expression to ensure the consistency of the formalism in the emerging space-time.

IV. The Exact Solution of the Two Body Problem

We want to solve the problem of two sources coupled to the Chern Simons gravity by reducing it to an equivalent one body problem. We make use of this terminology in the same sense that, e.g., the familiar two body central force problem is mathematically cast into the form of an equivalent one body problem. The main difference is that in the present case this reduction is acheived not by a coordinate trannsformation but by means of the topological features of the theory. In a topological gauge theory all the gauge invariant obsevables can be expressed in terms of Wilson loops [1]. By definition, the Wilson loop for the connection given by Eq. 2 in the representation $\mathcal{R}$ of the algebra is given by

$$W_\mathcal{R}(C) = Tr_\mathcal{R} P \exp\left[i \oint_C A \right] \quad (19)$$

where P denotes path ordering. The paths characterizing different Wilson loops are distinguished from each other not by their local coordinates but by their homotopy classes. This together with gauge invariance require that we carry out the reduction in terms of Wilson loops. Moreover, we note that the Casimir invariants of a Poincaré state, which we identify as mass and spin, must be Wilson loops. Since the equivalent one body problem must also be a Poincaré state specified by its mass and spin, it can be regarded as arising from a single source endowed with two charges: a charge $\Pi^a = (\Pi^0, \vec{\Pi})$ and a charge $\Psi^a = (\Psi^0, \vec{\Psi})$, such that the Casimir invarinats of the corresponding state are given, respectively, by $\Pi \cdot \Pi = H^2$ and $\Pi \cdot \Psi = HS$. We identify H and S as the mass and spin of the one body source. Being gauge invariant observables, they can be expressed in terms of Wilson loops of the two body system.

To evaluate H and S, let $W_\mathcal{R}(C_0)$ be a Wilson loop enclosing the equivalent one body source. In a frame in which this source is at rest at the origin, this gives

$$W_R(C_0) = Tr_R P \exp[-i (HJ_0 + SP_0)] \quad (20)$$

Since the dynamics generated by the reduced one body formalism must be identical to that of the original two body system, $W_R(C_0)$ must be equal to a Wilson loop, $W_R(C_{12})$, enclosing the two sources with charges (p_1^a, j_1^a) and (p_2^a, j_2^a), respectively. Thus, we require that [12-14]

$$W_R(C_0) = W_R(C_{12}) \qquad (21)$$

The path C_{12} can be chosen uniquely by the requirement that these Wison loops correctly represent the asymptotic observables of the emerging space-time [11,14].

It remains to explicitly evaluate $W_R(C_0)$ and

$$W_R(C_{12}) = Tr_R P \exp[i \oint_{C_{12}} (\omega^a J_a + e^a P_a)] \qquad (22)$$

Since the Poincaré algebra is not semi-simple, it turns out to be technically more convenient to evaluate these Wilson loops, $\hat{W}_R(C)$, for the anti-de Sitter algebra SO(2,2) $\sim$ SU(1,1)$\times$ SU(1,1) and then obtain their Poincaré limit by a group contraction.

To evaluate $\hat{W}_R(C_0)$, let the source representing the equivalent one body problem be endowed with anti-de Sitter charges

$$\hat{\Pi}^a = (\hat{\Pi}^0, \vec{\hat{\Pi}}) \; ; \quad \hat{\Psi}^a = (\hat{\Psi}^0, \vec{\hat{\Psi}}) \qquad (23)$$

These charges are subject to the requirement that after group contraction, $\hat{\Pi}^a \to \Pi^a$ and $\hat{\Psi}^a \to \Psi^a$ such that, we get

$$\Pi \cdot \Pi = H^2 \; ; \quad W = \Pi \cdot \Psi \qquad (24)$$

Then, with $\hat{Z}_\pm^a = \hat{\Pi}^a \pm \hat{\Psi}^a$, we obtain

$$\hat{W}_R(C_0) = 4 \cos \frac{|\hat{Z}_+|}{2} \cos \frac{|\hat{Z}_-|}{2} \qquad (25)$$

where

$$|\hat{Z}_\pm| = \left[(\hat{\Pi}^2 + \hat{\Psi}^2) \pm 2\hat{\Pi} \cdot \hat{\Psi}\right]^{\frac{1}{2}} \qquad (26)$$

The Casimir invariants of the anti-de Sitter group are given by

$$\hat{C}_1 \sim \hat{\Pi}^2 + \hat{\Psi}^2 \; ; \quad \hat{C}_2 \sim \hat{\Pi} \cdot \hat{\Psi} \qquad (27)$$

After group contraction

$$\hat{C}_1 \to C_1 = \Pi^2; \quad \hat{C}_2 \to C_2 = \Pi \cdot \Psi \qquad (28)$$

It follows that

$$|\hat{Z}_\pm| \to |Z_\pm| = \left[\Pi^2 \pm 2\Pi \cdot \Psi\right]^{\frac{1}{2}} = \left[H^2 \pm 2HS\right]^{\frac{1}{2}} \qquad (29)$$

Finally, we must compute the Wilson loop $W_R(C_{12})$ where C_{12} is a simple loop enclosing the two sources with charges (p_1^a, j_1^a) and (p_2^a, j_2^a), respectively. Again, for technical reasons it is advantageous to use the anti-de Sitter group and obtain $W_R(C_{12})$ by group contraction. Proceeding along the same lines as in the computation of $W_R(C_0)$, one obtains, after group contraction,

$$\begin{aligned} W_R(C_{12}) = 4 &\left[\cos \frac{|z_{1+}|}{2} \cos \frac{|z_{2+}|}{2} - \frac{z_{1+} \cdot z_{2+}}{|z_{1+}||z_{2+}|} \sin \frac{|z_{1+}|}{2} \sin \frac{|z_{2+}|}{2}\right] \\ \times &\left[\cos \frac{|z_{1-}|}{2} \cos \frac{|z_{2-}|}{2} - \frac{z_{1-} \cdot z_{2-}}{|z_{1-}||z_{2-}|} \sin \frac{|z_{1-}|}{2} \sin \frac{|z_{2-}|}{2}\right] \end{aligned} \qquad (30)$$

where
$$z^a_{k\pm} = p^a_k \pm j^a_k \qquad (31)$$

Substituting into Eq. 25 the results obtained in Eqs. 29 and 30, we get

$$\cos\{\frac{1}{2}\left[H^2 + 2HS\right]^{\frac{1}{2}}\} \cos\{\frac{1}{2}\left[H^2 - 2HS\right]^{\frac{1}{2}}\} = \frac{1}{4}W_R(C_{12}) \qquad (32)$$

To evaluate H and S from these expressions in terms of the invariants of the two body system, we note that they are two independent Casimir invariants. Therefore, the expression for H must be the same regardless of whether the two sources carry spin or not. Therefore, setting the spins of the two sources in Eqs. 30 and 32 equal to zero, we obtain

$$\cos\frac{H}{2} = \cos(\frac{m_1}{2})\cos(\frac{m_2}{2}) - \frac{p_1 \cdot p_2}{m_1 m_2}\sin(\frac{m_1}{2})\sin(\frac{m_2}{2}) \qquad (33)$$

Therefore, Eqs. 30, 32, and 33 provide a complete solution to the two body dynamics for particles with spin in terms of an equivalent one body problem. They can, in principle, be solved for S by substituting for H from Eq. 33.

V. The Physical Space-Time

Up to this point, we have constructed a Chern Simons gauge theory coupled to sources on $R\times\Sigma$ (x-space) which as we emphasized is not space-time. In the process, we have made use of such suggestive notions as the Poincaré group, the Hamiltonian, and Wilson loops. Now we want to show how the notion of a metrical space-time emerges from this formalism and what our gauge invariant observables correspond to in such a space-time. It is clear that the identification of quantities such as momenta and coordinates of physically realizable particles can only be made in a metrical space-time. For clarity, I will carry out this demonstration for spinless sources.

It has been noted [10,11] that for two or more sources the couplings to the Chern Simons action given by Eqs. 14 and 15 are not unique to the extent that the choice of phase space variables for such a system is not unique. This nonuniqueness is reminiscent of the freedom of choosing Majorana or Dirac couplings for neutrinos. We shall make use of this freedom to choose a set of phase space variables suitable for comparison with 't Hooft 's work [5]. To this end, let (p_1,q_1), and (p_2,q_2) be the phase space dynamical variables for the two particles, where p^a_1 and p^a_2 are the Poincaré "charges". Then consider the combinations

$$P^a = p^a_1 + p^a_2 \quad ; p^a = p^a_1 - p^a_2 \qquad (34)$$

Let Q^a and q^a be the corresponding canonically conjugate variables. This suggests the following action for the two particle (source) system coupled to gravity:

$$I_s = \int_{C_Q} d\tau \; \eta_{ab} \; P^a[\partial_\tau Q^b + T^\mu(e^b_\mu + \epsilon^b{}_{cd}Q^c\omega^d_\mu)] + \lambda_1[P^2 - m_1^2 - m_2^2 - \Delta]$$
$$+ \int_{C_q} d\tau \; \eta_{ab} \; p^a[\partial_\tau q^b + t^\mu(e^b_\mu + \epsilon^b{}_{cd}q^c\omega^d_\mu)] + \lambda_2[p^2 - m_1^2 - m_2^2 + \Delta] \qquad (35)$$

where Δ is the numerical invariant $2p_1 \cdot p_2$. In this action which is invariant under the Poincaré gauge transformations, the paths C_Q and C_q correspond not to the individual particle trajectories but to the source trajectories specified by the charges P^a and p^a, respectively. Q^a and q^a may be thought of as one dimensional fields defined, respectively, over the trajectories C_Q and C_q in $R\times\Sigma$.

As we have seen in the previous sections, the gauge invariant observables of this 2+1 dimensional topological gauge theory can be expressed in terms of Wilson loops.

To obtain some clues to the physical interpretation of such observables, let us consider some representative Wilson loops. Let

$$s = P \cdot P = (p_1 + p_2)^2 \equiv (E_1 + E_2)^2 - (\vec{p}_1 + \vec{p}_2)^2 \tag{36}$$

$$p^2 = p \cdot p = (p_1 - p_2)^2 \equiv (E_1 - E_2)^2 - (\vec{p}_1 - \vec{p}_2)^2 \tag{37}$$

The quantity s in Eq. 36 should not be confused with the spin, s, of previous sections. The simplest non-trivial Wilson loops are those enclosing only one source. For the spinless source with the charge P^a, it is easy to show that the corresponding Wilson loop is proportional to $cos(\sqrt{s}/2)$, so that the quantity s in Eq. 36 is a gauge invariant observable. Similarly, the Wilson loop around the other source will identify the quantity p in Eq. 37 as another gauge invariant observable.

The quantities s and p are kinematical invariants which we have identified with the Casimir invariants (or their eigenvalues) which label the Poincaré states. To understand the dynamics of the two body problem, we must consider a Wilson loop enclosing both sources. Following the same line of argument as in the previous section, we find that for the source action given by Eq. 35 the Casimir invariant H is now given by

$$\cos\frac{H}{2} = \cos\frac{\sqrt{s}}{2}\cos\frac{|p|}{2} - \frac{P \cdot p}{|p|\sqrt{s}}\sin\frac{\sqrt{s}}{2}\sin\frac{|p|}{2} \tag{38}$$

As expected, the exact Hamiltonian, H, is again a highly non-linear non-polynomial function of the invariants s and p. Since it is a gauge invariant quantity, it can be evaluated in any convenient gauge. In particular, in the gauge in which $\vec{Q} = 0$ it takes the form

$$\cos\frac{H}{2} = \cos[\frac{E_1 + E_2}{2}]\cos(\frac{|p|}{2}) - \sin[\frac{E_1 + E_2}{2}]\sin(\frac{|p|}{2}) \cdot [\frac{m_1^2 - m_2^2}{|p|(E_1 + E_2)}] \tag{39}$$

By requiring that $\vec{Q}$ remain at the origin, we have fixed four of the six Poincaré transformations. We can still make spatial rotations and translations along q^0. So consider the transformation

$$\vec{q}' = \vec{q}[\exp \tau^0 J_0] \tag{40}$$

where τ_0 is a spacial rotation given by,

$$\tau^0 = (1 - H/2\pi)\phi \tag{41}$$

Here H is numerical value of the exact Hamiltonian given by Eq. 39. This transformation which is clearly not 2π periodic leaves s, p, and hence H, invariant. But it is easy to see that the transformed coordinates $\vec{q}'$ acquire a phase under the rotation $\phi \rightarrow \phi + 2\pi$:

$$\vec{q}'(\phi + 2\pi) = [\exp{(2\pi - H)J_0}]\vec{q}'(\phi) \tag{42}$$

That is, they satisfy the matching conditions for the coordinates on a cone characterized by the deficit angle $\beta = H$. To have a well defined scattering problem it is also necessary [5] that

$$H < 2\pi/G \tag{43}$$

Here, we have reinserted the gravitational constant G which up to now had been set equal to unity. It thus follows that the general reduction of the two-body problem to an equivalent one-body problem always leads, in a particular gauge, to the motion of the relative coordinate on a cone. We know from the analysis of metrical general relativity [2-4] that point sources generate conical space-times. For a single source, the deficit angle of the cone is determined by the energy (mass), E, of the source. We must therefore identify the quantity H with the total gravitational energy of the two body system. It generates a cone over whcih the relative coordinate of the reduced two body system moves. Despite their similarities, this cone should not be confused with the conical space of the test particle approximation. As is clear from Eq. 39, our cone receives contributions from both sources.

It is thus clear that it is not the manifold $R \times \Sigma$ (x- space) but the q-space, M_q, from which the classical space-time is manufactured. Once the spatial part of q^a is identified with the cone, relativistic invariance requires that q^0 be identified with the "classical time". This is to be contrasted with the problem of time in metrical general relativity. In the gauge theory approach, the choice of gauge essentially fixes the time. It is remarkable that this geometrical picture emerges in full generality. Here we have a concrete example of how the dynamics and topology of a gauge theory can uniquely lead to a metrical space-time. The quantity H characterizing this space-time also supplies, as it should, the boundary term which is necessary for the consistency of the canonical formalism in the metrical theory.

The manner in which we arrived at this conical geometry is also consistent with the notion that a metrical theory of gravity corresponds to a broken phase of a Poincaré gauge theory. The picture is clearly gauge dependent. The requirement that the canonical coordinate $\vec{Q}$ remain fixed at the origin explicitly breaks the Poincaré gauge symmetry. The left over symmetry consists of invariances under time translations and spacial rotations, in agreement with asymptotic invariances of the corresponding metrical gravity.

Having shown how the classical space-time emerges from the Chern Simons gauge theory of the Poincaré group, we can now make contact with the work of 't Hooft [5]. In Eq. 39 if we set $m_1 = m_2$, it follows that $p = 2|\vec{p}_1|$. So, for small momenta such that $\cos(p/2) \approx 1$, we get

$$H \approx E_1 + E_2 \; ; \quad \mod 2\pi \qquad (44)$$

The ambiguity can be removed by requiring that H vanish when E_1 and E_2 vanish. This coincides with the free Hamiltonian of 't Hooft [5]. The conditions under which we have obtained it seem reasonable since for small momenta the back reactions of the particles on each other are small. We note that for the approximate expression in Eq. 44 to be applicable, the masses m_1 and m_2 need not be exactly equal.

We conclude this section by noting that the geometric structure of the metrical space-time which was manufactured in this section is independent of the choice of the phase space variables. The numerical values of its asymptotic observables such as H and S, however, do depend on this choice. The situation here is again similar to the distinction between Majorana and Dirac neutrinos. The physically correct choice for the phase space variables can be made, in principle, via cosmic string scattering experiments. This is the subject of the next section.

VI. Application to Particle Scattering

The results obtained in section V demonstrate the important role played by the equivalent one body formalism and the choice of gauge in arriving at the correct metrical space-time geometry. These two features play equally important roles in obtaining the two paricle scattering amplitude. We have seen that from the point of view of the Chern Simons gauge theory of the Poincaré group the very notion of space-time is gauge dependent. This means that the gauge invariant properties of the space-time theory can be studied in other, possibly more convenient, gauges of the Chern Simons theory. One such gauge invariant quantity is the scattering amplitude. To compute this quantity, it turns out to be more convenient to work in the manifold M_q rather than the cone which was obtained from it by a singular gauge transformation. In M_q, the equivalent one body source (Poincaré state) characterized by the pair of invariants H and S is located at the origin. So from the point of view of the scattered particle, the topology of M_q looks like that of Minkowski space with the origin cut out of it. Thus the exact scattering problem becomes *mathematically* identical to that of a test particle scattering from a fixed source. The scattering occurs because the topology of M_q is not trivial. This description amounts to approaching the two particle scattering in two steps. In the first step one writes down an effective one body action which generates the metrical manifold M_q (or equivalently a cone) with the values of H and S given by Eqs. 33 and 32, repectively. The two particle scattering can then be described by evaluating the quantities H and S in the rest frame of one of the particles, say (p_1, j_1), and then considering the scattering of the other particle, (p_2, j_2), from the source (H, S).

The quantization of this theory is accomplished by replacing the Poisson bracket between p and the corresponding conjugate variable q with a commutator and then constructing a Hilbert space of square integrable wave functions $\Psi(q)$.

To compute the scattering amplitude, let, $|i> \equiv |i, in>$ represent the initial state, and $|f> \equiv |f, out>$ the final state. Then, in standard notation, we have

$$|f, out> = S(f,i)|i, in> \qquad (45)$$

Since S(f,i) is a gauge invariant operator, it can be expressed in terms of Wilson loop operators in M_q. The paths of such loop operators differ from one another by the number of times they wrap around the source at the origin. Since they are all allowed *a priori*, we must sum over all of them with equal weight. Indexing these paths by an integer $n = 0, \pm 1, \pm 2, ...,$, we have

$$S(f,i) = \sum_n P \exp\left[-i \oint_{c_n} (e^a p_a + \omega^a j_a)\right] \qquad (46)$$

where the paths become closed by the addition of an n = 0 path to each of them. It is remarkable that in this formalism the S-matrix can be specified without an explicit reference to time or time ordering.

Using non-abelian Stoke's theorem, we have, in the rest frame of the first particle,

$$\oint_{c_n} (e^a p_a + \omega^a j_a) = \int_{\sigma_n} (F^a[e] p_a + F^a[\omega] j_a) \qquad (47)$$

$$= n \int_\sigma (F^a[e] p_a + F^a[\omega] j_a) = n (H j_2^0 + S p_2^0) \qquad (48)$$

Therefore,

$$|f> = \sum_{n=-\infty}^{\infty} \exp[in (H j_2^0 + S p_2^0)]|i> \qquad (49)$$

Following the steps given reference [14], we find that the exact expression for the scattering amplitude for particles of any spin is given by

$$f(\phi) = \frac{e^{\frac{i\phi}{2\pi\alpha}(Hs_2 + E_2 S)}}{2\alpha\sqrt{-2\pi i p_2}}$$

$$\times \left[e^{-i\pi(Hs_2 + E_2 S)} \cot\left(\frac{\phi - \pi}{2\alpha}\right) - e^{+i\pi(Hs_2 + E_2 S)} \cot\left(\frac{\phi + \pi}{2\alpha}\right) \right] \qquad (50)$$

In the test particle approximation, where $m_1 \gg m_2$ and $|\vec{p}_2|$ is small, it follows from Eqs. 32 and 33 that $H \approx m_1$ and $S \approx s_1$. Then, the phases in our Eq. 50 reduce to forms similar to the approximate results obtained by deSousa Gerbert and Jackiw [7] for spin one-half. Even in this approximation, however, our expression for the scattering amplitude is not identical to that given in reference [7]. The differences seem to be related to the choices of the initial wave function.

VII. Extension to Supersymmetric Theories

The analysis of the Chern Simons gauge theory of the Poincaré group coupled to sources presented in the previous sections can be extended to supergravity theories [18]. In 2+1 dimensions, simple and extended supergravity theories can also be cast into the form of a Chern Simons gauge theory of the corresponding supergroup [1,15]. Just as in the case of pure gravity, to maintain gauge invariance with respect to the full supergroup, it is necessary that the matter coupling be accomplished by the introduction of (super)sources. Taking our clue from the above developments, we take the supersources to be irreducible representations of an N- extended super Poincaré group. For a given N, these representations consist of a finite number of Poincaré states related to each other by the action of the supersymmetric charges. One of the remarkable features of the representations of the Poincaré group in 2+1 dimensions is that their spin values are not limited to integers and half integers and can be any real number. This feature allows us to construct irreducible representations of an extended super Poincaré group from Poincaré states carrying any spin (in particular, fractional) spin. Within such a supermultiplet, the spin of each Poincaré state still differs from the others by multiples of 1/2 unit. Two important conclusions can be drawn from the structure of such supermultiplets. One is that it is possible to construct representations of supersymmetry on states which are not fermions and bosons in the usual sense. The other is that such supermultiplets can be coupled to the corresponding Chern Simons supergravity in a super Poincaré gauge invariant way. This appears to permit the construction of unbroken supersymmetric theories in which the excited states are not ordinary fermions and bosons.

This work was supported in part by the Department of Energy under contract number DOE-FG02-84ER40153.

References

1. E. Witten, *Nucl. Phys.* **B311** (1988) 46 and **B323** (1989) 113

2. S.Deser, R.Jackiw, G. 't Hooft, *Ann. of Phys.* (N.Y.) **152** (1984) 220

3. S. Giddings, J. Abbott and K. Kuchar, *Gen. Rel. Grav.* **16**(1984) 751

4. J.R. Gott and M. Alpert, *Gen. Rel. Grav.* **16** (1984) 751

5. G. 't Hooft, *Comm. Math. Phys.* **117** (1988) 685

6. S.Deser and R.Jackiw, *Comm. Math. Phys.* **118** (1988) 495

7. P. de Sousa Gerbert and R. Jackiw, *Comm. Math. Phys.* **124** (1989) 229

8. S. Carlip, *Nucl. Phys.* **B324** (1989) 106

9. K. Koehler, F. Mansouri, C. Vaz, L. Witten, *Nucl. Phys.* **B348** (1990) 373

10. M.K. Falbo-Kenkel and F. Mansouri, *Mod. Phys. Lett.* **A7** (1992) 2173

11. M.K. Falbo-Kenkel and F. Mansouri, *Jour. Math. Phys.* **34** (1993) 139

12. F. Mansouri and M.K. Falbo-Kenkel, *Mod. Phys. Lett.* **A8** (1993) 2503

13. M.K. Falbo-Kenkel and F. Mansouri, *Phys. Lett.* **B309** (1993) 28

14. F. Mansouri, "Manufacturing Space-time from Gauge Theories" in *Particles and the Universe*, ed. by Z. Horvath, *et. al.*, World Scientific, 1994

15. A. Achucarro and P.K. Townsend, *Phys. Lett.* **B180** (1986) 89

16. S.W. MacDowell and F. Mansouri *Phys. Rev. Lett.* **38** (1977) 739

17. F. Mansouri, *Phys. Rev.* **D16** (1977) 2456

18. S. Kim and F. Mansouri, *in preparation*

THEOREMS ON ESTIMATING PERTURBATIVE COEFFICIENTS IN QUANTUM FIELD THEORY AND STATISTICAL PHYSICS

Mark A. Samuel[1] and Stephen D. Druger[2]

[1]Department of Physics
Oklahoma State University
Stillwater, OK 74078
Email address: physmas@mvs.ucc.okstate.edu
[2]Department of Physics and Astronomy
Northwestern University
Evanston, IL 60208-3112
Email address:druger@nwu.edu

ABSTRACT

We present rigorous proofs for several theorems on using Padé approximants to estimate coefficients in Perturbative Quantum Field Theory and Statistical Physics. As a result, we find new trigonometric and other identities where the estimates based on this approach are exact. We discuss hypergeometric functions, as well as series from both Perturbative Quantum Field Theory and Statistical Physics.

I. INTRODUCTION

Recently, we proposed[1-5] a method of estimating coefficients in Perturbative Quantum Field Theory and Statistical Physics with error bars for each estimate. The method makes use of Padé approximants and yields a Padé approximant approximation (PAP). There are many good references for Padé approximants, such as for example Refs. 6-10. We begin by defining the Padé approximant

$$[N/M] = \frac{a_0 + a_1 x + a_2 x^2 + \ldots + a_N x^N}{1 + b_1 x + b_2 x^2 + b_3 x^3 + \ldots + b_M x^M} \qquad (1.1)$$

to the series

$$S = S_0 + S_1 x + \ldots + S_{N+M} x^{N+M} \qquad (1.2)$$

where we set

$$[N/M] = S + O(x^{N+M+1}) . \qquad (1.3)$$

We have written a computer program that solves Eq. (1.3) numerically and then predicts the coefficient of the next term S_{N+M+1}. It works for arbitrary N and M. Furthermore, we have derived algebraic formulae for the [N/1], [N/2], [N/3], [N/4], [N/5], and [N/6] PAP's, where N is arbitrary.

To illustrate the method, consider the simple example

$$\frac{\ln(1+x)}{x} = 1 - \frac{x}{2} + \frac{x^2}{3} - \frac{x^3}{C} . \qquad (1.4)$$

We write the [1/1] Padé as

$$[1/1] = \frac{a_0 + a_1 x}{1 + b_1 x} . \qquad (1.5)$$

It is easy to show that $a_0 = 1$, $b_1 = 2/3$, $a_1 = 1/6$, and $C = 9/2$. We can see that the prediction for C is close to the correct value $C = 4$. For $x = 1$, we get $[1/1] = 7/10$, close to the correct result, $\ln 2 = 0.6931$. This is much better than the partial sum

$$1 - \frac{1}{2} + \frac{1}{3} = \frac{5}{6} = 0.8333 . \qquad (1.6)$$

By going to higher order, it is easy to show that

$$[1/2] = \frac{1 + \frac{x}{2}}{1 + x + \frac{x^2}{6}} \qquad (1.7)$$

and for $x = 1$ we obtain

$$m[1/2] = \frac{9}{13} = 0.6923 \qquad (1.8)$$

very close to ln 2. The PAP is $7/36 = 0.1944$ very close to the correct value of $1/5$.

As discussed in our recent article, the error bars are obtained by taking the reciprocals

$$r_n = \frac{1}{S_n} \tag{1.9}$$

and finding the PAP for r_{n+1}, and then taking the reciprocal. Then we consider the differences

$$t_n = r_{n+1} - r_n \tag{1.10}$$

and find the PAP for t_n. We then have

$$r_{n+1} = r_n + t_n \tag{1.11}$$

and then take the reciprocal

$$S_{n+1} = \frac{1}{r_{n+1}} . \tag{1.12}$$

Our error bar is calculated from the difference between the results from Eqs. (1.9) and (1.12).

II. THEOREMS

We consider the general series

$$S = \sum_n S_n x^n . \tag{2.1}$$

A. Sums of geometric series:

If S_n is a sum of M geometric series then the [N/M] PAP for $N \geq M-1$ is exact.
Proof: For $M = 2$,

$$S_n = ar^n + bs^n \tag{2.2}$$

and

$$S = \sum_{n=0}^{\infty} S_n x^n = \frac{a}{1-rx} + \frac{b}{1-sx} \tag{2.3}$$

so that

$$S = \frac{(a+b)-(as+br)x}{(1-rx)(1-sx)} \tag{2.4}$$

and the [1/2] and higher PAP's are exact.

To prove the general case we use mathematical induction. Assume that the theorem is true for the [M-1/M] PAP. Now for M → M+1, we have

$$\frac{P_{M-1}}{Q_M} + \frac{g}{1-ax} = \frac{P_{M-1}+gQ_M}{Q_M(1-ax)} \tag{2.5}$$

and the [M/M+1] PAP is exact. Here

$$S_n = ar^n + bs^n + \ldots + gw^n \tag{2.6}$$

and P_M and Q_M are polynomials of degree M.

B. Signs of geometric series

For

$$S_n = (-1)^{{}^nC_m} ar^n \tag{2.7}$$

the [m-1/m] PAP is exact, where ${}^nC_m$ denotes the binomial coefficient n!/[m! (n-m)!].
Proof: The proof is easily obtained by recognizing that the series in Eq. (2.7) is just a sum of geometric series and using Theorem A.

C. A sufficient condition for PAP's to be accurate:

If

$$g(n) = \frac{d^2 \ln S_n}{dn^2} \tag{2.8}$$

then a sufficient (but not necessary) condition for the PAP's to be accurate is

$$\lim_{n \to \infty} g(n) = 0 . \tag{2.9}$$

Proof: A_n and ε_n satisfy[2]

$$A_n \equiv 1 + \epsilon_n \equiv \frac{S_n S_{n+2}}{(S_{n+1})^2} \qquad (2.10)$$

and hence

$$A_n = 1 + \epsilon_n = e^{g(n)} . \qquad (2.11)$$

The % error is expressible in terms of the ϵ_n and if $g(n) \to 0$ then $\epsilon_n \to 0$. As shown in Ref. 2, $\epsilon_n \to 0$ implies that the PAP is accurate.

D. A generalization of Theorem C:

If, in addition to the conditions of Theorem C for S_n we generalize to a series

$$T = \sum_n T_n x^n \qquad (2.12)$$

where

$$T_n = (-1)^{{}^nC_m} S_n \qquad (2.13)$$

then the [m-1/m] and higher PAP's will be accurate.
Proof:

$$A_n' = 1 + \epsilon_n' = (-1)^{{}^nC_{m-2}} A_n \qquad (2.14)$$

where A_n is given by Eq. (2.10). Then we use Theorems B and C to prove Theorem D. For further details, see Ref. 2.

E. Polynomials of the nth degree:

For $S_n = P_n$ where P_n is a polynomial of degree n, the [N/M] PAP's are exact, where $M = n+1$ and $N \geq M-1$.
Proof: By differentiating

$$S = \sum_n x^n = (1-x)^{-1} \qquad (2.15)$$

and multiplying by x, we obtain

$$\sum_n (an+b)x^n = \frac{(a-b)x+b}{(1-x)^2} \qquad (2.16)$$

and the [N/2] is exact for $N \geq 1$. Now by induction we can easily obtain the desired result.
It should be emphasized that in all cases once the [N/M] PAP's are exact for $N \geq M-1$, the results remain exact for all higher order PAP's [N'/M'] for $M' > M$ and $N' > M'-1$.

III. SOME NEW TRIGONOMETRIC IDENTITIES

If we consider the series given by

$$S_n = \sin[(n+1)\theta + \delta] \qquad (3.1)$$

where θ and δ are arbitrary we will prove that the [N/M] PAP's are exact for $M \geq 2$ and $N \geq M-1$. This leads to new trigonometric identities corresponding to each of the [N/2], [N/3], [N/4], etc., PAP's.

From Eq. (3.1) it can easily be shown that

$$S = \sum_{n=0}^{\infty} S_n x^n$$
$$= \frac{x\cos(\theta+\delta)\sin\theta + \sin(\theta+\delta)(1-x\cos\theta)}{1-2x\cos\theta+x^2} . \qquad (3.2)$$

Hence the [N/M] PAP's are exact for $M \geq 2$, $N \geq M-1$. With $\delta = 0$, Eq. (3.2) becomes

$$S = \frac{\sin\theta}{1+x^2-2x\cos\theta} \qquad (3.3)$$

and, hence, the [0/2] PAP is exact.
Similarly for

$$S_n = \cos[(n+1)\theta + \delta] \qquad (3.4)$$

we can obtain

$$S = \frac{\cos(\theta+\delta)(1-x\cos\theta) - x\sin(\theta+\delta)\sin\theta}{1-2x\cos\theta+x^2} . \qquad (3.5)$$

In this case, however, if we set $\delta = 0$, we obtain

$$S = \frac{\cos\theta - x}{1 - 2x\cos\theta + x^2}, \quad (3.6)$$

and the [0/2] PAP is not exact!

Now, for each [N/M] PAP that is exact we will find a trigonometric identity. We begin with M = 2. The [1/2] PAP is given by $S_4 = N/D$ where

$$N = 2S_1 S_2 S_3 - S_0 S_3^2 - S_2^3 \quad (3.7)$$

and

$$D = S_1^2 - S_0 S_2 \quad (3.8)$$

with the S_n given by Eq. (3.1). Thus we have the identity

$$N - S_4 D \equiv 0 \quad (3.9)$$

which becomes

$$2\sin(2\theta + \delta)\sin(3\theta + \delta)\sin(4\theta + \delta)$$
$$-\sin(\theta + \delta)\sin^2(4\theta + \delta) - \sin^3(3\theta + \delta) \quad (3.10)$$
$$-\sin(5\theta + \delta)[\sin^2(2\theta + \delta) - \sin(\theta + \delta)\sin(3\theta + \delta)] = 0$$

Now one can step up in n, $S_n \to S_{n+1}$, and obtain another identity. This procedure can be repeated indefinitely. One can also step down in n, $S_n \to S_{n-1}$, where we set $S_{-1} = 0$. This gives a simpler identity, which can be obtained from simple known identities, for the [0/2] PAP. But we must set $\delta = k\pi$, $k = 0, 1, 2, \dots$, yielding

$$2\sin\theta \sin(2\theta)\sin(3\theta) - \sin^3(2\theta) - \sin(4\theta)\sin^2\theta = 0. \quad (3.11)$$

One can also use Eq. (3.9) to obtain the same identities for $\cos[(n+1)\theta + \delta]$. However, in this case there is no [0/2] identity for $\delta = 0$.

We now turn to M = 3. The [2/3] PAP is given by

$$S_6 = A/B \quad (3.12)$$

where

$$A = 2S_2^2 S_3 S_5 - 2S_1 S_3^2 S_5 + 2S_0 S_3 S_4 S_5$$

$$-2S_1 S_2 S_4 S_5 + S_1^2 S_5^2 - S_0 S_2 S_5^2 + S_2^2 S_4^2 \qquad (3.13)$$

$$-3S_2 S_3^2 S_4 + 2S_1 S_3 S_4^2 - S_0 S_4^3 + S_3^4$$

and

$$B = S_2^3 - 2S_1 S_2 S_3 + S_0 S_3^2 - S_0 S_2 S_4 + S_1^2 S_4 . \qquad (3.14)$$

Again we use Eqs. (3.1) and (3.4), but this time there are two identities in each case

$$A = 0 \quad \text{and} \quad B = 0 . \qquad (3.15)$$

The first part of Eq. (3.15) yields new identities, but the second part gives only previous identities for M = 2. This is in accordance with a theorem presented in an earlier paper.[2] We can again step up in n, $S_n \to S_{n+1}$, and obtain more identities for both $\sin[(n+1)\theta+\delta]$ and $\cos[(n+1)\theta+\delta]$. This process may be repeated as many times as desired. Now we may very easily also step down in n, $S_n \to S_{n-1}$, where we set $S_{-1} = 0$. This gives identities for the [1/3]. In this case the identity obtained results from

$$A - B S_6 = 0 \qquad (3.16)$$

for both sine and cosine, with $\delta \ne n\pi$. For the sine case, if $\delta = k\pi$, we obtain

$$A = 0 \quad \text{and} \quad B = 0 . \qquad (3.17)$$

If we step down once more to the [0/3] PAP and set $\delta = k\pi$, then we obtain an identity for sine, but not for cosine.

Although this process can be continued indefinitely for M = 4, 5, 6, ..., the formulae become increasingly complicated, as will soon be seen. So we present results for only one more value of M, namely M = 4. The [3/4] PAP is given by

$$S_8 = C/D \qquad (3.18)$$

where

$$C = 2(2S_2S_3S_4^2S_7 - S_3^3S_4S_7 - S_1S_4^3S_7$$

$$-S_2^2S_4S_5S_7 + S_1S_3S_4S_5S_7 - S_3^2S_4^2S_6 + 2S_1S_4^2S_5S_6$$

$$+S_2^2S_4S_6^2 - S_1S_3S_4S_6^2 - S_2S_3S_4S_5S_6 - S_2S_4^3S_6$$

$$+S_3^2S_4S_5^2 + S_2S_4^2S_5^2 - 2S_3S_4^3S_5 - S_1S_4S_5^3 + S_2S_3^2S_5S_7$$

$$+S_0S_4^2S_5S_7 + S_1S_2S_5^2S_7 - S_0S_3S_5^2S_7 - S_2^2S_4S_5S_7$$

$$-S_1S_3S_4S_5S_7 - S_2^2S_3S_6S_7 - S_0S_3S_4S_6S_7 - S_1^2S_5S_6S_7$$ (3.19)

$$+S_0S_2S_5S_6S_7 + S_1S_2S_4S_6S_7 + S_1S_3^2S_6S_7 - S_1S_2S_3S_7$$

$$-S_3^3S_5S_6 - S_1S_3S_4S_6^2 - S_1S_2S_5S_6^2 + S_1S_3S_5^2S_6$$

$$+S_0S_3S_5S_6^2 - S_2S_3S_5^3) - 3S_0S_4S_5^2S_6 + S_2^3S_7^2 + S_0S_3^2S_7^2$$

$$+S_1^2S_4S_7^2 - S_0S_2S_4S_7^2 + S_2S_3^2S_6^2 + S_1^2S_6^3 - S_0S_2S_6^3$$

$$+S_3^2S_4S_5^2 + S_2^2S_5^2S_6 + S_4^5 + S_2S_4^2S_5^2 + S_3^2S_4^2S_6$$

$$-S_2S_4^3S_6 + S_0S_4^2S_6^2 + S_0S_5^4$$

and

$$D = 2(S_1S_3^2S_5 + S_1S_2S_4S_5 - S_1S_3S_4^2$$

$$-S_2^2S_3S_5 - S_1S_2S_3S_6 - S_0S_3S_4S_5) + 3S_2S_3^2S_4$$ (3.20)

$$-S_3^4 - S_1^2S_5^2 + S_1^2S_4S_6 + S_0S_4^3 + S_0S_2S_5^2$$

$$+S_0S_3^2S_6 - S_0S_2S_4S_6 + S_2^3S_6 - S_2^2S_4^2 .$$

The identities for the [3/4] PAP are

$$C = 0 \quad \text{and} \quad D = 0 . \tag{3.21}$$

for arbitrary δ in both the sine and cosine cases. Again we may step up $S_n \to S_{n+1}$, etc., and obtain new identities. We may also step down to the [2/4] PAP. For the [1/4] PAP the identity is obtained for arbitrary δ from

$$C - DS = 0 \tag{3.22}$$

for arbitrary δ. For $\delta = k\pi$ we obtain for the sine case the identities

$$C = 0 \quad \text{and} \quad D = 0 \tag{3.23}$$

For the [0/4], for $\delta = k\pi$, sine works but not cosine.

We believe these identities are new, except for the [0/2] PAP. We would be interested in hearing from anyone who believes any of these identities are already known.

IV. THE GENERALIZED HYPERGEOMETRIC FUNCTION

The hypergeometric function $_kF_m$ represents a large number of elementary functions. Thus we can consider PAP's for many functions at once. We will see that the PAP's are accurate for arbitrary k and m and a large number of parameters a, b, c, ... For many examples of how numerous mathematical functions can be expressed in terms of the hypergeometric function $_2F_1$ or the confluent hypergeometric function $_1F_1$ see, for example, the books by Arfken,[12] by Abramowitz and Stegun,[13] and by Gradshteyn and Rhyzik.[14]

Consider the hypergeometric series given by

$$S_n = \frac{(a)_n (b)_n}{(c)_n n!} \tag{4.1}$$

where

$$(a)_n = a(a+1)...(a+n+1) \tag{4.2}$$

and hence

$$_2F_1(a,b,c;x) = \sum_{n=0}^{\infty} S_n x^n . \tag{4.3}$$

For the [N/2] PAP the percentage error is given by 100p where

$$p = \frac{\epsilon_N^2/\epsilon_{N-1} - \epsilon_{N+1}(1+\epsilon_N)^2}{(1+\epsilon_N)^2(1+\epsilon_{N+1})}; \quad N \geq 1 . \tag{4.4}$$

It can be shown for the $_2F_1$ hypergeometric function that

$$p \sim \frac{-2B(1+B)}{N^4} \tag{4.5}$$

where

$$B = c+1-a-b \tag{4.6}$$

and, hence, the PAP's quickly become accurate as $N \to \infty$. For ${}_1F_1(a,c;x)$

$$p \sim +\frac{2}{N^2} \tag{4.7}$$

and for ${}_2F_0(a,b,c;x)$

$$p \sim -\frac{2}{N^2}. \tag{4.8}$$

For the general case ${}_kF_m$, if $k \neq m+1$,

$$p \sim -\frac{2A}{N^2} \tag{4.9}$$

where

$$A = k-(m+1), \tag{4.10}$$

and if $k = m+1$, then

$$p \sim \frac{-2B(1+B)}{N^4} \tag{4.11}$$

where

$$B = 2+k^2-2k+m-km+\sum_{i=1}^{m} c_i - \sum_{i=1}^{k} a_i. \tag{4.12}$$

In general if

$$\varepsilon_n \sim A/N \tag{4.13}$$

$$p \sim -2A^2/N^2, \tag{4.14}$$

and if

$$\varepsilon_n \sim B/N^2, \tag{4.15}$$

$$p \sim -2B(1+B)/N^4. \tag{4.16}$$

To check the behavior for $M \neq 2$ we have written a computer program that scans over a, b, c values (skipping over integers) and evaluates the corresponding PAP's. The parameters a, b, and c vary from -5.0 to 5.0 in steps of 0.125. For each [N/M] PAP, the fractional error p is evaluated, and the maximum and minimum values of p listed as TESTMAX and TESTMIN, respectively. The results for $_2F_1$, $_1F_1$, and $_2F_0$ are presented in Tables I, II, and III, respectively. It can be seen that the TESTMIN and TESTMAX values decrease rapidly in going to progressively higher order. We have listed only diagonal PAP's, but other Padé's were also computed and gave very good results.

V. OTHER EXAMPLES OF EXACT PAP's

Other examples can be found in which PAP's are exact. Any series whose sum is a rational fraction of two polynomials

$$S = \sum_{n=0}^{\infty} S_n x^n = \frac{P_{N_0}(x)}{Q_{M_0}(x)} \tag{5.1}$$

will be exact for the [N/M] PAP where $N \geq N_0$ and $M \geq M_0$. Some examples include

$$S_n = (2n+1); \quad N_0 = 1, \ M_0 = 2 \tag{5.2}$$

$$S_n = (n+1)^2; \quad N_0 = 1, \ M_0 = 3 \tag{5.3}$$

$$S_n = (2n+1)^2; \quad N_0 = 2, \ M_0 = 3 \tag{5.4}$$

$$S_n = (a+nd); \quad N_0 = 1, \ M_0 = 2 \tag{5.5}$$

TABLE I. Padé estimates for $_2F_1$.

[N/M]	TESTMIN	TESTMAX
[4/4]	0.649×10^{-8}	0.649
[5/5]	0.105×10^{-9}	34.0
[6/6]	0.103×10^{-11}	2.57
[7/7]	0.100×10^{-13}	7.0
[8/8]	0.107×10^{-15}	0.494
[9/9]	0.136×10^{-17}	0.687×10^{-2}
[10/10]	0.173×10^{-19}	0.332×10^{-3}
[11/11]	0.284×10^{-21}	0.151×10^{-4}
[12/12]	0.442×10^{-23}	0.215×10^{-5}
[13/13]	0.757×10^{-25}	0.294×10^{-6}
[14/14]	0.151×10^{-26}	0.386×10^{-7}
[15/15]	0.652×10^{-30}	0.492×10^{-8}
[16/16]	0.197×10^{-29}	0.611×10^{-9}
[17/17]	0.705×10^{-30}	0.720×10^{-10}
[18/18]	0.192×10^{-29}	0.847×10^{-11}

TABLE II. Padé estimates for $_1F_1$.

[N/M]	TESTMIN	TESTMAX
[4/4]	0.536×10^{-5}	0.246×10^{-6}
[5/5]	0.170×10^{-6}	6.721
[6/6]	0.195×10^{-6}	35.4
[7/7]	0.251×10^{-6}	4.56
[8/8]	0.124×10^{-7}	0.456
[9/9]	0.878×10^{-8}	0.102×10^{-1}
[10/10]	0.835×10^{-9}	0.114×10^{-2}
[11/11]	0.168×10^{-8}	0.697×10^{-3}
[12/12]	0.148×10^{-9}	0.260×10^{-3}
[13/13]	0.394×10^{-10}	0.270×10^{-3}
[14/14]	0.419×10^{-11}	0.676×10^{-5}
[15/15]	0.157×10^{-11}	0.119×10^{-5}
[16/16]	0.530×10^{-13}	0.844×10^{-6}
[17/17]	0.393×10^{-13}	0.121×10^{-6}
[18/18]	0.984×10^{-14}	0.201×10^{-7}
[19/19]	0.259×10^{-14}	0.217×10^{-8}
[20/20]	0.158×10^{-14}	0.128×10^{-8}

TABLE III. Padé estimates for $_2F_0$.

[N/M]	TESTMIN	TESTMAX
[4/4]	0.146×10^{-4}	127.3
[5/5]	0.118×10^{-4}	45.3
[6/6]	0.458×10^{-6}	0.304
[7/7]	0.259×10^{-5}	3.00
[8/8]	0.986×10^{-8}	0.307×10^{-1}
[9/9]	0.262×10^{-8}	0.247×10^{-1}
[9/10]	0.528×10^{-8}	0.227×10^{-1}
[10/9]	0.572×10^{-9}	0.298×10^{-2}
[10/10]	0.590×10^{-8}	0.152×10^{-1}
[11/11]	0.165×10^{-8}	0.774×10^{-1}
[13/13]	0.132×10^{-10}	0.573×10^{-4}
[13/14]	0.298×10^{-10}	0.227×10^{-3}
[14/13]	0.214×10^{-10}	0.776×10^{-3}
[16/16]	0.136×10^{-11}	0.151×10^{-5}
[17/17]	0.377×10^{-13}	0.105×10^{-6}
[18/18]	0.272×10^{-13}	0.275×10^{-7}
[19/19]	0.349×10^{-14}	0.729×10^{-8}
[20/20]	0.296×10^{-16}	0.198×10^{-8}

$$S_n = (n+1); \quad N_0 = 0, \; M_0 = 2 \qquad (5.6)$$

$$S_{n+1} - S_n = (n+2), \; S_0 = 1; \quad N_0 = 0, \; M_0 = 3 \qquad (5.7)$$

and

$$S_n = 1; \quad N_0 = 0, \; M_0 = 1. \qquad (5.8)$$

VI. NON-SINGLET MOMENTS OF DEEP INELASTIC STRUCTURE FUNCTIONS IN QCD

In this section we make use of some recent results of Larin, van Rittergen, and Vermeseren.[15] They have calculated the next-to-next leading QCD approximations for non-singlet moments of deep inelastic structure functions, in the leading twist approximation, for the moments N = 2, 4, 6, 8 of the non-singlet deep inelastic structure function F_L. They have calculated the three-loop anomalous dimensions of the corresponding non-singlet operators and the three-loop coefficient functions of the structure factor F_L, in the leading twist and massless quark approximation.

We present our results in Tables IV-XI. In each case, we estimate the $O(\alpha_s^3)$ coefficient and compare our estimate with the now-known Larin et al. result. We neglect the term that depends on the sum of the quark charges Σq_f, since the term is small in all cases of interest. We present our results for N_f = 3, 4, 5, where N_f is the number of quark flavors. We then present our estimates for the next (unknown) $O(\alpha_s^4)$ coefficients, in each case.

In Table IV we present results for $C_{L,2}$. It is seen that, for N_f = 3, 4, 5, our estimates are within the error-bars, for the $O(\alpha_s^3)$ terms and we estimate the next (unknown) $O(\alpha_s^4)$ terms. Table V shows the results for $C_{L,4}$, Table VI for $C_{L,6}$, and Table VII for $C_{L,8}$.

In Tables VIII-XI we present our results for the anomalous dimensions γ_2, γ_4, γ_6, and γ_8. Here again, in each case, our estimates are within the error bars of the Larin et al. results for $O(\alpha_s^3)$ and we estimate the next (unknown) $O(\alpha_s^4)$ term. For further details on how we obtain our error bars, see Ref. 11.

VII. EXAMPLES FROM STATISTICAL PHYSICS

In this section we consider two examples[16] from statistical physics. They are the low temperature ferromagnetic susceptibility coefficients in the Ising Model. Table XII gives the results for the honey-combed (hc) lattice and the results for the square (sq) lattice are shown in Table XIII.

It can be seen that the results are excellent and that the % error decreases in going to higher order. In all cases the estimates are within 2σ of the exact results for the known coefficients and the next (unknown) coefficient is predicted. For the hc lattice the estimate is 538,596,000 ± 10,500 and for the square lattice it is 185,000,000 ± 55,000,000.

VIII. CONCLUSIONS

We have proved several theorems on using Padé approximants to estimate coefficients in Perturbative Quantum Field Theory and Statistical Physics. These theorems give sufficient conditions for the PAP method of estimating the next term in a series expansion to work.

TABLE IV. Padé estimates for $C_{L,2}$.

Estimate	Error	Exact	|Estimate - Exact|
$\underline{N_f = 3}$			
1046	1046	2230	1184
82,812	205,498		
$\underline{N_f = 4}$			
837	837	2313	1476
82,233	32,915		
$\underline{N_f = 5}$			
652	652	2420	1768
80,203	13,522		

TABLE V. Padé estimates for $C_{L,4}$.

Estimate	Error	Exact	|Estimate - Exact|
$\underline{N_f = 3}$			
1376	668	1473	137
56,946	29,021		
$\underline{N_f = 4}$			
1106	553	1166	60
39,468	19,846		
$\underline{N_f = 5}$			
897	449	881	16
25,076	4004		

TABLE VI. Padé estimates for $C_{L,6}$.

Estimate	Error	Exact	|Estimate - Exact|
$N_f = 3$			
2305	1153	1433	872
41,989	17,795		
$N_f = 4$			
2018	1009	1159	859
27,443	12,654		
$N_f = 5$			
1750	875	905	845
15,894	8434		

TABLE VII. Padé estimates for $C_{L,8}$.

Estimate	Error	Exact	|Estimate - Exact|
$N_f = 3$			
1437	719	1985	548
124,711	78,967		
$N_f = 4$			
1226	613	2043	817
130,095	139,494		
$N_f = 5$			
1031	516	2118	1087
133,699	814,125		

TABLE VIII. Padé estimates for γ_2.

Estimate	Error	Exact	|Estimate - Exact|
$N_f = 3$			
424	212	448	24
5159	2270		
$N_f = 4$			
358	179	306	52
2607	636		
$N_f = 5$			
298	149	162	136
677	238		

TABLE IX. Padé estimates for γ_4.

Estimate	Error	Exact	|Estimate - Exact|
$N_f = 3$			
636	318	762	126
8606	3146		
$N_f = 4$			
517	259	503	14
4953	387		
$N_f = 5$			
410	205	239	175
954	297		

TABLE X. Padé estimates for γ_6.

Estimate	Error	Exact	|Estimate - Exact|
$N_f = 3$			
744	372	946	202
10,676	4001	—	—
$N_f = 4$			
596	298	621	25
5245	1266	—	—
$N_f = 5$			
464	232	290	174
1201	348	—	—

TABLE XI. Padé estimates for γ_8.

Estimate	Error	Exact	|Estimate - Exact|
$N_f = 3$			
833	417	1081	248
12,225	4629	—	—
$N_f = 4$			
662	331	709	47
6018	2552	—	—
$N_f = 5$			
510	255	330	180
1401	393	—	—

TABLE XII. Padé estimates for the low temperature ferromagnetic susceptibility coefficients in the Ising model (hc lattice).

Estimate	Error	Exact	|Estimate - Exact|
8749	818	8792	43
35,682	120	35,622	60
143,333	447	143,079	254
569,470	950	570,830	1360
2,264,740	631	2,264,649	91
8,942,853	2031	8,942,436	417
35,159,776	8724	35,169,616	9840
137,838,225	5787	137,839,308	1083
538,596,320	10,430	——	——

TABLE XIII. Padé estimates for the low temperature ferromagnetic susceptibility coefficients in the Ising model (sq lattice).

Estimate	Error	Exact	|Estimate - Exact|
449	138	416	33
2715	830	2791	76
18,699	1592	18,296	403
118,069	35,392	118,016	53
751,928	146	752,008	80
4747×10^3	1410×10^3	4746×10^3	$O(10^3)$
2973×10^4	721×10^4	2973×10^4	$O(10^3)$
$18,502 \times 10^4$	5494×10^4	——	——

In addition, we have presented new trigonometric identities which we obtained as a result of the PAP being exact. We have also considered the generalized hypergeometric function, for which the method works. As a result, many series are dealt with at the same time, since the hypergeometric function can represent many elementary functions merely by changing the parameters.

We have considered several series from QCD. These are for the non-singlet moments of deep inelastic structure functions. We have also considered two series from statistical physics. These are the low temperature ferromagnetic susceptibility coefficients in the Ising model.

In all cases, the method works beautifully! Thus the information needed for estimating the next term in perturbative series is, in fact, contained in the lower-order results.

ACKNOWLEDGEMENTS

One of us (MAS) thanks the theory groups at the Stanford Linear Accelerator Center and at the Argonne National Laboratory for their kind hospitality. He also thanks the following people for very helpful discussions: David Atwood, Bill Bardeen, Richard Blankenbecler, Eric Braaten, Stan Brodsky, Guowen Li, Helen Perk, Jacques Perk, Tom Rizzo, Davison Soper, George Sudarshan, Levan Surguladze, Alan White, and Cosmos Zachos. This work was supported by the U.S. Department of Energy under Grant No. DE-FG02-94ER40852.

REFERENCES

1. M. A. Samuel, G. Li, and E. Steinfields, *Phys. Rev.* **D48**, 869 (1993).

2. M. A. Samuel, G. Li, and E. Steinfields, "*On Estimating Perturbative Coefficients in Quantum Field Theory, Condensed Matter Theory, and Statistical Physics*", *Phys. Rev. E*, to be published (May, 1995).

3. M. A. Samuel, G. Li, and E. Steinfields, *Phys. Lett.* **B323**, 188 (1994); M. A. Samuel and G. Li, "*Estimating Perturbative Coefficients in High Energy Physics and Condensed Matter Theory*", International Journal of Theoretical Physics (1994).

4. M. A. Samuel and G. Li, *Phys. Lett.* **B331**, 114 (1994).

5. M. A. Samuel and G. Li, "*On the R and R_τ Ratios at the Five-Loop Level of Perturbative QCD*", SLAC-PUB-6370, October (1993).

6. J. Zinn-Justin, *Physics Reports* **1**, 55 (1971).

7. J. Nutall, *J. Math. Anal.* **31**, 147 (1970).

8. G. A. Baker, Jr. **Essentials of Padé Approximants**, (Academic, New York, 1975).

9. C. Bender and S. Orzag, **Advanced Mathematical Methods for Scientists and Engineers**, (McGraw-Hill, New York, 1978).

10. C. Chlouber, G. Li, and M. A. Samuel, "*Padé Approximants - Type I and Type II - and Their Application*", Oklahoma State University Research Note 265, February (1992).

11. M. A. Samuel, "*On Estimating Perturbative Coefficients in Quantum Field Theory and Statistical Physics*", Oklahoma State University Research Note 290, May (1994).

12. G. Arfken, **Mathematical Methods of Physics**, (Academic, New York, 1985).

13. M. Abramowitz and I.A. Stegun, **Handbook of Mathematical Functions**, (U.S. Gov't Printing Office, Wash., D.C., 1964).

14. I. S. Gradshteyn and Ryzhik, **Tables of Integrals, Series, and Products**, (Academic, New York, 1980).

15. S. A. Larin, T. van Ritbergen, and J. A. M. Vermaseren, "*The Next-Next-to-Leading QCD Approximation for Nonsinglet Moments of Deep Inelastic Structure Functions*", NIKHEF-H-93-29, December (1993).

16. C. Domb, **Ising Model in Phase Transitions and Critical Phenomena**, vol. 3, ed. by C. Domb and M. S. Green, (Academic, New York, 1974).

SECTION VII

SPIN PHYSICS AT HIGH ENERGY

HIGHLIGHTS OF THE SPIN '94 MEETING*

A. D. Krisch
Randall Laboratory of Physics, University of Michigan
Ann Arbor, Michigan 48109-1120 USA

This lecture will discuss some highlight of the joint meeting of the 11th International Symposium on High Energy Spin Physics and the 8th International Symposium on Polarization Phenomena in Nuclear Physics. This 15-22 September 1994 SPIN '94 meeting at Indiana University provided an excellent opportunity for nuclear and high energy spin physicists to learn something from each other.

Professor Ternov gave an excellent historical talk related to electron spin physics. In 1963, along with Professor Sokolov, he discovered the Sokolov-Ternov effect of self-polarization,[1] where electrons and positrons become polarized along the accelerator magnets' vertical field direction because of their different spin-up and spin-down synchrotron radiation rates. This self-polarization has recently become very important to our field. In the 1960's, self-polarization seemed a clever abstraction, which was only interesting to theorists. Now HERA and LEP, two of the world's largest electron facilities, both operate with polarized beams using the Sokolov-Ternov self-polarization effect. It was a pleasure and an honor to have Professor Ternov lecture at this Symposium.

Some early proton-proton spin effects are shown Figure 1; this figure is especially appropriate for this meeting, which includes both high energy and nuclear spin physicists. It displays the spin-spin correlation parameter for $90°_{cm}$ proton-proton elastic scattering from the lowest up to the highest measured energy;[2] Professor Haeberli helped me to make this compilation. When I started studying spin around 1970, most people were quite sure that there would be no spin effects at high energy; this graph certainly does not support that belief.

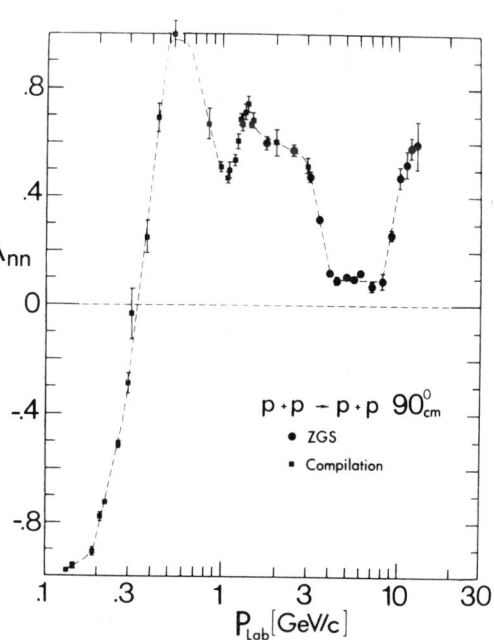

FIGURE 1. Momentum Dependence of Proton-Proton Elastic A_{nn} at $90°_{cm}$.

Unified Symmetry: In the Small and in the Large 2
Edited by B. N. Kursunoglu *et al.*, Plenum Press, New York, 1995

Even after this large two-spin effect was discovered at the ZGS,[3] many people said, "Perhaps there are two-spin effects when the beam and the target are both polarized, but, surely there will be no one-spin effects at high energy." The talks by Professors Devlin and Pondrom[4] on inclusive hyperon polarization and hyperon magnetic moments referred to this perturbative QCD prediction that A should go to zero at high energy and high $P_\perp$ for all hadronic reactions. Figure 2 shows the hyperon polarization plotted against transverse momentum at 12 GeV, at 400 GeV, and at 2000 GeV; this data certainly does not support the A = 0 prediction. Moreover, it seems to me that 2000 GeV is a fairly high energy.

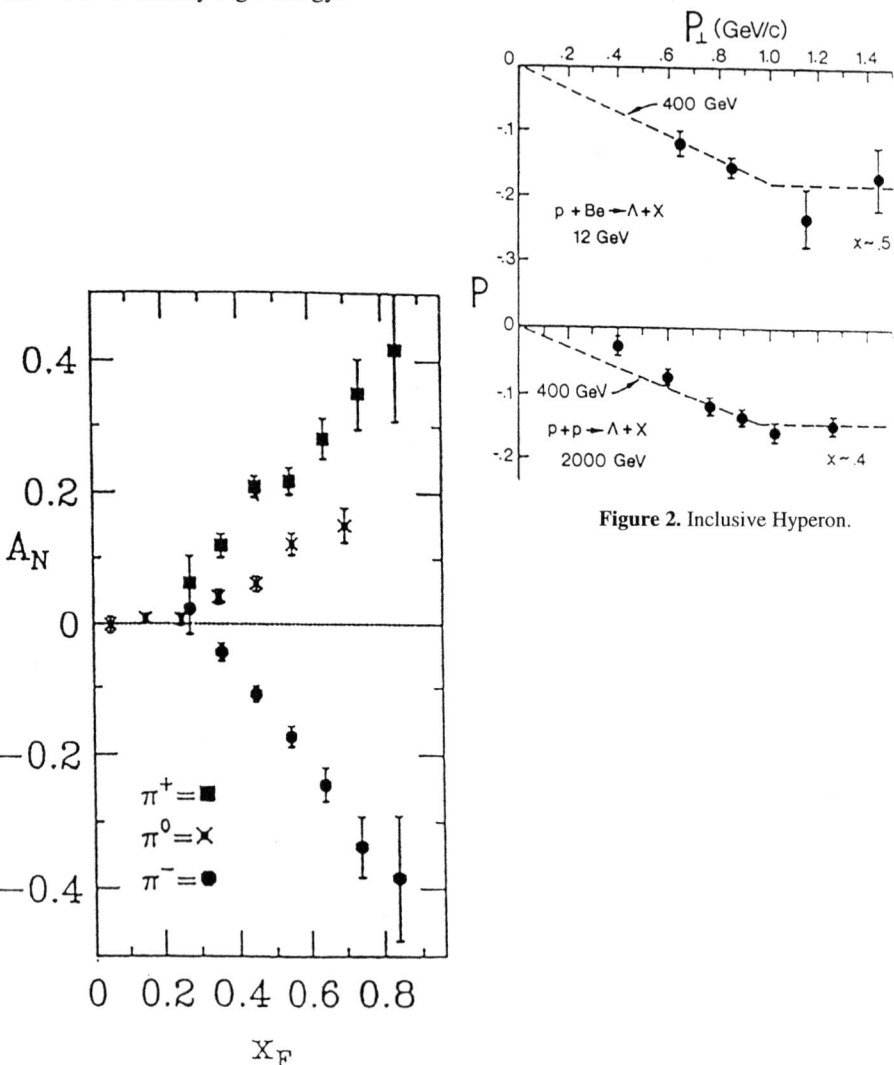

Figure 2. Inclusive Hyperon.

Figure 3. Fermilab E-704 Left-Right Asymmetry in Inclusive Pion Production.

Some interesting results[5] from E-704 at Fermilab are shown in Figure 3; they used the 200 GeV polarized proton beam produced by hyperon-decay to measure the left-right asymmetry in inclusive pion production. Notice that, at large Feynman-x, there are certainly large left-right asymmetries which should not exist, according to PQCD.

In these experiments, $P_\perp$ was only about 1 GeV/c; some people said that perturbative QCD would surely force one-spin asymmetries to go to zero at higher $P_\perp$. Figure 4 shows some elastic one-spin data from the AGS and CERN. The prediction of perturbative QCD for elastic scattering in this region was again that A should be zero; the AGS data[6] certainly do not agree with this A = 0 prediction at $P_\perp^2$ = 7 (GeV/c)^2. Thus, the predictions of the PQCD theory of hadronic interactions do not agree with these four hadronic spin experiments. I was quite amused by the earlier quoted comment of Bjorken that perhaps theorists on Program Committees should ban spin experiments to help protect PQCD.

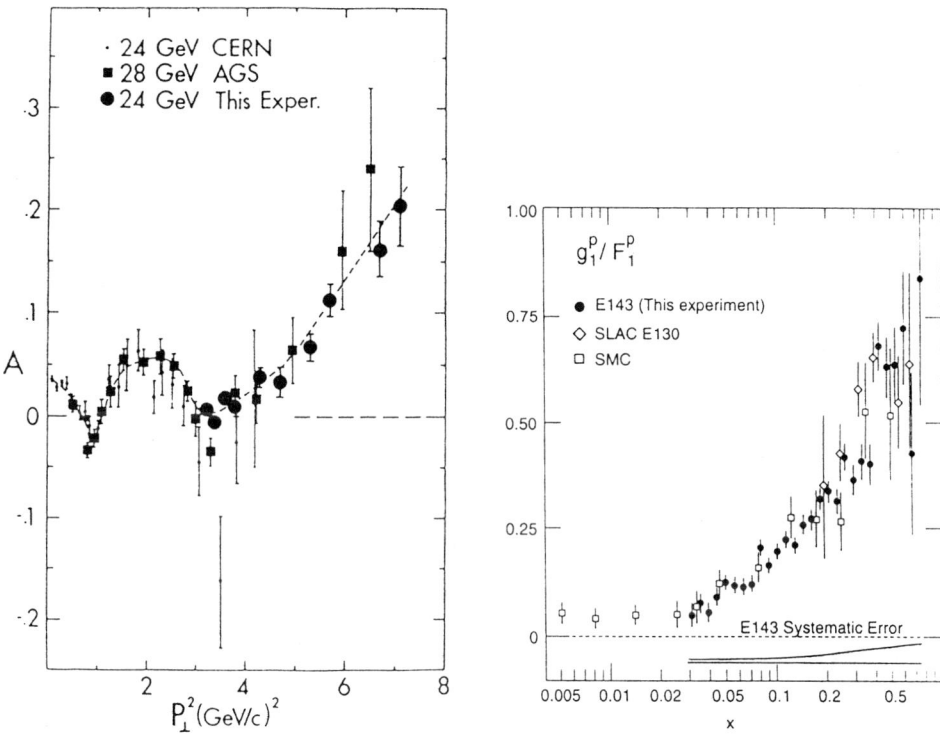

FIGURE 4. Spin Asymmetry in p-p Elastic Scattering at High-$P_\perp$.

FIGURE 5. "Spin Crisis" Form Factor Ratio Plotted Against x.

The "Spin Crisis" has been our most recent excitement; we heard many comments about this in the talks of Thomas, R. Voss, Windmolders, Day, Jackson, Soffer, and Anselmino.[7] We heard from Professor Thomas that perhaps now there is not so much of a spin crisis. I then asked him, "What changed? the data, the theory, or the definition of 'crisis'?" Apparently, there have been some small changes in the data, but Figure 5 shows that the recent SMC data, using a CERN polarized muon beam on a polarized proton target, agrees rather well with the SLAC polarized electron data from experiments E-130 and E-143. Turning to the theory, perhaps there was a bit of overconfidence in the validity of some sum rules and some extrapolations to very small values of x, which had not been experimentally tested; now they have been tested. But probably the biggest change was in our definition of the word "crisis"; the cancellation of the SSC helped us to better understand the word crisis. In any case, the data now indicates that probably each proton does not contain three simple quarks which carry most of the proton's spin; Professor Prescott told me that the best present theoretical estimate is that the quarks and anti-quarks together carry about $\frac{1}{3}$ of the proton's spin.

Three different workshops were summarized at the Symposium. Professor Anderson reported on the May 1993 Workshop on Polarized Ion Sources and Polarized Gas Targets in Madison, Wisconsin;[8] Professor Mori[9] also reviewed this subject. Many exciting results were presented, but I will only mention Figure 6 which shows the intense atomic beam sources at Heidelberg and Wisconsin. The present source technology is very impressive, especially the high-gradient permanent magnet sextupoles of $4 \text{ T} \cdot \text{cm}^{-1}$. These new polarized sources work very well; they produce intensities of $4 \cdot 10^{16}$ per second with a polarization of over 80%. The situation is now very different from 1970 when Hilton Glavish from New Zealand sold us ZGS people the world's first commercial polarized proton source; it cost about $250,000 and produced 6 μA of polarized protons. The progress in polarized ion sources has been outstanding.

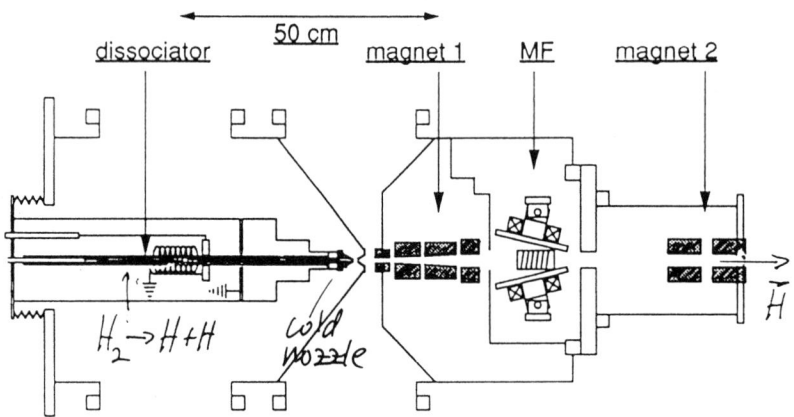

FIGURE 6. Wisconsin-Heidelberg-Marburg-Munich Source of Polarized Hydrogen Atoms.

Professor Nurushev discussed the 5th International Workshop on High Energy Spin Physics (SPIN '93).[10] These Workshops, which he organizes in the odd-numbered years at Protvino, are really small symposia which allow young physicists from the former Soviet Union to hear what is happening in spin physics. Spin physics is an area of great activity in the former Soviet Union, but the currency exchange problems make travel to foreign scientific meetings very difficult. I thank Professor Nurushev for organizing these valuable workshops; they have helped to keep physics healthy in these difficult times.

Werner Meyer reviewed the 7th Workshop on Polarized Target Materials and Techniques,[11] which had 49 participants from many institutions. One highlight was the successful Virginia-Basel-SLAC polarized target, shown in Figure 7, which is now being used in fixed-target experiments at SLAC. Its clever arrangement of magnetic fields allows a longitudinal polarization. The target uses frozen ND_3 or NH_3; some recent results are shown in Figure 8: the deuteron polarization was over 30% and the proton polarization was over 70% in an intense beam of over 10^{11} electrons per second. This impressive polarized target works very well in these extreme conditions.

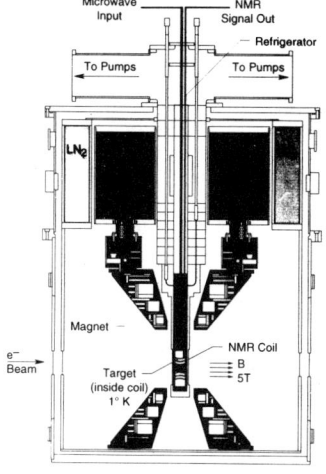

FIGURE 7. Virginia-Basel-SLAC Polarized Target.

FIGURE 8. ND3 Polarization with Beam on.

A major highlight of the Symposium was the polarization work at the large electron facilities: SLC, HERA, and LEP. Dr. Placidi[12] gave a very nice talk on polarization at LEP; his Figure 9 shows the transverse polarization obtained at LEP in August 1993 plotted against time. Note that the polarization reached about 55%; this clearly demonstrates the success of the Sokolov-Ternov self-polarization mechanism.[1] This electron polarization was used to calibrate the LEP energy, which then provided a precise calibration of the Z mass. This was a rather significant contribution to high energy physics.

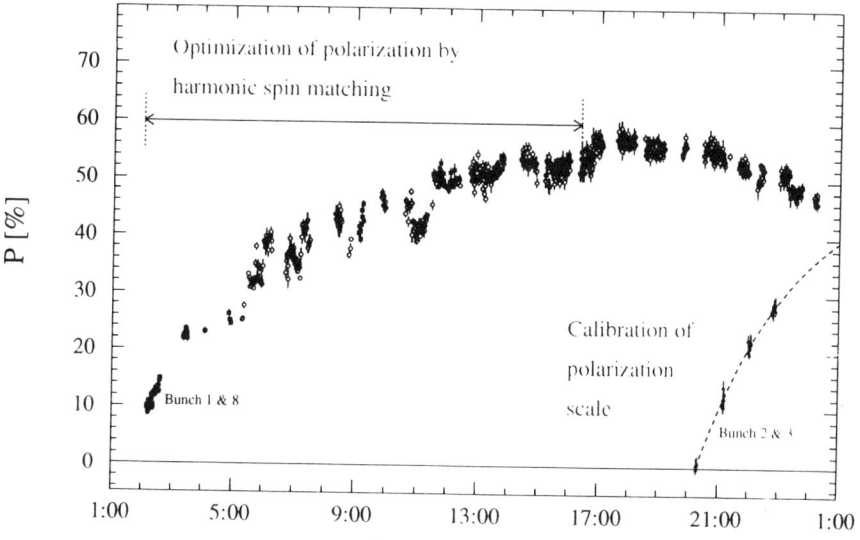

FIGURE 9. High Transverse Beam Polarization in LEP Plotted against Hours.

271

In the HERA rings at DESY, there has been a strong emphasis on polarization. G. Voss, Barber, and Jackson[13] each discussed various aspects of polarization at HERA. The HERA ring is shown in Figure 10 along with the H-1 experiment, the Zeus experiment, the HERMES experiment, and the spin direction at each experiment. The polarimeter is also shown.

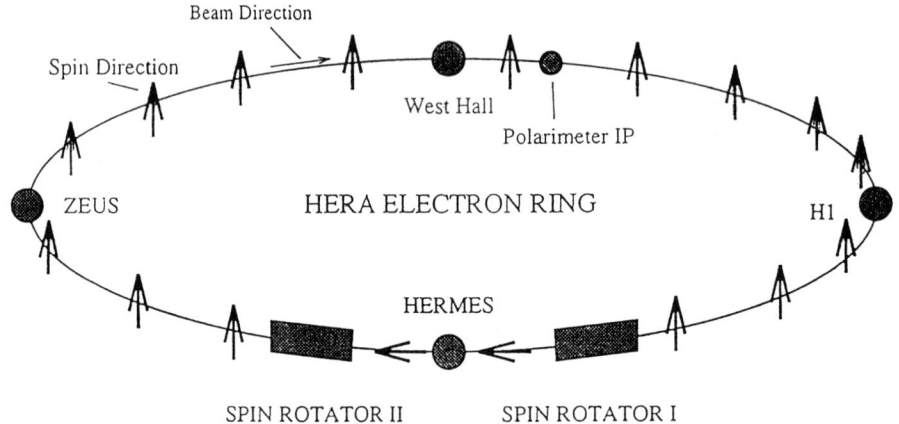

FIGURE 10. HERA Electron Ring.

There is only time to discuss a few parts of the HERA polarization program; I will first show the HERA polarimeter in Figure 11. The incoming laser light interacts with the polarized electrons, and then the analyzer detects the scattered photons. Measuring the differences in the event rate for each spin direction gives a precise determination of the electron beam polarization. This polarimeter contains some very impressive new technology.

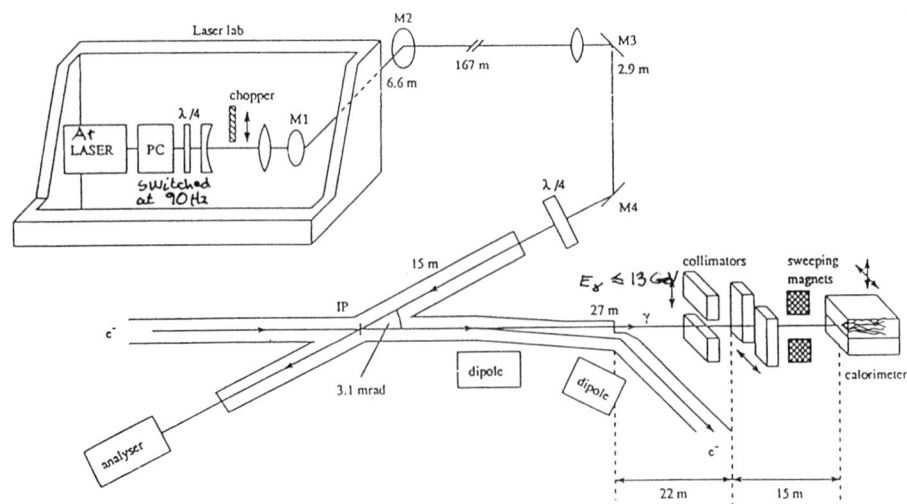

FIGURE 11. HERA Polarimeter.

The beam polarization measured at HERA is plotted against time in Figure 12; the transverse polarization reached about 70% and the longitudinal polarization reached about 55% in about one hour. This was another success of the Sokolov-Ternov self-polarization mechanism.[1] Also plotted is something that I named the "Soergel Limit." In a comment[14] at the 1990 Bonn Spin Symposium, Professor Soergel said something like, "We decided that HERMES is approved if the HERA polarization reaches 50%. If it does not reach 50%, it is not approved." This seemed a wise thing for a Director to do; it eliminated the need for a decision and encouraged polarizers to work harder. There were about 300 witnesses to his comment, so apparently DESY decided to approve HERMES.

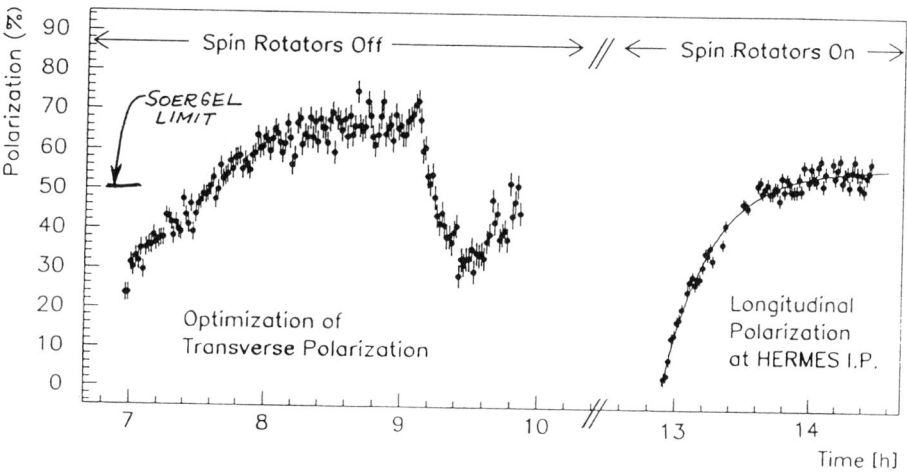

FIGURE 12. Transverse and Longitudinal Polarization at HERA.

The HERMES experiment uses a "storage cell" internal target which is a bit different than a Jet; this clever storage cell idea was probably first proposed by Professor Haeberli. As shown in Figure 13, the storage cell has an incoming polarized jet, but the jet does not pass through the beam. Instead it is trapped inside the storage cell, which is open only at the ends so that the proton-polarized hydrogen atoms can only escape slowly. Therefore, most of them remain in the storage cell for some time; this increases the target thickness by a factor of 10 to 100. This storage cell is another indication that hard-working, clever, and persistent physicists can solve most problems.

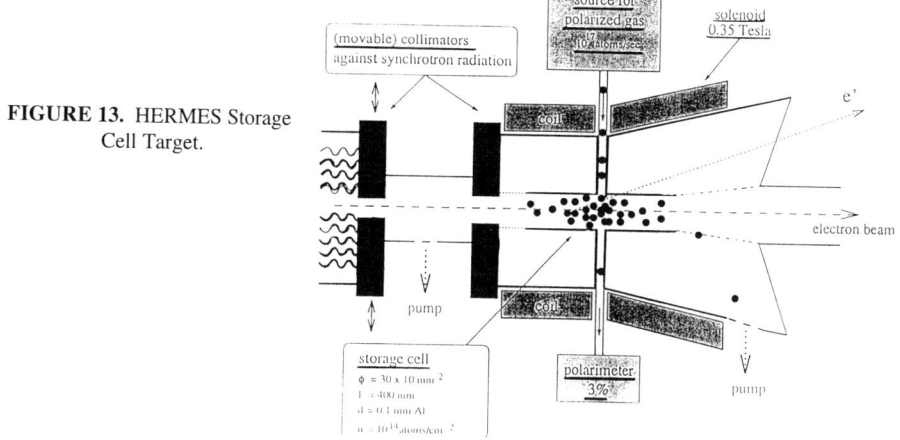

FIGURE 13. HERMES Storage Cell Target.

SLAC now has a very nice facility called SLC. There were talks about SLAC and SLC from Woods, Steiner, and others.[15] In recent years, SLAC, which is certainly one of the largest electron facilities in the world, has become almost exclusively a polarized electron facility. The hardware items associated with the SLAC polarized beam are shown in Figure 14. The polarized electrons are first produced in a polarized source; then one must maintain the polarization in the accumulator rings. Next the polarized electrons are accelerated in the LINAC and pass through SLC's somewhat complex non-planar arcs. Finally, one must measure the polarization near the interaction region point. So far there has been no attempt to polarize the positrons; that would be an exciting goal. Perhaps someday, we will figure out how to polarize positrons and even antiprotons.

During the past few years, the SLAC polarized source team[15] made great progress with the polarized electron source, which is shown in Figure 15. The electrons are emitted by the gallium-arsenide cathode. Then the polarized laser pumps them into the proper spin-polarization state; the laser polarizes the electrons very well. In 1992, the polarization reached 40% for fixed-target running and 22% at SLC. In 1993, it reached 85% and 63%. In 1994, the polarization reached 85% for fixed-target running and 80% at SLC. These polarizations are impressive.

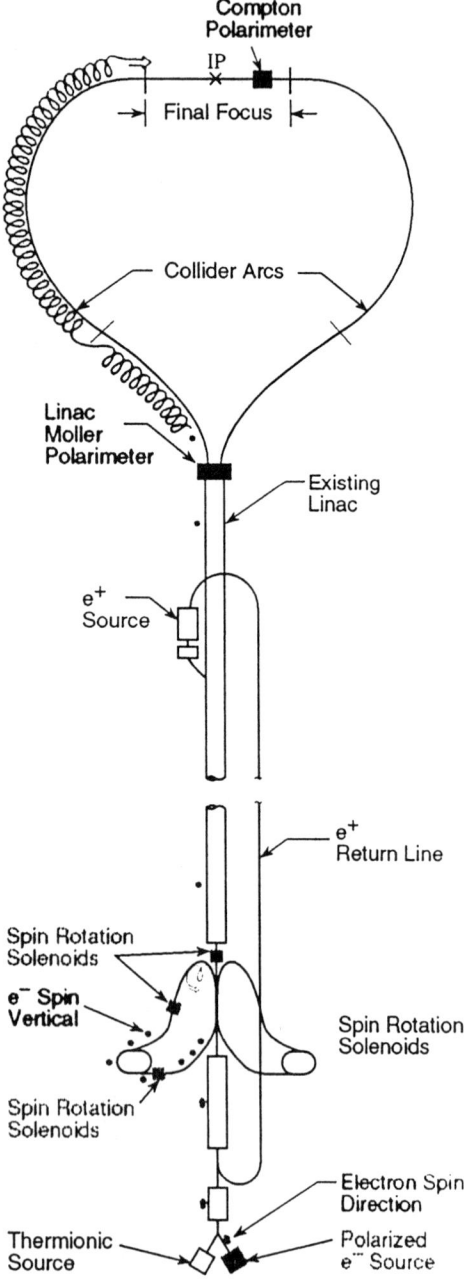

FIGURE 14. Polarization Hardware at SLAC.

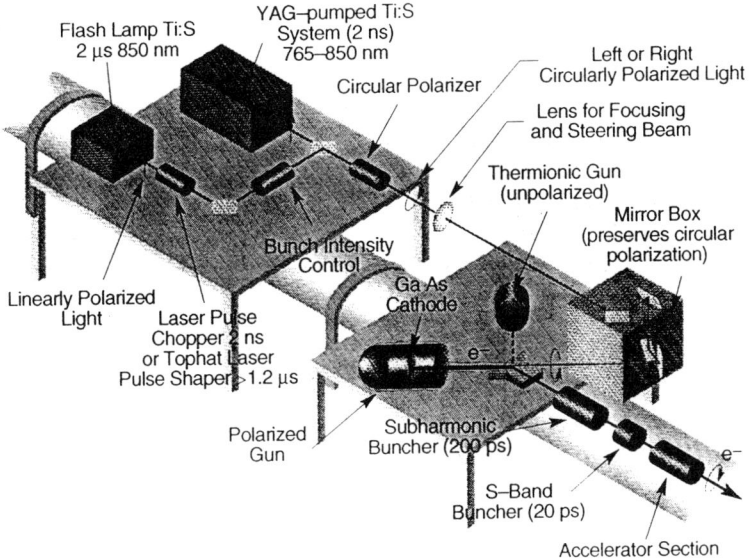

FIGURE 15. Polarized Electron Source at SLAC.

A very significant physics result from SLAC is shown in Figure 16, where $\sin^2$ of the Weinberg angle is plotted for various experiments. The two parallel horizontal lines represent the error bars for the average of all the LEP data. The first square point on the left represents the data from SLD at SLC. All of the LEP data taken together still has a better error than the SLAC data, but only about 2.5 times better. Note that there may be some difference between the SLAC and LEP results. Using my earlier definition, I am not yet ready to call this difference a crisis, but it certainly seems interesting. Perhaps in the future, we should use the word "crisis" more sparingly and use "interesting" more often.

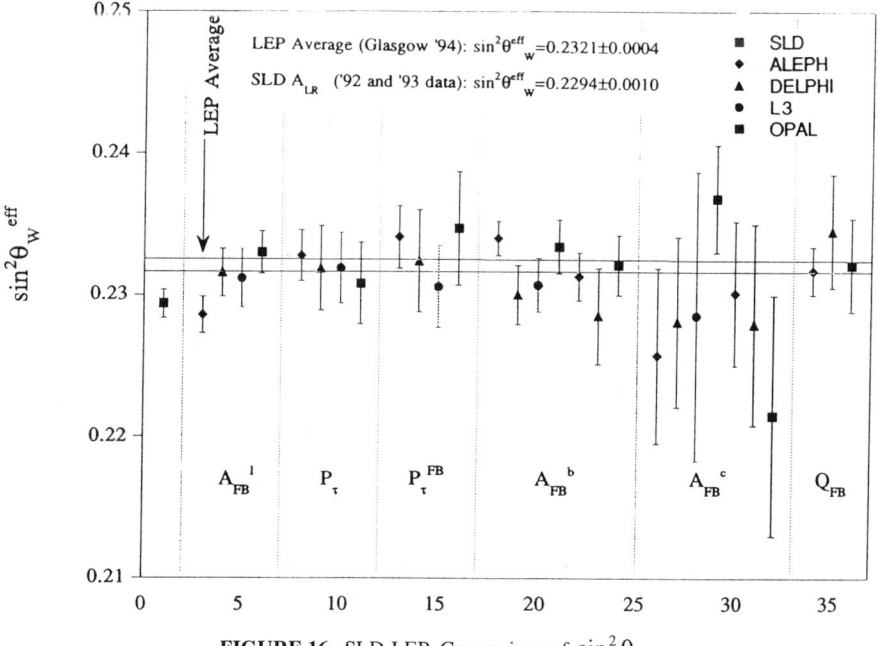

FIGURE 16. SLD-LEP Comparison of $\sin^2\theta_W$.

Now, I will discuss the proton facilities in slightly more detail; Professor Heller asked for this detail because Professor Ado was unable to give the requested lecture on polarized beams at Fermilab and UNK. I will begin with the development and testing of Siberian snakes. The Siberian snake is an extraordinarily clever idea which was invented by Derbenev and Kondratenko;[16] moreover, it works.

The first test of the Siberian snake was at the IUCF Cooler Ring, which is shown in Figure 17; to some people it looks like an accelerator, but we Siberian snake people think of it as an experiment.[17] Polarized protons are injected into the Cooler Ring from the IUCF cyclotrons. The kicker magnets are sometimes used for injection, and sometimes used as rf dipoles. Notive the Siberian snake, the polarimeter, the rf solenoid, and the cooling magnets which we have also used to create imperfection magnetic fields for our experiments. Thus, we are using many existing hardware items for extra jobs.

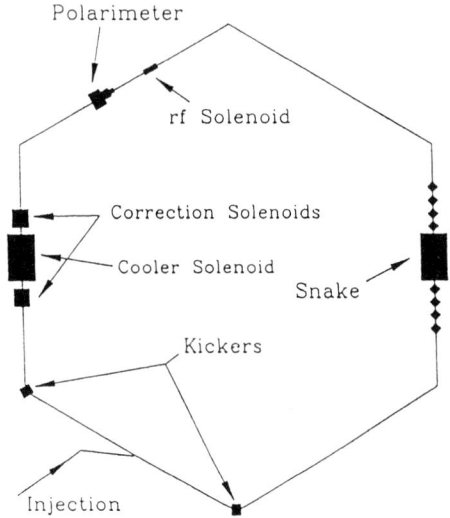

FIGURE 17. Siberian Snake at the IUCF Cooler Ring.

The Siberian snake[17] is shown in Figure 18; its main element is an aging and temperamental superconducting solenoid of two Tesla-meters. The four skew and four normal quadrupoles do nothing to the spin, but they correct the solenoid's rather strong focusing and beam twisting of the 100 or 200 MeV beam. On each turn around the ring, the snake rotates the spin by 180°. This makes any depolarizing effects, that may occur during one turn around the ring, cancel themselves during the next turn. The snake forces all the problems to cancel themselves; it is a very clever idea.[16]

FIGURE 18. Siberian Snake at the Cooler Ring.

Many different aspects of Siberian snakes have now been tested. Recently we accelerated polarized protons from 95 to 140 MeV through the $G\gamma = 2$ imperfection depolarizing resonance at 108 MeV.[18] The measured polarization is plotted against the imperfection field integral in Figure 19; the circles show the polarization with a partial Siberian snake turned on, while the squares show the polarization with no snake. Clearly, with no snake, the polarization drops when the imperfection field is large; however, with the snake on, there is no depolarization. Thus the Siberian snake overcomes the depolarization during acceleration through an imperfection depolarizing resonance.

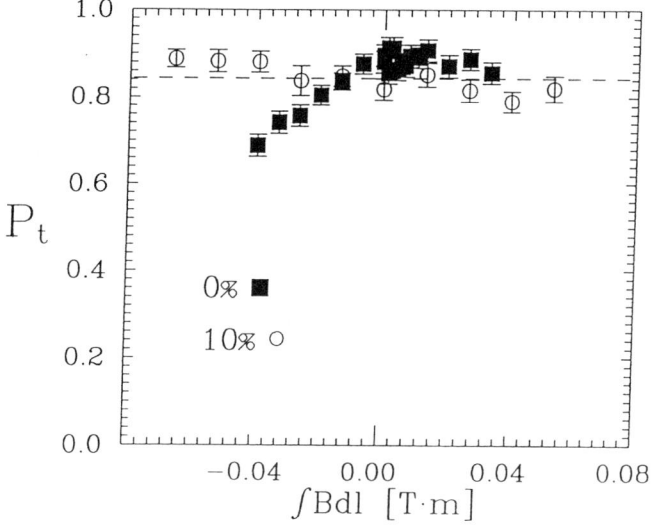

FIGURE 19. Acceleration through a Depolarizing Resonance at $G\gamma = 2$.

Since the last SPIN Symposium in 1992, we also studied the adiabatic turn-on of a Siberian snake at an energy where the spin tune $G\gamma$ is a half integer; this occurs at 370 MeV at the IUCF Cooler Ring. Professor Courant had proposed that adiabatic turn-on should allow switching from one type of polarization correction to another during an acceleration cycle. The data[19] demonstrated that, within our measured precision of 2%, there was no polarization loss when the snake was turned on and off at these magic energies. At other energies, there is some depolarization during adiabatic turn-on.

Several speakers emphasized the importance of flipping the spin of a stored polarized beam from up to down to discriminate against systematic errors. Dr. Phelps from Michigan has been much involved with our recent spin-flipping studies using an RF solenoid. The studies[20] showed that, with very careful tuning, there was no polarization loss within our error of $\pm 0.05\%$. This suggests that one can flip the spin perhaps 100 to 1,000 times without a significant loss of polarization. This is very important for reducing systematic errors. Professor Cameron seems to support further spin-flipping studies because there are several approved experiments at the IUCF Cooler Ring that need this capability.

High energy proton polarized accelerators need Siberian snakes because Professor Ternov's self-polarization formula[1] does not work very well for most proton accelerators. At a Coral Gables meeting, Kent Terwilliger, R.R. Wilson, and I once calculated that self-polarization would work rather well for proton storage rings at about 70 TeV. However, until Congress funds a 70 TeV SSC, we proton polarizers must either painfully correct each resonance or use Siberian snakes.

Professor Vigdor[21] gave a very nice talk about Indiana's proposed Light Ion Spin Synchrotron (LISS), which would use Siberian snakes to overcome the depolarizing resonances during the acceleration of polarized protons and other light ions to about 20 GeV. This proposal seems rather natural since Indiana is quite familiar with Siberian snakes. LISS is shown in Figure 20; some of its parameters are: stored polarized beams of about 20 GeV for protons and other light ions, a very high luminosity of 10^{33}cm^{-2}s^{-1}, and long straight sections. LISS could provide a very exciting spin physics program with strong components in both High Energy Physics and Nuclear Physics. The successful SPIN '94 Meeting at IUCF should help to focus the scientific program for LISS.

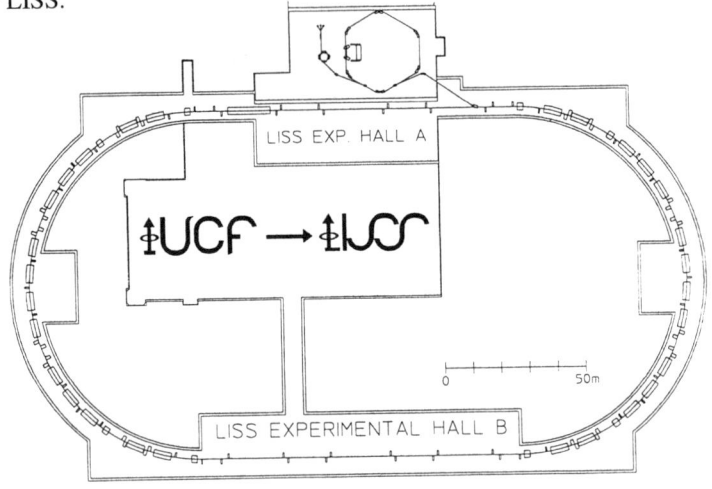

FIGURE 20. IUCF Light Ion Spin Synchrotron.

The Michigan-IUCF Siberian snake experiments had no competition anywhere in the world for six years. This is no longer true; now there is another Siberian snake in the AGS, which Huang and Roser[22] discussed. Figure 21 shows a photograph of this partial warm Siberian snake; it is a rather large warm solenoid because it is quite difficult to ramp superconducting solenoids. The partial Siberian snake is rampable along with the AGS energy; it can operate as a 5% partial snake up to about 25 GeV.

FIGURE 21. AGS Partial Siberian Snake.

Figure 22 shows some results[22] from the April 1994 AGS polarized beam run. The upper graph shows spin-flipping with the partial Siberian snake. The lower graph plots the AGS vertical polarization against $G\gamma$, which is the spin tune. The maximum $G\gamma$ of about 22.5 corresponds to about 11 GeV. The snake was only a partial snake, which should overcome the imperfection depolarizing resonances but not the three intrinsic depolarizing resonances; note the significant polarization loss at each intrinsic resonance. However, the partial snake did overcome the many imperfection resonances with no observable depolarization; thus, the complex system of 96 correction dipoles[23] was not needed to painfully correct them. The AGS pulsed quadrupoles were not used because their large 20 MW power supplies had no maintenance for six years; they should allow a higher polarization in the next AGS polarized beam run.

The Siberian snake's ability to accelerate polarized protons in the AGS, without weeks of tune-up time, may revive the AGS polarized beam physics program; moreover, it could provide polarized protons for RHIC. Dr. Makdisi[24] reviewed the RHIC polarized beam program. His Figure 23 shows an overview of the polarization hardware; note the AGS's partial Siberian snake, its pulsed quadrupoles, its polarimeters, and the four full Siberian snakes and two polarimeters in RHIC, which also has eight spin rotators for the STAR and PHENIX detectors. With Siberian snakes, RHIC could be the world's highest energy polarized proton-proton collider.

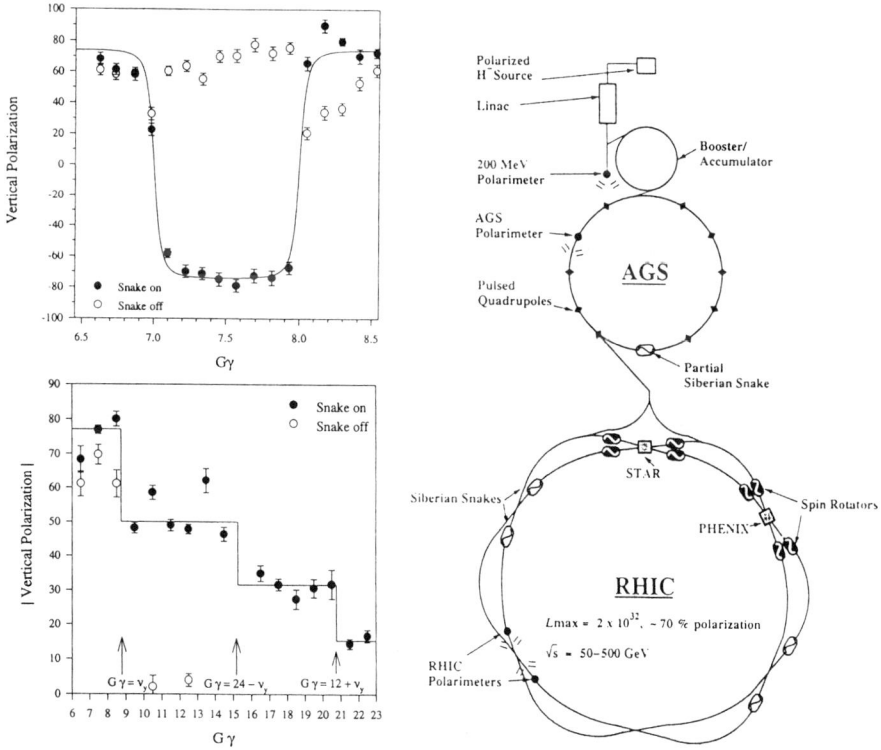

FIGURE 22. AGS Partial Snake Data.   **FIGURE 23.** Polarized Protons at RHIC.

I hope that they find a way to accelerate polarized beams at RHIC; it could provide an excellent program of spin physics. The main goal of the first Workshop on High Energy Spin Physics[25], held in Ann Arbor in 1977, was to accelerate polarized protons in ISABEL. Now, ISABEL is long gone; its successor, SSC, is also gone. However, one may still get polarized protons in the old ISABEL tunnel. Spin physicists may not be fast, but we are persistent.

Fermilab is also somewhat interested in polarized protons. In 1991 Fermilab first commissioned the SPIN Collaboration to do detailed studies of how to accelerate polarized protons in the various rings at Fermilab. Figure 24 shows the first page of the 1 August 1994 Polarized Tevatron Progress Report,[26] which includes detailed drawings, budgets, and schedules on how to accelerate polarized protons in both the Main Injector and the Tevatron. Figure 25 shows the new polarized hardware required in the various Fermilab rings and injectors. Two possibilities are being considered for the polarized source: an atomic beam-type polarized source (ABS) and an optically pumped polarized ion source (OPPIS). There is a competition with equal funding going to the ABS team at IUCF and the OPPIS team at TRIUMF. Both teams have made good progress; whoever develops the best source may get to build a polarized source for Fermilab.

The source's polarized H^- ions would be first accelerated by an RFQ, which might be built by Professor Teplyakov's team at IHEP-Protvino. There would be no depolarization in the transfer line or LINAC. The Booster would have a polarimeter, two modest pulsed quadrupoles, and one partial Siberian snake. The Main Injector would have two polarimeters and two full helical Siberian snakes. Professor Ado and Dr. Ludmirsky at IHEP-Protvino built a 10% scale model of a Main Injector helical dipole snake; this working model was successfully tested at both Protvino and Michigan.

August 1, 1994

PROGRESS REPORT

Acceleration of Polarized Protons to 1 TeV in the Fermilab Tevatron

SPIN Collaboration

Michigan, Indiana, Fermilab,
IHEP-Protvino, JINR-Dubna, Moscow, INR-Moscow
KEK
TRIUMF

ABSTRACT

The SPIN collaboration has been studying the acceleration of a polarized proton beam in the Fermilab Tevatron. This Progress Report briefly describes some physics goals for a polarized proton beam near 1 TeV. It also contains a preliminary estimate of the budget and schedule for the Polarized Tevatron project. The Report describes in some detail the hardware and commissioning procedures needed to accelerate polarized protons in the Tevatron and to perform polarized proton experiments; some highlights are:

- Twenty new 6 T superconducting dipoles would replace thirty-six existing 4.4 T Tevatron ring dipoles to make empty spaces in the Tevatron for six Siberian snakes and four Collider-detector spin rotators. This critical-path item now needs some guidance from the Fermilab management.
- Nine Siberian snakes and other minor hardware should allow full polarization of about 75% to be maintained and manipulated in the Booster, Main Injector, and Tevatron.
- Recent progress in ABS and OPPIS polarized ion sources (160 μA and 120 μA) should allow a polarized Collider luminosity above 10^{31} cm^{-2}s^{-1}. We are now supporting R & D for both source types which might lead to a polarized luminosity of 2×10^{31} to 10^{32} cm^{-2}s^{-1} by 1998. The source's three-year fabrication time makes it a critical-path item.

FIGURE 24. SPIN Collaboration Report.

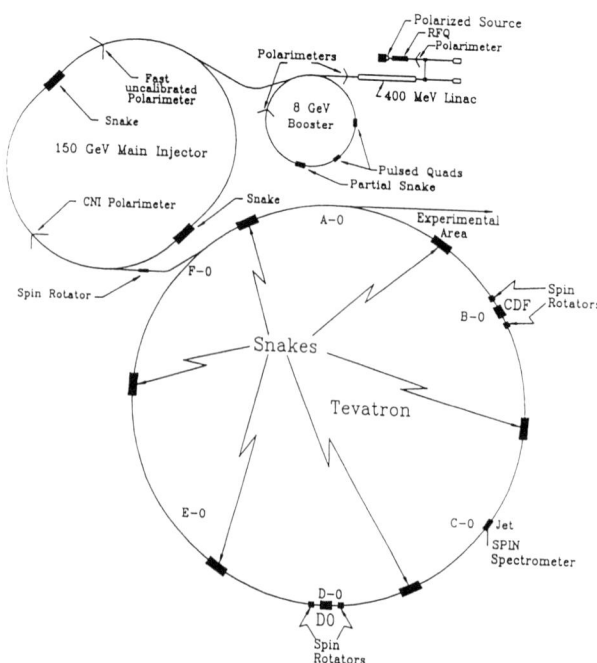

FIGURE 25. Polarized Tevatron.

There might be two types of fixed-target spin physics experiments. In the 120 GeV extracted beam area, there could be an experimental program using the Michigan solid polarized target, which has a polarization of over 90% in a $10^{11} s^{-1}$ beam intensity. The C0 internal target area may have a Mark-III polarized jet, similar to the Mark-II jet for UNK (see Figure 30).

The most complex part of the Polarized Tevatron project came from the discovery that there were no places to put the Tevatron Siberian snakes with the proper symmetry. We decided to make spaces for the six Tevatron snakes by removing at each snake point four existing 6-m-long 4.4 Tesla dipoles. We could then install, in the resulting 24-m-long space, two 8.8-m-long 6 Tesla dipoles; this would create a 6.4-m space for one 5-m-long Siberian snake. These 6 Tesla dipoles might be slightly modified HERA-type dipoles, UNK-2-type dipoles, or new Fermilab-designed dipoles. We are now interacting with many high-field dipole experts.

Fermilab has been funding these polarized studies and R&D, but they certainly have not approved the required $25 to $30 million; we expect a decision by mid-1995. We are developing a schedule so that the Polarized Tevatron installation, which requires removing 36 existing Tevatron ring magnets, could occur during the Main Injector installation now scheduled for February 1998.

In early 1994, Dr. Stanfield asked us to interact with the CDF and D0 people about this Polarized Tevatron project. We had several meetings, which at first were slightly painful because the collider-detector people and the polarized beam people have not talked much to each other for the last few decades. However, a good idea came out of this "painful" interaction. During a Polarized Tevatron meeting, Professor Weerts from Michigan State University showed Figure 26, where the inclusive cross-section for jet production is plotted against transverse energy. I had seen this graph at many seminars, but I never before thought of it as being related to polarization. I suddenly saw that at a huge transverse energy (or $P_\perp$) of 100, 200, or 300 GeV, there is still a very high event rate. I remembered that perturbative QCD predicted that A must eventually be zero, but essentially ignored all existing spin experiments because they were not at "a high enough energy or high enough $P_\perp$". However, Figure 26 shows high-event-rate data at $\sqrt{s}$ of almost 2 TeV and $P_\perp$ of 100 or 200 GeV/c. These inclusive Jets could easily test with great precision the perturbative QCD prediction that A must be zero. Weerts said that high precision measurements at $P_\perp = 100$ GeV/c can be made in a few hours. If A is not zero at $\sqrt{s} = 2$ TeV and $P_\perp = 100$ GeV/c, then it may be difficult to say where PQCD is useful.

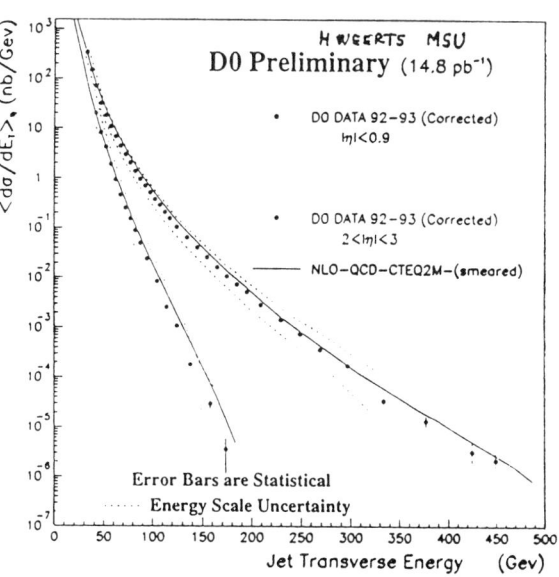

FIGURE 26. Inclusive Jet Cross Section.

The UNK proton accelerator is now being built at IHEP in Protvino, Russia; IHEP is sometimes called Serpukhov, which is a city ten miles away. IHEP's major activity now is to build the UNK-1 accelerator and to get the NEPTUN and NEPTUN-A polarization experiments running as soon as possible. This huge new facility is shown in Figure 27. There is also an informal proposal to develop a 70 GeV polarized proton beam at U-70 by Professor Ado and his student, Anferov, by making spaces for the snakes in the existing U-70 ring by installing some higher field warm magnets.

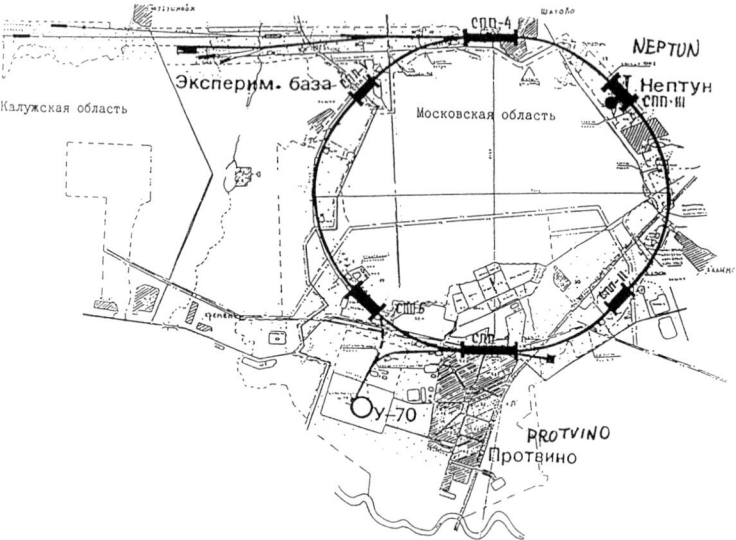

FIGURE 27. UNK and U-70 in Protvino, Russia.

In March 1994, IHEP successfully extracted a beam from the 70 GeV accelerator U-70 and transported it through the 2.7 km transfer line to the UNK tunnel.[27] The extraction and transfer efficiency was above 90%. When all of its magnets are installed, the UNK tunnel will look like this transfer line, which uses UNK-1's standard 6-m-long warm dipoles, its quadrupoles, and its standard vacuum system. Figure 28 shows the transfer line. They have not yet started installing the UNK-1 magnets in the main 21-km UNK tunnel. Although about 1500 of the 2200 dipoles and all of the 500 quadrupoles are already on site, they need the tunnel air conditioning to install magnets.

FIGURE 28. Transfer Line from the U-70 Accelerator into the UNK Tunnel.

It has been very challenging to build UNK in these complex times in Russia. However, UNK still progresses at a fairly good rate, and there is good progress in some areas. Apparently it was decided to concentrate all efforts during the next few years on finishing and then operating the 400 to 600 GeV UNK-1 accelerator and the polarized internal jet target experiments NEPTUN and NEPTUN-A. NEPTUN is led by Professor Solovianov, and I lead NEPTUN-A. The large SS-3 underground hall, shown in Figure 29, is essentially finished; it was dedicated last fall. The Mark-II polarized jet target and the NEPTUN and NEPTUN-A spectrometers are shown along with the underground electronics hall. There may also be a second upstream jet target built by Professor Pilipenko at Dubna with several smaller spectrometers. Thus, for several years, the 21 km circumference UNK complex may be completely devoted to polarization experiments until the 3 TeV superconducting UNK-2 ring operates.

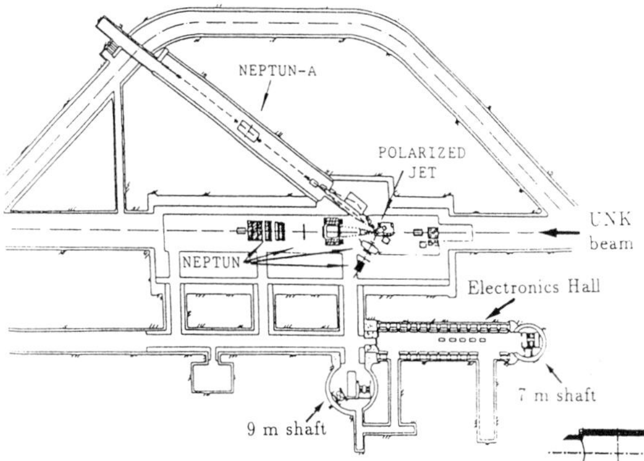

FIGURE 29. NEPTUN and NEPTUN-A Internal Polarized Jet Experiments.

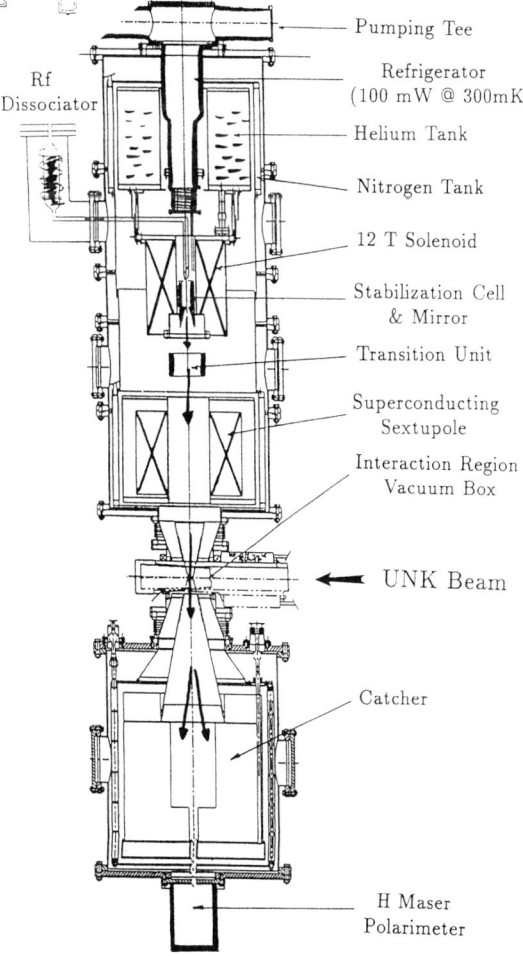

Figure 30 shows the ultra-cold spin-polarized atomic hydrogen jet[28] which is being built at Michigan; Dr. Luppov is heading this effort. This jet uses the ultra-cold technique which was discussed at several of the sessions. We plan to take this Mark-II spin-polarized proton jet to Protvino and use it as the internal target for the NEPTUN and NEPTUN-A experiments, which will measure A in 400 GeV proton-proton elastic and inelastic collisions. Our goal luminosity with this jet and UNK's 10^{19} s^{-1} circulating protons is 10^{32} s^{-1} cm^{-2}.

Figure 30. Michigan Mark-II Ultra-Cold Spin-Polarized H Jet.

I will end with a few comments on behalf of the International High Energy Spin Physics Committee. Figure 31 demonstrates an interesting situation where five different Committees organized the two Symposia that met together in the Indiana Memorial Union; about 340 physicists attended. Despite the five Committees we had a marvelous scientific meeting.

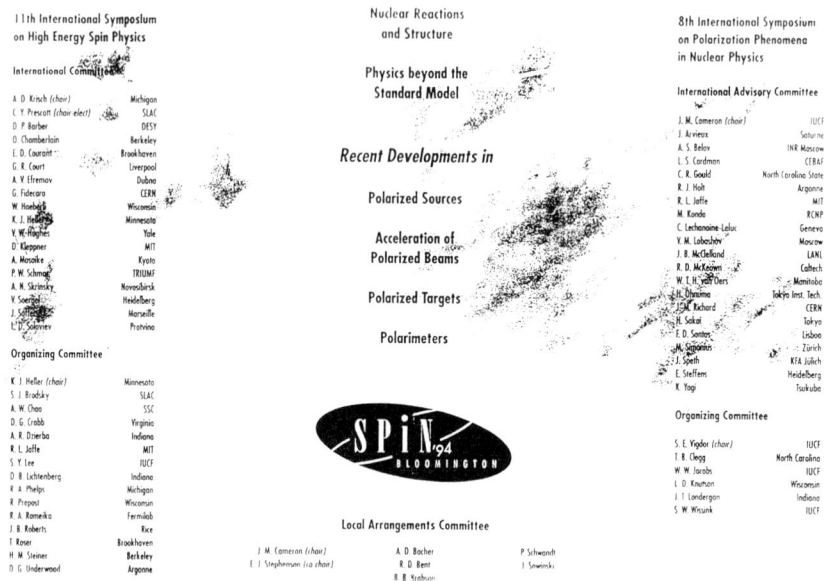

FIGURE 31. SPIN '94 Committees.

At its 20 September 1994 Meeting, the International Committee for High Energy Spin Physics Symposia approved several future Symposia and Workshops:

The 12th International High Energy Spin Physics Symposium will be held during 9-14 September 1996 in Amsterdam. Professor C.W. de Jager will be the Chair of the Organizing Committee.

The 5th Workshop on Polarized Ion Sources and Gas Targets. It will occur during 6-9 June 1995 in Cologne. Professor H.P.g. Schieck will chair the Workshop.

The 6th Protvino Spin Workshop; it will take place during 18-23 September 1995 in Protvino. Professor S.B. Nurushev will again chair this Workshop.

The 8th Workshop on Solid Polarized Targets was also approved. It will be at TRIUMF in either May or June 1996; the Chair will be Professor P. Delheij.

Four very distinguished International Committee Members retired at the end of 1994: Owen Chamberlain, Vernon Hughes, Daniel Kleppner, and Akiro Masaike. I am pleased that there are four distinguished new Committee Members: John Cameron, Hiroyasu Ejiri, William Happer, and Yoshi Mori. These additions represent an effort to make the Committee even more international and to encourage more intermediate energy and nuclear physicists to participate in the Amsterdam High Energy Spin Physics Symposium in 1996. This was my last Symposium as Chair; it was a rewarding, 20-year term. Our new Chair is Charles Prescott of SLAC. The trial partial merger of the two Spin Symposia was exciting, exhausting and very successful.

REFERENCES

* This work was supported by a Research Grant from the U.S. Department of Energy.
1. I.M. Ternov, *Proc. of 11th International Symposium on High Energy Spin Physics*, eds. K.J. Heller and S. Smith (AIP, New York 1995), to be published. A.A. Sokolov and I.M. Ternov, Sov. Phys. Dokl. **8**, 1203 (1964).
2. A.D. Krisch, Z. Phys. **C46**, S133 (1990).
3. E.A. Crosbie *et al.*, Phys. Rev. **D23**, 600 (1981).
4. T.J. Devlin and L.G. Pondrom, *Proc. of 11th Inter. Symposium on High Energy Spin Physics,* eds. K.J. Heller and S. Smith (AIP, New York 1995), to be published.
5. A. Yokosawa, *Proc. of 10th Int. Symp. on High Energy Spin Physics*, eds. T. Hasegawa *et al.*, p. 93 (Univ. Acc. Press Inc., Tokyo 1993).
6. D.G. Crabb *et al.*, Phys. Rev. Letters **65**, 3241 (1990).
7. A. Thomas, R. Voss, R. Windmolders, D. Day, H. Jackson, J. Soffer and M. Anselmino, *Proc. of 11th International Symposium on High Energy Spin Physics*, eds. K.J. Heller and S. Smith   (AIP, New York 1995), to be published.
8. *Proc. of Workshop on Polarized Ion Sources and Polarized Gas Targets*, L.W. Anderson and W. Haeberli eds., AIP Conf. Proc. **293** (AIP, New York 1993).
9. Y. Mori, *Proc. of 11th International Symposium on High Energy Spin Physics*, eds. K.J. Heller and S. Smith (AIP, New York 1995), to be published.
10. *SPIN '93 Proceedings*, S.B. Nurushev *et al.* eds., (IHEP, Protvino 1995).
11. Bad Honneff Workshop Proceedings, hopefully to be published; W. Meyer, *Proc. of 11th   International Symposium on High Energy Spin Physics*, eds. K.J. Heller and S. Smith (AIP, New York 1995), to be published.
12. M. Placidi, *Proc. of 11th International Symposium on High Energy Spin Physics*, eds. K.J. Heller and S. Smith (AIP, New York 1995), to be published.
13. G. Voss, D. Barber, and H. Jackson, *Proc. of 11th International Symposium on High Energy Spin Physics*, eds. K.J. Heller and S. Smith (AIP, New York 1995), to be pub.
14. V. Soergel, Unpublished comment at plenary session of 9th Int. Symp. on High Energy Spin Physics, Bonn 1990.
15. M. Woods and H.B. Steiner, *Proc. of 11th International Symposium on High Energy Spin Physics*, eds. K.J. Heller and S. Smith (AIP, New York 1995), to be published.
16. Ya.S. Derbenev and A.M. Kondratenko, Part. Accel **8**, 115 (1978).
17. A.D. Krisch *et al.*, Phys. Rev. Letters **63**, 1137 (1989).
18. B.B. Blinov *et al.*, Phys. Rev. Letters **73**, 1621 (1994).
19. R.A. Phelps *et al.*, Phys. Rev. Letters **72**, 1479 (1993).
20. D.D. Caussyn *et al.*, Phys Rev. Letters **73**, 2857 (1994).
21. S.E. Vigdor, *Proc. of 11th International Symposium on High Energy Spin Physics*, eds. K.J. Heller and S. Smith (AIP, New York 1995), to be published.
22. H. Huang and T. Roser, *Proc. of 11th International Symposium on High Energy Spin Physics*, eds. K.J. Heller and S. Smith (AIP, New York 1995), to be published; H. Huang *et al.*, Phys. Rev. Letters **73**, 2982 (1994).
23. F.Z. Khiari *et al.*, Phys. Rev. **D39**, 45 (1989).
24. Y. Makdisi, *Proc. of 11th International Symposium on High Energy Spin Physics*, eds. K.J. Heller and S. Smith (AIP, New York 1995), to be published.
25. *Proc. of 1977 Ann Arbor Workshop on Higher Energy Proton Polarized Beams*, eds. A.D. Krisch and A.J. Salthouse, AIP Conf. Proc. **42** (AIP, New York 1978).
26. *Polarized Tevatron Progress Report*, unpublished University of Michigan Report UM HE 94-15 (August 1, 1994).
27. CERN Courier, June 1994, pp. 18-20.
28. V.G. Luppov *et al.*, Phys. Rev. Letters **71**, 2405 (1993).

ACCELERATING POLARIZED BEAMS AT THE AGS

Thomas Roser

AGS Department
Brookhaven National Laboratory
Upton, N.Y. 11973, USA

INTRODUCTION

The acceleration of polarized beams in circular accelerators is complicated by the presence of numerous depolarizing resonances. During acceleration, a depolarizing resonance is crossed whenever the spin precession frequency equals the frequency with which spin-perturbing magnetic fields are encountered. There are two main types of depolarizing resonances corresponding to the possible sources of such fields: imperfection resonances, which are driven by magnet errors and misalignments, and intrinsic resonances, driven by the focusing fields.

The resonance conditions are usually expressed in terms of the spin tune ν_s, which is defined as the number of spin precessions per revolution. For an ideal planar accelerator, where orbiting particles experience only the vertical guide field, the spin tune is equal to $G\gamma$[1], where $G = 1.7928$ is the anomalous magnetic moment of the proton and γ is the relativistic Lorentz factor. The resonance condition for imperfection depolarizing resonances arise when $\nu_s = G\gamma = n$, where n is an integer. Imperfection resonances are therefore separated by only 523 MeV energy steps. The condition for intrinsic resonances is $\nu_s = G\gamma = kP \pm \nu_y$, where k is an integer, ν_y is the vertical betatron tune and P is the superperiodicity. For the AGS, $P = 12$ and $\nu_y \approx 8.8$. For most of the time during the acceleration cycle, the precession direction, or stable spin direction, coincides with the main vertical magnetic field. Close to a resonance, the stable spin direction is perturbed away from the vertical direction by the resonance driving fields. When a polarized beam is accelerated through an isolated resonance, the final polarization can be calculated analytically[2] and is given by

$$P_f/P_i = 2e^{-\frac{\pi|\epsilon|^2}{2\alpha}} - 1,$$

where P_i and P_f are the polarizations before and after the resonance crossing, respectively, ϵ is the resonance strength obtained from the spin rotation of the driving fields, and α is the

E880 Partial Snake Test at the AGS

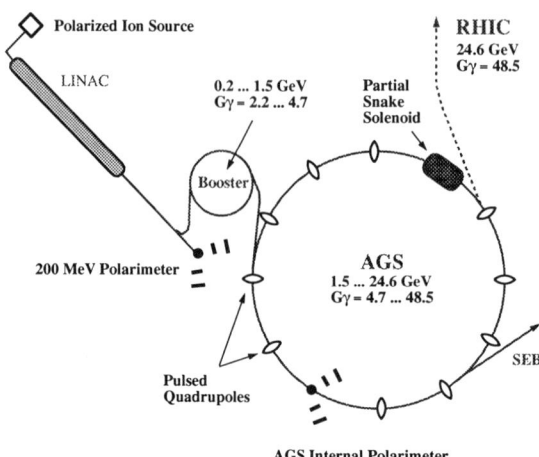

Figure 1: Layout of the AGS accelerator complex showing the location of the Partial Siberian Snake, the pulsed quadrupoles, and the AGS internal polarimeter

change of the spin tune per radian of the orbit angle. When the beam is slowly ($\alpha \ll |\epsilon|^2$) accelerated through the resonance, the spin vector will adiabatically follow the stable spin direction resulting in spin flip. However, for a faster acceleration rate partial depolarization or partial spin flip will occur. Traditionally, the intrinsic resonances are overcome by using a betatron tune jump, which effectively makes α large, and the imperfection resonances are overcome with the harmonic corrections of the vertical orbit to reduce the resonance strength ϵ[3]. At high energy, these traditional methods become difficult and tedious.

By introducing a 'Siberian Snake' [4], which is a 180° spin rotator of the spin about a horizontal axis, the stable spin direction remains unperturbed at all times as long as the spin rotation from the Siberian Snake is much larger than the spin rotation due to the resonance driving fields. Therefore the beam polarization is preserved during acceleration. Such a spin rotator can be constructed by using either solenoidal magnets or a sequence of interleaved horizontal and vertical dipole magnets producing only a local orbit distortion. Since the orbit distortion is inversely proportional to the momentum of the particle, such a dipole snake is particularly effective for high-energy accelerators, e.g. energies above about 30 GeV. For lower-energy synchrotrons, such as the Fermilab booster and the Brookhaven AGS with weaker depolarizing resonances, a partial snake[5], which rotates the spin by less than 180°, is sufficient to keep the stable spin direction unperturbed at the imperfection resonances.

AGS PARTIAL SIBERIAN SNAKE TESTS

Two polarized beam test runs of experiment E-880 at the AGS have recently demonstrated the feasibility of polarized proton acceleration using a 5% partial Siberian Snake. During the first run[6] in April 1994 it was shown that a 5% Snake is sufficient to avoid depolarization due to the imperfection resonances without using the harmonic correction method. Fig. 2 shows the evolution of the beam polarization as the beam energy and therefore $G\gamma$ is

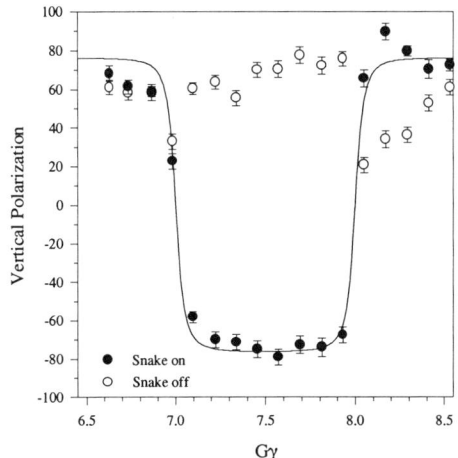

Figure 2: The measured vertical polarization as a function of the spin tune $G\gamma$ for a 10% snake is shown with and without a snake. Note here that partial depolarization at $G\gamma = 8$ is avoided by using a 10% snake. The solid line is the predicted energy dependence of the polarization.

increased. As predicted the polarization reverses the sign whenever $G\gamma$ is equal to an integer.

Fig. 3 shows the achieved polarization as a function of beam energy. It shows that no polarization was lost at the imperfection resonances. The only polarization loss occurred at the location of the intrinsic resonances for which the pulsed quadrupoles are required for the tune jump method. During the first run the pulsed quadrupoles were not available. During the second run in December 1994 it was shown that it is possible to use the tune jump method in the presence of the partial Snake. A new record energy for accelerated polarized beam of $25\,GeV$ was reached with about 12 % beam polarization left. Again no polarization was lost due to the imperfection resonances and depolarization from most intrinsic resonances was avoided with the tune jump quadrupoles. However, as can be seen from Fig. 3, significant amount of polarization was lost at $G\gamma = 0 + \nu_y$, $12 + \nu_y$ and $G\gamma = 36 + \nu_y$. The first two of these three resonances were successfully crossed previously and it will require further study to explain the unexpected polarization loss. The strength of the tune jump quadrupoles is not sufficient to jump the last resonance. We attempted to induce spin flip at this resonance but were only partially successful. During the next study run the method of inducing spin flip at intrinsic resonances will be further investigated[7].

TOWARDS A POLARIZED PROTON COLLIDER

With the successful tests of Siberian Snakes the stage is set for the acceleration of polarized proton beams to much higher energies to be used in collider experiments to explore spin effects at the highest energies attainable. Polarized protons from the Brookhaven AGS will be injected into the two RHIC rings to allow for up to $\sqrt{s} = 500\,GeV$ collisions with both beams polarized. With full Siberian Snakes all depolarizing resonances should be avoided

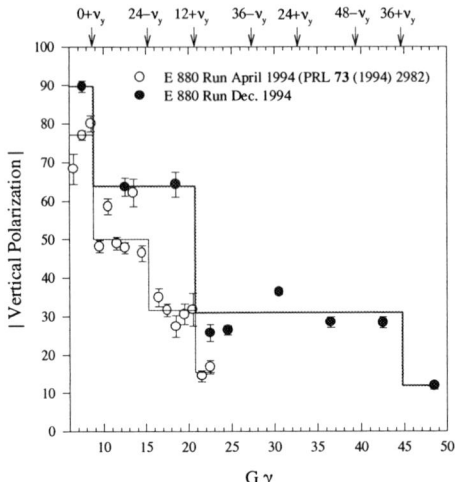

Figure 3: The measured absolute value of the vertical polarization at $G\gamma = n + \frac{1}{2}$ up to $G\gamma = 48.5$ which corresponds to an energy fo $25\,GeV$. The partial depolarization is due to intrinsic spin resonances at $G\gamma$ values indicated at the top of the figure. The results from the Dec. 1994 run are preliminary.

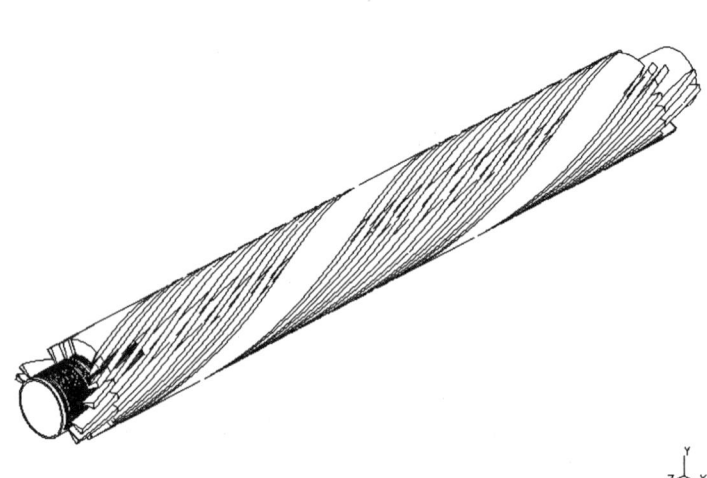

Figure 4: Aluminum former for the helical dipol magnet prototype under construction at BNL

since the spin tune is a half-integer independent of energy. However, if the spin disturbance from small horizontal fields is adding up sufficiently between the Snakes depolarization can still occur. This is most pronounced when the spin rotation from all the focusing filed add up coherently which is the case at the strongest intrinsic resonances. At RHIC two Snakes can still cope with the strongest intrinsic resonance whereas at Tevatron energies six Snakes would be needed.

Of particular interest is the design of the Siberian Snakes (two for each ring) and the spin rotators (four for each collider experiment) for RHIC. Proposed by V. Ptitsin and Yu. Shatunov from BINP[8], it is based on helical dipole magnet modules each having a full 360 degree helical twist. Using helical magnets minimizes orbit excursions which is most important at injection energy. Fig. 4 shows the aluminum former for the prototype helical dipole magnet now under construction at Brookhaven. The construction of a large bore high field helical dipole presents a formidable challenge for present superconducting magnet technology.

ACKNOWLEDGMENT

This work was performed under the auspices of the U.S. Department of Energy.

REFERENCES

[1] L.H. Thomas, Phil. Mag. **3**, 1 (1927); V. Bargmann, L. Michel, V.L. Telegdi, Phys. Rev. Lett. **2**, 435 (1959).
[2] M. Froissart and R. Stora, Nucl. Instr. Meth., **1**, 297 (1960).
[3] T. Khoe et al., Part. Accel. **6**, 213 (1975); J.L. Laclare et al., J. Phys. (Paris), Colloq. **46**, C2-499 (1985); H. Sato et al., Nucl. Inst. Meth., Phys. Res. Sec A**272**, 617 (1988); F.Z. Khiari, et al., Phys. Rev. D**39**, 45 (1989).
[4] Ya.S. Derbenev et al., Part. Accel. **8**, 115 (1978).
[5] T. Roser, AIP Conf. Proc. No. 187, ed. K.J. Heller p.1442 (AIP, New York, 1988).
[6] H.Huang et al., Phys. Rev. Lett. 73, 2982 (1994)
[7] T. Roser, in Proc. of the 10th Int. Symp. on High Energy Spin Physics, Nagoya, Japan, p. 429 (1992).
[8] V.I.Ptitsin and Yu.M.Shatunov, Helical Spin Rotators and Snakes, Proc. Third Workshop on Siberian Snakes and Spin Rotators (A.Luccio and T.Roser Eds.) Upton, NY, Sept. 12-13,1994, Brookhaven National Laboratory Report BNL-52453, p.15

Polarized Proton Beams at Fermilab

R. A. Phelps

Randall Laboratory of Physics
University of Michigan
Ann Arbor, Michigan 48109

In recent years interest in spin phenomena in high energy particle physics had increased significantly. Polarized beams are being planned or used at SLAC[1], HERA[2], RHIC[3] and LEP[4]. In this paper, I will discuss a study our University of Michigan group, along with IHEP-Protvino, JINR-Dubna, Indiana University, Fermilab, INR-Moscow, Moscow State University, KEK, and TRIUMF, has performed for the last 3 years on the acceleration of polarized protons in the Fermilab Tevatron[5, 6]. I will not discuss the possibility of polarizing the Tevatron antiproton beam.

Our interest in polarized proton beam acceleration stems from our interest in polarized proton-proton elastic scattering. With a beam and target both polarized normal to the scattering plane, one measures the 2 spin parameter A_{nn} defined by:

$$A_{nn} = \frac{d\sigma(\uparrow,\uparrow) + d\sigma(\downarrow,\downarrow) - 2d\sigma(\uparrow,\downarrow)}{\sum_{i,j} d\sigma(i,j)}. \tag{1}$$

Here the target and beam polarization are denoted by the arrows or i and j, and $d\sigma$ is the elastic differential cross section for the polarization state indicated.

If only the beam or target is polarized normal to the scattering plane, one measures the analyzing power, A, defined by:

$$A = \frac{d\sigma(\uparrow) - d\sigma(\downarrow)}{d\sigma(\uparrow) + d\sigma(\downarrow)}. \tag{2}$$

Note that if one has beam and target polarized, a measurement of A is obtained for free along with A_{nn} by averaging the data over one of the polarizations.

The quantities A and A_{nn} were once thought to vanish for high energy and high $P_\perp^2$. The existing high energy data for A_{nn}, shown in fig. 1, clearly shows that this is not the case. Experiments at the Brookhaven AGS with a polarized proton target and an unpolarized proton beam[7] showed that A not only does not go to 0 at high energy, but actually increases for $P_\perp^2 = 7\ (GeV/c)^2$. These A data along with data from an

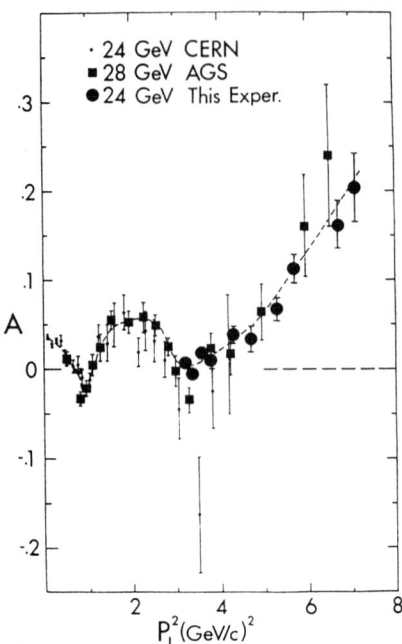

Figure 1: Plot of A against $P_\perp^2$ for $p_\uparrow + p \to p + p$.

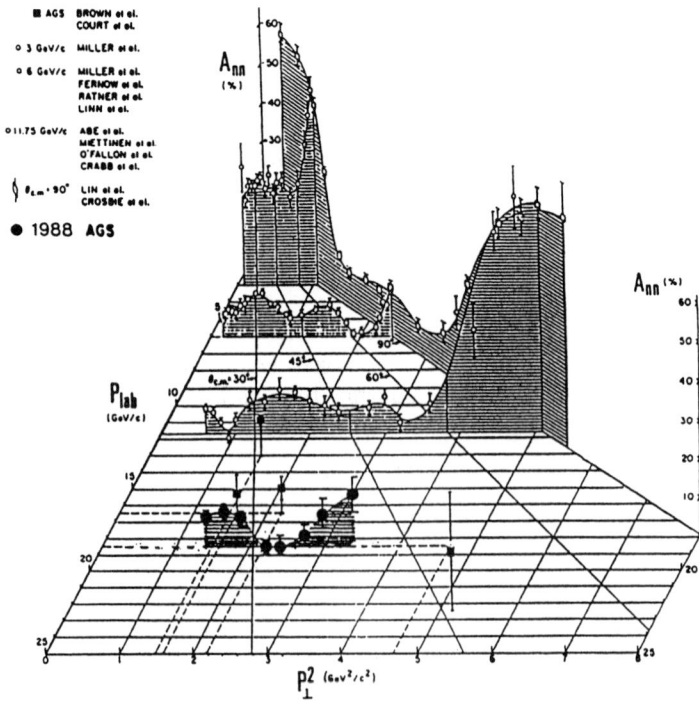

Figure 2: All high energy A_{nn} data for proton-proton elastic scattering plotted against $P_\perp^2$ and P_{lab}.

earlier CERN experiment[8] are shown in fig.2. We would like to see if these non-zero spin effects persist at energies in the TeV range.

One could also arrange transverse or longitudinal polarized protons to collide with unpolarized antiprotons. The D0 and CDF detectors could be used for the following studies: QCD tests by looking for non-zero inclusive jet production asymmetries, asymmetries from weak interactions, and polarized structure function measurements. Polarized protons might also give control of W production background for top quark studies. The advantage in performing these experiments at Fermilab is that these collider detectors would require little modification and are well understood with regard to systematics. These studies are outlined in ref. [10].

When polarized protons circulate in a storage ring, their spins will precess around the vertical guide field of the ring with a frequency given by $f_s = f_c G \gamma$, where f_c is the circulation frequency, $G=1.792847$ is the proton's anomalous magnetic moment in nuclear magnetons, and $\gamma = E/m_p$. The quantity $\nu_s = G\gamma$, the number of 360° precessions in one revolution, is commonly called the spin tune. If the spins are initially vertical, for most energies they will continue to remain vertical; the polarization in the vertical direction will be preserved. Protons also encounter non-vertical magnetic fields, kicking the spin away from the vertical direction. These kicks are usually uncorrelated, except at certain energies where the kick frequency becomes synchronized with f_s; then kicks will add up coherently from turn to turn and the polarization will be destroyed. These "depolarizing resonances" can be due to two effects: 1) the closed orbit is distorted due to imperfections in magnet alignment and field uniformity, causing the protons to pass through the focusing quadrupoles off axis vertically (imperfection resonances), seeing non-vertical fields at a frequency that is an integer multiple of f_c, and 2) the protons experience vertical oscillations in the "potential well" caused by the focusing quadrupoles (intrinsic resonances) and see non-vertical quadrupole fields with this oscillation frequency; this vertical oscillation frequency in terms of the circulation frequency is called the vertical betatron tune, ν_y. The conditions for depolarizing resonances are then:

$$\nu_s = n + m\nu_y, \qquad (3)$$

where n and m are integers. $m = 0$ gives the imperfection resonances and $m = 1$, $n = kP$ (k=integer, P=ring periodicity) gives the strongest intrinsic resonances. Since $\nu_s = G\gamma$ in equation 3, if the polarized beam is accelerated to high energies, many of these depolarizing resonances will be crossed, and the polarization will not survive. The beam depolarization from resonance crossing depends on the strength of the non-vertical magnetic fields and the acceleration rate.

At the Brookhaven AGS, the polarization of the accelerated proton beam was maintained by correcting each resonance individually during the acceleration cycle[9]. Since the imperfection resonances alone occur every 523 MeV, the Tevatron will have over 1,000 resonances; this long correction process would be impossible. However, there is a device, called a Siberian Snake[11], which can eliminate all depolarizing resonances simultaneously with little or no adjustments.

A Siberian Snake is a set of magnets in a circular accelerator which rotates the spin of the proton by an angle ϕ around some non-vertical direction without affecting the particle orbit outside the snake (it is "optically transparent"). A short calculation gives the following formula for the spin tune in terms of $G\gamma$ and ϕ:

$$cos(\pi\nu_s) = cos(\pi G\gamma)cos(\phi/2). \qquad (4)$$

From eq. 4 one finds that $\phi = \pi$ (full Siberian Snake) gives $\nu_s = \frac{1}{2}$ for all energies. Since $\nu_s \neq G\gamma$, the depolarizing resonances given by equation 3 are no longer encountered, and the polarization will survive acceleration.

A Siberian Snake made up of a 2 T·m superconducting solenoid and quadrupole beam correctors was installed in the Indiana University Cyclotron Facility (IUCF) Cooler Ring to test this principle[12, 13]. The result of one of the earliest and most important tests is shown in fig. 3, which is a plot of the beam polarization, both with and without the snake, when the strength of an imperfection resonance was varied (the resonance strength was proportional to $\int_{long} Bdl$ of a set of solenoids whose currents were varied). This data was taken at 104 MeV, an energy very close to the $G\gamma = 2$ imperfection resonance. Fig. 3 shows that without the snake, polarization survives only in a very narrow range around $\int_{long} Bdl = 0$, but with the snake, the polarization is insensitive to the resonance strength. We have since performed many more experiments with a Siberian Snake at IUCF to understand its effect on a polarized beam[14, 15, 16, 17, 18].

To successfully accelerate polarized protons in the Tevatron, modifications will have to be made to all stages of the Fermilab complex. A picture of these modification is shown in fig. 4. I will briefly discuss most of these modifications.

The acceleration of 400 MeV to 8 GeV in the Booster is similar to the Brookhaven AGS. The non-vertical fields required for a snake can be produced using either dipoles or solenoids. For a solenoid, the spin precession is proportional to $p^{-1} \int Bdl$, hence at energies of a few GeV, the field required for a full snake becomes impractically large. For dipoles, the field required is independent of energy, but at 400 MeV the dipole-generated orbit excursions inside the snake would be large, requiring very large magnets. We envision avoiding depolarization in the same manner as that employed in the AGS. The Booster has a single strong intrinsic resonance at $\nu_s = 0 + \nu_y$. This resonance can be crossed by "tune jumping"; as the resonance is approached, the vertical betatron tune is shifted rapidly by a set of fast pulsed quadrupoles. The resonance condition is then past through quickly and polarization is maintained. The imperfection resonances are avoided by using a "partial snake", a solenoid which rotates the spin by about 7.2° per revolution. This single magnet will strengthen all the imperfection resonances so that during acceleration the spin is completely flipped through the vertical direction. A similar device has been successfully tested at the AGS[19].

Resonance strengths grow with energy; the 8 GeV to 150 GeV Main Injector has enough strong resonances that the individual intrinsic resonance correction and partial snake scheme becomes impossible. The injection energy is high enough so that a Siberian Snake made of dipoles gives acceptably small orbit distortions inside the snake if one is clever with the snake design. Two full Siberian Snakes are required to overcome the depolarizing resonances in the Main Injector. To reduce the orbit excursions to an acceptable level, we propose to use "helical" snakes; each would be a warm dipole with pole faces having a twist in the beam direction. Orbit excursions are minimized in this type of magnet. These snakes should be placed 180° apart in azimuth to stabilize the spin tune at 1/2 during acceleration. They would be 13m long conventional magnets with a total $\int Bdl$ of about 22 T·m.

The resonances in the Tevatron are even stronger and more numerous than the Main Injector. We estimate that 6 symmetrically located snakes will be required to avoid depolarization in the Tevatron. Unlike the Main Injector, these snakes can be superconducting, hence will be much shorter than the Main Injector snakes. Our group has several designs for Tevatron snakes, one such snake uses 5 dipoles with lengths from 0.6 to 0.9m with various field orientations and magnitudes for a total length of 4.3m and $\int Bdl$ of 20.8 T·m. Even with shorter snakes, space in the Tevatron is in short supply. To accommodate the snakes, we propose to replace some existing sets

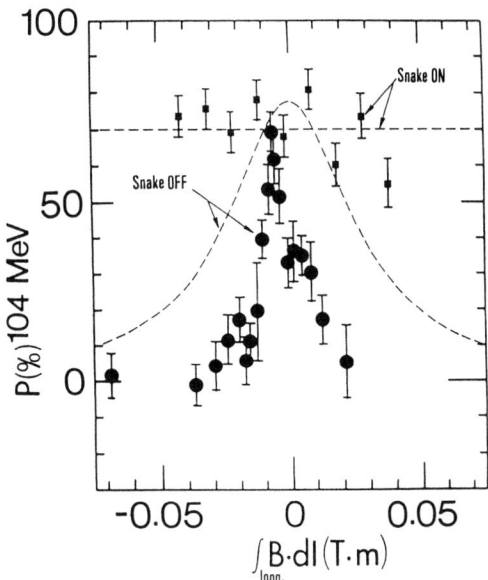

Figure 3: Siberian Snake overcoming depolarization effects from an imperfection resonance at the Indiana University Cyclotron Facility.

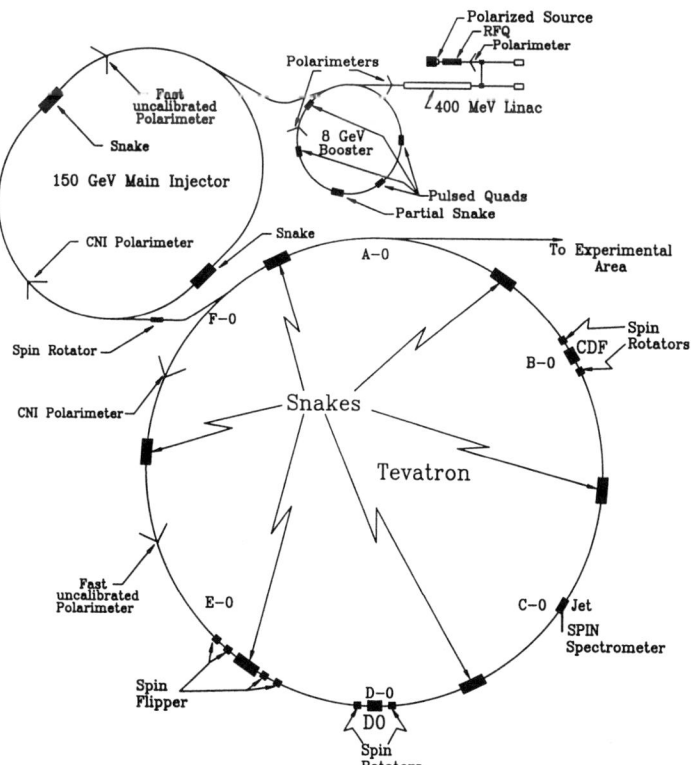

Figure 4: Modifications to the Fermilab complex for a polarized proton beam.

of 4 Tevatron 4.4 T dipoles with shorter strings of 3-6 T dipoles on each side of the snakes to increase the straight section length available for each snake. With 6 snakes, the spin direction would be everywhere vertical; to have longitudinal polarization at the D0 and CDF colliders, one would need a spin rotator preceeding each collider to orient the spin longitudinally and then another after to rotate the spin back to the vertical direction. This technique has been used successfully in the HERA electron ring[20] for longitudinal polarization at the HERMES experiment. These spin rotators are basically "half snakes", and would also require extra space that is presently not available. We therefore propose to replace Tevatron dipoles near the colliders with 6 T dipoles to make space available there also. Finding a suitable 6 T replacement for Tevatron dipoles is a difficult task; there are no exiting superconducting dipoles which satisfy all the requirements. If design and manufacture time becomes an issue, it may be necessary to initiate a "phased" installation of hardware; all hardware up to the Main Injector could be constructed and installed during the Main Injector shutdown in 1997-1998, to provide a 120 GeV extracted polarized proton beam[5]. The Tevatron snakes, spin rotators and 6 T dipoles could be installed at a later date.

One question unpolarized proton users are interested in is the impact a polarized to unpolarized beam changeover (or vice versa) would have on the running schedule. Will it take a large amount of potential running time to implement a changeover? Fortunately, Siberian Snakes requires little or no tuning, and are designed to be optically transparent; they should have little impact on beam tuning. There will be adjustments of the pulsed quadrupoles in the 8 GeV Booster to jump the intrinsic resonance; this could be done while the collider is still running unpolarized beam, since the Booster only works part time to fill the collider. Hence there should be little time required to tune up the polarized beam after unpolarized running. Going from polarized to unpolarized proton running is easier still; the Siberian Snakes are turned off and the beam is tuned up with the unpolarized source.

The intensity of the polarized proton beam depends on the intensity of the polarized source and the level of beam stacking in the Main Injector. A conservative estimate of the luminosity for $p_\uparrow - \bar{p}$ collisions is $1.2 \, 10^{31} \text{cm}^{-2}\text{s}^{-1}$. This is about 12% of the expected luminosity for unpolarized $p - \bar{p}$ collisions in the Main Injector era. This assumes a polarized proton source of intensity $150\mu A$, and 60% efficiency for "slip stacking" and RF gymnastics in the Main Injector. This polarized proton source intensity should be easily achieved; atomic beam source (ABS) and optically pumped polarized ion source (OPPIS) groups in the SPIN collaboration now discuss intensities in the near future approaching 1mA. An internal ultra-cold polarized atomic hydrogen jet target[21] of thickness 10^{13}cm^{-2} would give $p_\uparrow - p_\uparrow$ collisions for A_{nn} measurements in the Tevatron at a luminosity of $2 \, 10^{30}\text{cm}^{-2}\text{s}^{-1}$. The same assumptions on the source and stacking give a Main Injector 120 GeV extracted polarized proton beam intensity of $1.4 \, 10^{11}\text{s}^{-1}$. With the University of Michigan solid polarized target[22] of thickness $2 \, 10^{23}\text{cm}^{-2}$, we could do an A_{nn} measurement at 120 GeV with a luminosity of $3 \, 10^{34}\text{cm}^{-2}\text{s}^{-1}$.

Studies of the SPIN collaboration show that accelerating a polarized proton beam to the highest available energy in the Tevatron is not only feasible but could be implemented in a few years. Since the spin of the proton is the only other initial condition to control in high energy hadron collisions besides the energy, it makes sense for experiments to start controlling it.

This work was supported by grants from the U.S. Department of Energy and the U.S. National Science Foundation.

References

[1] Polarization at SLAC. By M. Woods (SLAC), SLAC-PUB-6694, Jan 1995. 17pp. Presented at 11th International Symposium on High Energy Spin Physics and the 8th International Symposium on Polarization Phenomena in Nuclear Physics (SPIN 94), Bloomington, IN, 15-22 Sep 1994.

[2] D.P. Barber et al., Nucl. Instrum. Meth. **A329**, 79 (1993).

[3] RHIC Spin Proposal M. Beddo et al.; Brookhaven National Lab proposal (1992), (unpublished).

[4] K. Knudsen et al., Phys. Lett. **B270**, 97 (1991).

[5] Acceleration of Polarized Protons to 120 and 150 GeV in the Fermilab Main Injector, SPIN Collaboration, Univ. of Michigan Report (March 1992), (unpublished).

[6] Progress Report: Acceleration of Polarized Protons to 1 TeV in the Fermilab Main Tevatron, SPIN Collaboration, Univ. of Michigan Report (August 1994), (unpublished).

[7] P.R. Cameron et al., Phys. Rev. **D32**, 3070 (1985);
P.H. Hansen et al., Phys. Rev. Lett. **50**, 802 (1983);
D.C. Peaslee et al., Phys. Rev. Lett. **51**, 2359 (1983);
D.G. Crabb et al., Phys. Rev. Lett. **65**, 3241 (1990).

[8] J. Antille et al., Nucl. Phys. **B185**, 1 (1981).

[9] F. Z. Khiari et al., Phys. Rev. **D39**, 45 (1989).

[10] H. Weerts et al.,, Polarization Physics at the Tevatron Collider, Expression of Interest by Michigan State University (1994), (unpublished).

[11] Ya.S. Derbenev and A. M. Kondratenko, Part. Accel. **8**, 115 (1978).

[12] A. D. Krisch et al., Phys. Rev. Lett. **63**, 1137 (1989).

[13] J. E. Goodwin et al., Phys. Rev. Lett. **64**, 2779 (1990).

[14] M.G. Minty et al., Phys. Rev. **D44**, R1361 (1991).

[15] V. A. Anferov et al., Phys. Rev. **A46**, R7383 (1992).

[16] R. Baiod et al., Phys. Rev. Lett. **70**, 2557 (1993).

[17] R.A. Phelps et al., Phys. Rev. Lett. **72**, 1479 (1994).

[18] B.B. Blinov et al., Phys. Rev. Lett. **73**, 1621 (1994).

[19] H. Huang et al., Phys. Rev. Lett. **73**, 2982 (1994).
Also see T. Roser, these proceedings.

[20] D.P. Barber et al., Phys. Lett. **B343**, 436 (1995).

[21] W.A. Kaufman et al., Nucl. Instrum. Meth. **A335**, 17 (1993).

[22] D.G. Crabb et al., Phys. Rev. Lett. **64**, 2627 (1990).

INDEX

angular momentum 93

b quark 168
big bang 53,58,64
bispinors 178
black holes 197,200
Bose-Einstein correlations 99
BSRT 175

causality 15
Center for Theoretical Studies 8
Chern-Simons Gauge 220
chiral phase transition 11,20
closed time path formalism 11
condensation 79
cold dank matter 87
cosmology 111
critical behavior 225

deep inelastic scattering 256
de Sitter 122
dilation 111

electroweak interactions 178,203
Einstein equations 121
Fermi gas 146,160
fermion 213
fractals 98
Friedmann-Lemâitric equation 60,67

galaxy distribution 97
Gamons 74
gauge bosons 178
gauge symmetry 175
generalized gravitation 36
ghosts 177
gluons 11,17

Goldstone Boson 190
gravitational condensation 33
gravitational waves 113,125
gravity 29,119
GUTS 166

hadrons 99
halo particles 88
heavy ion collisions 21
HERP 272
HERMES 273
Higgs production 167
Hoja 8
Hubble 102
Hubble parameters 80,82
hypergeometic functions 250

inclusive neutrino reactions 153
inflation 111,112,119
interferometers 128,135
invariant momentum 99

KARMEN 151

LAMPF 139,161
Landau 204
Laser Interferometer 126
LIGO 125,133
LSND 139,151

magnetic charge 30,31,43
Michel spectrum 152
microwave background 54
monopole 41
muon 165,269
muon capture 158
muon colliders 165

oscillation 139

Penturbative coefficients 241
Planck Mass 122
Planck Scale 120
plasma 21
Poincavi group 229
polarized beams 287,293

QCD 107,262,268,295
quarks 11,17

radiative breaking 205
random phase approximation 162
reflector matrices 222
relativity 29

scaling 103,106
scattering 236
Schwinger 3,7
Sine-Gordon equation 213
solar neutrinos 153
spin 31,267
spin fields 175

standard model 166
standard quantum limit 133
steady state model 56
Stefan-Boltzmann Law 78
strings 111,193
super conformal fields 176
super strings 197
supersymmetry 30,187,238
SUSY 169,203,205
SUSY-GUT 169

tachyon 38,206,207
thermal distribution 199
thermal noise 133

unitarity 183

VIRGO 125

wavelet 97
weyl 179

Zeldovich 89